CAMBRIDGE LIBRARY COLLECTION
Books of enduring scholarly value

Life Sciences

Until the nineteenth century, the various subjects now known as the life sciences were regarded either as arcane studies which had little impact on ordinary daily life, or as a genteel hobby for the leisured classes. The increasing academic rigour and systematisation brought to the study of botany, zoology and other disciplines, and their adoption in university curricula, are reflected in the books reissued in this series.

The Natural History and Antiquities of Selborne, in the County of Southampton

Gilbert White (1720–1793) published his *Natural History and Antiquities* as one volume in 1789. Both works consist of a series of letters written by White to the barrister Daines Barrington (1727–1800) and the zoologist Thomas Pennant (1726–1798). The letters in *Natural History*, White's best-known work, contain detailed information about his observations of local flora, fauna and wildlife. White was a pioneer of the study of birds and animals in their natural habitats, rather than as specimens removed from their environments. His methods of observation enabled him to identify and record many previously unknown species. (He was the first, for example, to distinguish the chiffchaff from the warbler by differences in song.) The letters in *Antiquities* are concerned with the topographical, social and ancient history of Selborne. They include details of important Roman coin finds and are an indispensable source for the history of local churches and buildings.

Cambridge University Press has long been a pioneer in the reissuing of out-of-print titles from its own backlist, producing digital reprints of books that are still sought after by scholars and students but could not be reprinted economically using traditional technology. The Cambridge Library Collection extends this activity to a wider range of books which are still of importance to researchers and professionals, either for the source material they contain, or as landmarks in the history of their academic discipline.

Drawing from the world-renowned collections in the Cambridge University Library, and guided by the advice of experts in each subject area, Cambridge University Press is using state-of-the-art scanning machines in its own Printing House to capture the content of each book selected for inclusion. The files are processed to give a consistently clear, crisp image, and the books finished to the high quality standard for which the Press is recognised around the world. The latest print-on-demand technology ensures that the books will remain available indefinitely, and that orders for single or multiple copies can quickly be supplied.

The Cambridge Library Collection will bring back to life books of enduring scholarly value (including out-of-copyright works originally issued by other publishers) across a wide range of disciplines in the humanities and social sciences and in science and technology.

The Natural History and Antiquities of Selborne, in the County of Southampton

GILBERT WHITE

CAMBRIDGE UNIVERSITY PRESS

Cambridge, New York, Melbourne, Madrid, Cape Town,
Singapore, São Paolo, Delhi, Tokyo, Mexico City

Published in the United States of America by Cambridge University Press, New York

www.cambridge.org
Information on this title: www.cambridge.org/9781108138369

This edition first published 1789
This digitally printed version 2011

ISBN 978-1-108-13836-9 Paperback

Pl. I.

North

½ East view of SELBORNE, from the SHORT LY

Published Nov.ʳ 1. 1788, as the Act directs, by B. White & Son.

East view of SELBORNE, from the SHORT LY

Published Nov.ʳ 1. 1788, as the Act directs, by B. White & Son.

YTHE.

THE

NATURAL HISTORY

AND

ANTIQUITIES

OF

SELBORNE,

IN THE

COUNTY OF SOUTHAMPTON:

WITH

ENGRAVINGS, AND AN APPENDIX.

— — — " ego Apis Matinæ
" More modoque
" Grata carpentis — — — per laborem
" Plurimum," — — — — —   HOR.

" Omnia benè defcribere, quæ in hoc mundo, a Deo facta, aut Naturæ creatæ viribus
" elaborata fuerunt, opus eft non unius hominis, nec unius ævi. Hinc *Faunæ* & *Floræ*
" utiliffimæ; hinc *Monographi* præftantiffimi."   SCOPOLI ANN. HIST. NAT.

LONDON:

PRINTED BY T. BENSLEY;

FOR B. WHITE AND SON, AT HORACE'S HEAD, FLEET STREET.

M,DCC,LXXXIX.

ADVERTISEMENT.

THE Author of the following Letters takes the liberty, with all proper deference, of laying before the public his idea of *parochial history*, which, he thinks, ought to confift of natural productions and occurrences as well as antiquities. He is alfo of opinion that if ftationary men would pay fome attention to the diftricts on which they refide, and would publifh their thoughts refpecting the objects that furround them, from fuch materials might be drawn the moft complete county-hiftories, which are ftill wanting in feveral parts of this kingdom, and in particular in the county of *Southampton*.

And here he feizes the firft opportunity, though a late one, of returning his moft grateful acknowledgments to the reverend the Prefident and the reverend and worthy the Fellows of *Magdalen College* in the univerfity of *Oxford*, for their liberal behaviour in permitting their archives to be fearched by a member of their own fociety, fo far as the

A   evidences

evidences therein contained might refpect the parifh and priory of *Selborne*. To that gentleman alfo, and his affiftant, whofe labours and attention could only be equalled by the very kind manner in which they were beftowed, many and great obligations are alfo due.

Of the authenticity of the documents above-mentioned there can be no doubt, fince they confift of the identical deeds and records that were removed to the College from the Priory at the time of it's diffolution; and, being carefully copied on the fpot, may be depended on as genuine; and, never having been made public before, may gratify the curiofity of the antiquary, as well as eftablifh the credit of the hiftory.

If the writer fhould at all appear to have induced any of his readers to pay a more ready attention to the wonders of the Creation, too frequently overlooked as common occurrences; or if he fhould by any means, through his refearches, have lent an helping hand towards the enlargement of the boundaries of hiftorical and topographical knowledge; or if he fhould have thrown fome fmall light upon ancient cuftoms and manners, and efpecially on thofe that were monaftic; his purpofe will be fully anfwered. But if he fhould not have been fuccefsful in any of thefe his intentions,

tions, yet there remains this confolation behind—that thefe his purfuits, by keeping the body and mind employed, have, under Providence, contributed to much health and cheerfulnefs of fpirits, even to old age: and, what ftill adds to his happinefs, have led him to the knowledge of a circle of gentlemen whofe intelligent communications, as they have afforded him much pleafing information, fo, could he flatter himfelf with a continuation of them, would they ever be deemed a matter of fingular fatisfaction and improvement.

Selborne,
January 1ft, 1788.

GIL. WHITE.

THE
NATURAL HISTORY
OF
SELBORNE.

S. H. Grimm del. *D. Lerpinière sculp.*

—— where the Hermit hangs his straw-clad cell.

Τρηχει, αλλ' αγαθη κυροτροφος. υτι εγωγε
'Ης γαιης δυναμαι γλυκερωτερον αλλο ιδεσθαι.   Homeri Odyſſ.

Tota denique noſtra illa aſpera, & montuoſa, & fidelis, & ſimplex, & fautrix ſuorum regio.
Cicero Orat. pro Cn. Plancio.

THE
NATURAL HISTORY
OF
SELBORNE.

LETTER I.

TO THOMAS PENNANT, ESQUIRE.

THE parish of SELBORNE lies in the extreme eastern corner of the county of *Hampshire*, bordering on the county of *Sussex*, and not far from the county of *Surrey*; is about fifty miles south-west of *London*, in latitude 51, and near midway between the towns of *Alton* and *Petersfield*. Being very large and extensive it abuts on twelve parishes, two of which are in *Sussex*, viz. *Trotton* and *Rogate*. If you begin from the south and proceed westward the adjacent parishes are *Emshot, Newton Valence, Faringdon, Harteley Mauduit, Great Ward le ham, Kingsley, Hedleigh, Bramshot, Trotton, Rogate, Lysse,* and *Greatham*. The soils of this district are almost as various and diversified as the views and aspects. The high part to the south-west consists of a vast hill of chalk, rising three

B   hundred

hundred feet above the village; and is divided into a sheep down, the high wood, and a long hanging wood called *The Hanger*. The covert of this eminence is altogether beech, the most lovely of all forest trees, whether we consider it's smooth rind or bark, it's glossy foliage, or graceful pendulous boughs. The down, or sheep-walk, is a pleasing park-like spot, of about one mile by half that space, jutting out on the verge of the hill-country, where it begins to break down into the plains, and commanding a very engaging view, being an assemblage of hill, dale, wood-lands, heath, and water. The prospect is bounded to the south-east and east by the vast range of mountains called *The Sussex Downs*, by *Guild-down* near *Guildford*, and by the *Downs* round *Dorking*, and *Ryegate* in *Surrey*, to the north-east, which altogether, with the country beyond *Alton* and *Farnham*, form a noble and extensive outline.

At the foot of this hill, one stage or step from the uplands, lies the village, which consists of one single straggling street, three quarters of a mile in length, in a sheltered vale, and running parallel with *The Hanger*. The houses are divided from the hill by a vein of stiff clay (good wheat-land), yet stand on a rock of white stone, little in appearance removed from chalk; but seems so far from being calcarious, that it endures extreme heat. Yet that the freestone still preserves somewhat that is analogous to chalk, is plain from the beeches which descend as low as those rocks extend, and no farther, and thrive as well on them, where the ground is steep, as on the chalks.

The cart-way of the village divides, in a remarkable manner, two very incongruous soils. To the south-west is a rank clay, that requires the labour of years to render it mellow; while the gardens to the north-east, and small enclosures behind, consist of a

warm

warm, forward, crumbling mould, called *black malm*, which seems highly saturated with vegetable and animal manure; and these may perhaps have been the original site of the town; while the woods and coverts might extend down to the opposite bank.

At each end of the village, which runs from south-east to north-west, arises a small rivulet: that at the north-west end frequently fails; but the other is a fine perennial spring, little influenced by drought or wet seasons, called *Well-head*[a]. This breaks out of some high grounds joining to *Nore Hill*, a noble chalk promontory, remarkable for sending forth two streams into two different seas. The one to the south becomes a branch of the *Arun*, running to *Arundel*, and so falling into the *British* channel: the other to the north. The *Selborne* stream makes one branch of the *Wey*; and, meeting the *Black-down* stream at *Hedleigh*, and the *Alton* and *Farnham* stream at *Tilford-bridge*, swells into a considerable river, navigable at *Godalming*; from whence it passes to *Guildford*, and so into the *Thames* at *Weybridge*; and thus at the *Nore* into the *German* ocean.

Our wells, at an average, run to about sixty-three feet, and when sunk to that depth seldom fail; but produce a fine limpid water, soft to the taste, and much commended by those who drink the pure element, but which does not lather well with soap.

To the north-west, north and east of the village, is a range of fair enclosures, consisting of what is called a *white malm*, a sort of

---

[a] This spring produced, *September* 14, 1781, after a severe hot summer, and a preceding dry spring and winter, nine gallons of water in a minute, which is five hundred and forty in an hour, and twelve thousand nine hundred and sixty, or two hundred and sixteen hogsheads, in twenty-four hours, or one natural day. At this time many of the wells failed, and all the ponds in the vales were dry.

rotten

rotten or rubble ſtone, which, when turned up to the froſt and rain, moulders to pieces, and becomes manure to itſelf [b].

Still on to the north-eaſt, and a ſtep lower, is a kind of white land, neither chalk nor clay, neither fit for paſture nor for the plough, yet kindly for hops, which root deep into the freeſtone, and have their poles and wood for charcoal growing juſt at hand. This white ſoil produces the brighteſt hops.

As the pariſh ſtill inclines down towards *Wolmer-foreſt*, at the juncture of the clays and ſand the ſoil becomes a wet, ſandy loam, remarkable for timber, and infamous for roads. The oaks of *Temple* and *Blackmoor* ſtand high in the eſtimation of purveyors, and have furniſhed much naval timber; while the trees on the freeſtone grow large, but are what workmen call *ſhakey*, and ſo brittle as often to fall to pieces in ſawing. Beyond the ſandy loam the ſoil becomes an hungry lean ſand, till it mingles with the foreſt; and will produce little without the aſſiſtance of lime and turnips.

LETTER II.

TO THE SAME.

IN the court of *Norton* farm houſe, a manor farm to the north-weſt of the village, on the white malms, ſtood within theſe twenty years a *broad-leaved elm*, or *wych hazel*, *ulmus folio latiſſimo ſcabro* of *Ray*, which, though it had loſt a conſiderable leading

----

[b] This ſoil produces good wheat and clover.

bough

bough in the great ftorm in the year 1703, equal to a moderate
tree, yet, when felled, contained eight loads of timber; and, being too bulky for a carriage, was fawn off at feven feet above the
butt, where it meafured near eight feet in the diameter. This elm
I mention to fhew to what a bulk *planted elms* may attain; as this
tree muft certainly have been fuch from it's fituation.

In the centre of the village, and near the church, is a fquare
piece of ground furrounded by houfes, and vulgarly called *The
Pleftor* [c]. In the midft of this fpot ftood, in old times, a vaft oak,
with a fhort fquat body, and huge horizontal arms extending
almoft to the extremity of the area. This venerable tree, furrounded with ftone fteps, and feats above them, was the delight
of old and young, and a place of much refort in fummer evenings; where the former fat in grave debate, while the latter fro-
licked and danced before them. Long might it have ftood, had
not the amazing tempeft in 1703 overturned it at once, to the
infinite regret of the inhabitants, and the vicar, who beftowed
feveral pounds in fetting it in it's place again: but all his care
could not avail; the tree fprouted for a time, then withered and
died. This oak I mention to fhew to what a bulk *planted oaks*
alfo may arrive: and planted this tree muft certainly have been, as
will appear from what will be faid farther concerning this area,
when we enter on the antiquities of *Selborne.*

On the *Blackmoor* eftate there is a fmall wood called *Lofel's*, of a
few acres, that was lately furnifhed with a fet of oaks of a peculiar growth and great value; they were tall and taper like firs,
but ftanding near together had very fmall heads, only a little
brufh without any large limbs. About twenty years ago the

[c] Vide the plate in the antiquities.

bridge

bridge at the *Toy*, near *Hampton Court*, being much decayed, some trees were wanted for the repairs that were fifty feet long without bough, and would meafure twelve inches diameter at the little end. Twenty fuch trees did a purveyor find in this little wood, with this advantage, that many of them anfwered the defcription at fixty feet. Thefe trees were fold for twenty pounds apiece.

In the centre of this grove there ftood an oak, which, though fhapely and tall on the whole, bulged out into a large excrefcence about the middle of the ftem. On this a pair of ravens had fixed their refidence for fuch a feries of years, that the oak was diftinguifhed by the title of *The Raven-tree*. Many were the attempts of the neighbouring youths to get at this *eyry:* the difficulty whetted their inclinations, and each was ambitious of furmounting the arduous tafk. But, when they arrived at the fwelling, it jutted out fo in their way, and was fo far beyond their grafp, that the moft daring lads were awed, and acknowledged the undertaking to be too hazardous. So the ravens built on, neft upon neft, in perfect fecurity, till the fatal day arrived in which the wood was to be levelled. It was in the month of *February*, when thofe birds ufually fit. The faw was applied to the butt, the wedges were inferted into the opening, the woods echoed to the heavy blows of the beetle or mallet, the tree nodded to it's fall; but ftill the dam fat on. At laft, when it gave way, the bird was flung from her neft; and, though her parental affection deferved a better fate, was whipped down by the twigs, which brought her dead to the ground.

LETTER

Pl. III.

MYTILUS, Crista Galli.

Published 1 Nov.r 1788, as the Act directs, by B. White & Son.

LETTER III.

TO THE SAME.

THE foffil-fhells of this diftrict, and forts of ftone, fuch as have fallen within my obfervation, muft not be paffed over in filence. And firft I muft mention, as a great curiofity, a fpecimen that was plowed up in the chalky fields, near the fide of the *Down*, and given to me for the fingularity of it's appearance, which, to an incurious eye, feems like a petrified fifh of about four inches long, the cardo paffing for an head and mouth. It is in reality a *bivalve* of the *Linnæan Genus* of *Mytilus*, and the fpecies of *Crifta Galli*; called by *Lifter, Raftellum*; by *Rumphius, Oftreum plicatum minus*; by *D'Argenville, Auris Porci*, f. *Crifta Galli*; and by thofe who make collections *cock's comb*. Though I applied to feveral fuch in London, I never could meet with an entire fpecimen; nor could I ever find in books any engraving from a perfect one. In the fuperb *mufeum* at *Leicefter-houfe* permiffion was given me to exa- mine for this article; and, though I was difappointed as to the foffil, I was highly gratified with the fight of feveral of the fhells themfelves in high prefervation. This bivalve is only known to inhabit the *Indian* ocean, where it fixes itfelf to a *zoophyte*, known by the name *Gorgonia*. The curious foldings of the future the one into the other, the alternate flutings or grooves, and the curved form of my fpecimen being much eafier expreffed by the pencil than by words, I have caufed it to be drawn and engraved.

Cornua Ammonis are very common about this village. As we were cutting an inclining path up *The Hanger*, the labourers found

them

them frequently on that steep, just under the soil, in the chalk, and of a considerable size. In the lane above *Well-head*, in the way to *Emshot*, they abound in the bank in a darkish sort of marl; and are usually very small and soft: but in *Clay's Pond*, a little farther on, at the end of the pit, where the soil is dug out for manure, I have occasionally observed them of large dimensions, perhaps fourteen or sixteen inches in diameter. But as these did not consist of firm stone, but were formed of a kind of *terra lapidosa*, or hardened clay, as soon as they were exposed to the rains and frost they mouldered away. These seemed as if they were a very recent production. In the chalk-pit, at the north-west end of *The Hanger*, large *nautili* are sometimes observed.

In the very thickest strata of our freestone, and at considerable depths, well-diggers often find large *scallops* or *pectines*, having both shells deeply striated, and ridged and furrowed alternately. They are highly impregnated with, if not wholly composed of, the stone of the quarry.

LETTER IV.

TO THE SAME.

As in a former letter the *freestone* of this place has been only mentioned incidentally, I shall here become more particular.

This stone is in great request for hearth-stones, and the beds of ovens: and in lining of lime-kilns it turns to good account; for the workmen use sandy loam instead of mortar; the sand of which fluxes, [d] and runs by the intense heat, and so cases over the

---

[d] There may probably be also in the chalk itself that is burnt for lime a proportion of sand: for few chalks are so pure as to have none.

whole

whole face of the kiln with a ftrong vitrified coat like glafs, that it is well preferved from injuries of weather, and endures thirty or forty years. When chifeled fmooth, it makes elegant fronts for houfes, equal in colour and grain to the *Bath* ftone; and fuperior in one refpeçt, that, when feafoned, it does not fcale. Decent chimney-pieces are worked from it of much clofer and finer grain than *Portland*; and rooms are floored with it; but it proves rather too foft for this purpofe. It is a freeftone, cutting in all direçtions; yet has fomething of a grain parallel with the horizon, and therefore fhould not be *furbedded*, but laid in the fame pofition that it grows in the quarry e. On the ground abroad this fireftone will not fucceed for pavements, becaufe, probably fome degree of faltnefs prevailing within it, the rain tears the flabs to pieces f. Though this ftone is too hard to be açted on by vinegar; yet both the white part, and even the *blue rag*, ferments ftrongly in mineral acids. Though the white ftone will not bear wet, yet in every quarry at intervals there are thin ftrata of *blue rag*, which refift rain and froft; and are excellent for pitching of ftables, paths and courts, and for building of dry walls againft banks; a valuable fpecies of fencing, much in ufe in this village, and for mending of roads. This *rag* is rugged and ftubborn, and will not hew to a fmooth face; but is very durable: yet, as thefe ftrata are fhallow and lie deep, large quantities cannot be procured but at confiderable expenfe. Among the *blue rags* turn up fome blocks tinged

e To *furbed* ftone is to fet it edgewife, contrary to the pofture it had in the quarry, fays Dr. *Plot. Oxfordfh.* p. 77. But *furbedding* does not fucceed in our dry walls; neither do we ufe it fo in ovens, though he fays it is beft for *Teynton* ftone.

f " Fireftone is full of falts, and has no fulphur: muft be clofe grained, and have no " interftices. Nothing fupports fire like falts; faltftone perifhes expofed to wet and " froft." *Plot's Staff.* p. 152.

C   with

with a ſtain of *yellow* or *ruſt colour,* which ſeem to be nearly as laſting as the blue; and every now and then balls of a friable ſubſtance, like ruſt of iron, called *ruſt balls.*

In *Wolmer Foreſt* I ſee but one ſort of ſtone, called by the workmen *ſand,* or *foreſt-ſtone.* This is generally of the colour of ruſty
iron, and might probably be worked as iron ore; is very hard and
heavy, and of a firm, compact texture, and compoſed of a ſmall
roundiſh cryſtalline grit, cemented together by a brown, terrene,
ferruginous matter; will not cut without difficulty, nor eaſily ſtrike
fire with ſteel. Being often found in broad flat pieces, it makes
good pavement for paths about houſes, never becoming ſlippery
in froſt or rain; is excellent for dry walls, and is ſometimes uſed
in buildings. In many parts of that waſte it lies ſcattered on the
ſurface of the ground; but is dug on *Weaver's Down,* a vaſt hill
on the eaſtern verge of that foreſt, where the pits are ſhallow, and
the ſtratum thin. This ſtone is imperiſhable.

From a notion of rendering their work the more elegant, and
giving it a finiſh, maſons chip this ſtone into ſmall fragments about
the ſize of the head of a large nail; and then ſtick the pieces into
the wet mortar along the joints of their freeſtone walls: this em-
belliſhment carries an odd appearance, and has occaſioned ſtrangers
ſometimes to aſk us pleaſantly, "whether we faſtened our walls
" together with tenpenny nails."

LETTER

LETTER V.

TO THE SAME.

AMONG the fingularities of this place the two rocky hollow lanes, the one to *Alton*, and the other to the foreft, deferve our attention. Thefe roads, running through the malm lands, are, by the traffick of ages, and the fretting of water, worn down through the firft ftratum of our freeftone, and partly through the fecond; fo that they look more like water-courfes than roads; and are bedded with naked *rag* for furlongs together. In many places they are reduced fixteen or eighteen feet beneath the level of the fields; and after floods, and in frofts, exhibit very grotefque and wild appearances, from the tangled roots that are twifted among the ftrata, and from the torrents rufhing down their broken fides; and efpecially when thofe cafcades are frozen into icicles, hanging in all the fanciful fhapes of froft-work. Thefe rugged gloomy fcences affright the ladies when they peep down into them from the paths above, and make timid horfemen fhudder while they ride along them; but delight the naturalift with their various botany, and particularly with their curious *filices* with which they abound.

The manor of *Selborne*, was it ftrictly looked after, with all it's kindly afpects, and all it's floping coverts, would fwarm with game; even now hares, partridges, and pheafants abound; and in old days woodcocks were as plentiful. There are few quails, becaufe they more affect open fields than enclofures; after harveft fome few land-rails are feen.

The parifh of *Selborne*, by taking in fo much of the foreft, is a vaft diftrict. Thofe who tread the bounds are employed part of

three

three days in the bufinefs, and are of opinion that the outline, in all it's curves and indentings, does not comprife lefs than thirty miles.

The village ftands in a fheltered fpot, fecured by *The Hanger* from the ftrong wefterly winds. The air is foft, but rather moift from the effluvia of fo many trees; yet perfectly healthy and free from agues.

The quantity of rain that falls on it is very confiderable, as may be fuppofed in fo woody and mountainous a diftrict. As my experience in meafuring the water is but of fhort date, I am not qualified to give the mean quantity [g]. I only know that

		Inch.	Hund.
From May 1, 1779, to the end of the year there fell		28	37 !
From Jan. 1, 1780, to Jan. 1, 1781	- -	27	32
From Jan. 1, 1781, to Jan. 1, 1782	- -	30	71
From Jan. 1, 1782, to Jan. 1, 1783	- -	50	26 !
From Jan. 1, 1783, to Jan. 1, 1784	- -	33	71
From Jan. 1, 1784, to Jan. 1, 1785	- -	33	80
From Jan. 1, 1785, to Jan. 1, 1786	- -	31	55
From Jan. 1, 1786, to Jan. 1, 1787	- -	39	57

The village of *Selborne*, and large hamlet of *Oakhanger*, with the fingle farms, and many fcattered houfes along the verge of

[g] A very intelligent gentleman affures me (and he fpeaks from upwards of forty years experience) that the mean rain of any place cannot be afcertained till a perfon has mea-fured it for a very long period. " If I had only meafured the rain," fays he, " for the " four firft years, from 1740 to 1743, I fhould have faid the mean rain at *Lyndon* was " 16 1-hf. inch for the year; if from 1740 to 1750, 18 1-hf. inches. The mean rain be-" fore 1763 was 20 1-qr. from 1763 and fince 25 1-hf. from 1770 to 1780, 26. If only " 1773, 1774 and 1775, had been meafured, *Lyndon* mean rain would have been called " 32 inches."

the

the foreſt, contain upwards of ſix hundred and ſeventy inhabitants [h].

[h] A STATE *of the* Pariſh *of* SELBORNE, *taken* OCTOBER 4, 1783.

The number of tenements or families, 136.

The number of inhabitants in the ſtreet is 313 ⎱ Total 676; near five inhabitants to
In the reſt of the pariſh - - - 363 ⎰ each tenement.

In the time of the Rev. *Gilbert White,* Vicar, who died in 1727-8, the number of inhabitants was computed at about 500.

Average of baptiſms for 60 years.

From 1720 to 1729, both years incluſ. { Males 6,9 / Females 6,0 } 12,9 | From 1740 to 1749 incl. { M. 9,2 / F. 6,6 } 15,8 | From 1760 to 1769, incl. { M. 9,1 / F. 8,9 } 18,0

From 1730 to 1739, both years incluſ. { Males 8,2 / Females 7,1 } 15,3 | From 1750 to 1759 incl. { M. 7,6 / F. 8,1 } 15,7 | From 1770 to 1779, incl. { M. 10,5 / F. 9,8 } 20,3

Total of baptiſms of Males 515 ⎱ 980
Females 465 ⎰

Total of baptiſms from 1720 to 1779, both incluſive - - - 60 years - - - 980.

Average of burials for 60 years.

From 1720 to 1729, both years incluſ. { Males 4,8 / Females 5,1 } 9,9 | From 1740 to 1749, incl. { M. 4,6 / F. 3,8 } 8,4 | From 1760 to 1769, incl. { M. 6,9 / F. 6,5 } 13,4

From 1730 to 1739, both years incluſ. { Males 4,8 / Females 5,8 } 10,6 | From 1750 to 1759, incl. { M. 4,9 / F. 5,1 } 10,0 | From 1770 to 1779, incl. { M. 5,5 / F. 6,2 } 11,7

Total of burials of Males 315 ⎱ 640
Females 325 ⎰

Total of burials from 1720 to 1779, both incluſive - - - 60 years - - - 640.

Baptiſms exceed burials by more than one third.
Baptiſms of Males exceed Females by one tenth, or one in ten.
Burials of Females exceed Males by one in thirty.
It appears that a child, born and bred in this pariſh, has an equal chance to live above forty years.
Twins thirteen times, many of whom dying young have leſſened the chance for life.
Chances for life in men and women appear to be equal.

A TABLE *of the* Baptiſms, Burials, *and* Marriages, *from* January 2, 1761, *to* December 25, 1780, *in the* Pariſh *of* SELBORNE.

	BAPTISMS.			BURIALS.			MAR.		BAPTISMS.			BURIALS.			MAR.
	M.	F.	Tot.	M.	F.	Tot.			M.	F.	Tot.	M.	F.	Tot.	
1761.	8	10	18	2	4	6	3	1771.	10	6	16	3	4	7	4
1762.	7	8	15	10	14	24	6	1772.	11	10	21	6	10	16	3
1763.	8	10	18	3	4	7	5	1773.	8	5	13	7	5	12	3
1764.	11	9	20	10	8	18	6	1774.	6	13	19	2	8	10	1
1765.	12	6	18	9	7	16	6	1775.	20	7	27	13	8	21	6
1766.	9	13	22	10	6	16	4	1776.	11	10	21	4	6	10	6
1767.	14	5	19	6	5	11	2	1777.	8	13	21	7	3	10	4
1768.	7	6	13	2	5	7	6	1778.	7	13	20	3	4	7	5
1769.	9	14	23	6	5	11	2	1779.	14	8	22	5	6	11	5
1770.	10	13	23	4	7	11	3	1780.	8	9	17	11	4	15	3
	95	94	189	62	65	127	43		103	94	197	61	58	119	40
									95	94	189	62	65	127	43
									198	188	386	123	123	246	83

During this period of twenty years the births of males exceeded thoſe of females - - 10.
The burials of each ſex were equal.
And the births exceeded the deaths - - - 140.

We

We abound with poor; many of whom are fober and induf-trious, and live comfortably in good ftone or brick cottages, which are glazed, and have chambers above ftairs: mud buildings we have none. Befides the employment from hufbandry, the men work in hop gardens, of which we have many; and fell and bark timber. In the fpring and fummer the women weed the corn; and enjoy a fecond harveft in September by hop picking. For-merly, in the dead months they availed themfelves greatly by fpinning wool, for making of *barragons*, a genteel corded ftuff, much in vogue at that time for fummer wear; and chiefly manu-factured at *Alton*, a neighbouring town, by fome of the people called Quakers: but from circumftances this trade is at an end [i]. The inhabitants enjoy a good fhare of health and longevity; and the parifh fwarms with children.

LETTER VI.

TO THE SAME.

SHOULD I omit to defcribe with fome exactnefs the *foreft* of *Wol-mer*, of which three fifths perhaps lie in this parifh, my account of *Selborne* would be very imperfect, as it is a diftrict abounding with many curious productions, both animal and vegetable; and has often afforded me much entertainment both as a fportfman and as a naturalift.

[i] Since the paffage above was written, I am happy in being able to fay that the fpin-ning employment is a little revived, to the no fmall comfort of the induftrious houfe-wife.

The

The royal *foreſt* of *Wolmer* is a tract of land of about ſeven miles in length, by two and a half in breadth, running nearly from North to South, and is abutted on, to begin to the South, and ſo to proceed eaſtward, by the pariſhes of *Greatham*, *Lyſſe*, *Rogate*, and *Trotton*, in the county of *Suſſex*; by *Bramſhot*, *Hedleigh*, and *Kingſley*. This royalty conſiſts entirely of ſand covered with heath and fern; but is ſomewhat diverſified with hills and dales, without having one ſtanding tree in the whole extent. In the bottoms, where the waters ſtagnate, are many bogs, which formerly abounded with ſubterraneous trees; though Dr. *Plot* ſays poſitively [k], that "there never were any fallen trees hidden in the moſſes of the "ſouthern counties." But he was miſtaken: for I myſelf have ſeen cottages on the verge of this wild diſtrict, whoſe timbers conſiſted of a black hard wood, looking like oak, which the owners aſſured me they procured from the bogs by probing the ſoil with ſpits, or ſome ſuch inſtruments: but the peat is ſo much cut out, and the moors have been ſo well examined, that none has been found of late [l]. Beſides the oak, I have alſo been ſhewn pieces of foſſil-wood of a paler colour, and ſofter nature, which the inhabitants called

fir:

[k] See his Hiſt. of *Staffordſhire*.

[l] Old people have aſſured me, that on a winter's morning they have diſcovered theſe trees, in the bogs, by the hoar froſt, which lay longer over the ſpace where they were concealed, than on the ſurrounding moraſs. Nor does this ſeem to be a fanciful notion, but conſiſtent with true philoſophy. Dr. *Hales* ſaith, "That the warmth of the earth, at "ſome depth under ground, has an influence in promoting a thaw, as well as the change "of the weather from a freezing to a thawing ſtate, is manifeſt, from this obſervation, "viz. Nov. 29, 1731, a little ſnow having fallen in the night, it was, by eleven the next "morning, moſtly melted away on the ſurface of the earth, except in ſeveral places in "*Buſhy-park*, where there were *drains* dug and covered with earth, on which the ſnow "continued to lie, whether thoſe drains were full of water or dry; as alſo where *elm-pipes* "lay under ground: a plain proof this, that thoſe drains intercepted the warmth of the

"earth

fir: but, upon a nice examination, and trial by fire, I could difcover nothing resinous in them; and therefore rather suppose that
they were parts of a willow or alder, or some such aquatic tree.

This lonely domain is a very agreeable haunt for many sorts of
wild fowls, which not only frequent it in the winter, but breed
there in the summer; such as lapwings, snipes, wild-ducks, and,
as I have discovered within these few years, teals. Partridges in vast
plenty are bred in good seasons on the verge of this forest, into
which they love to make excursions: and in particular, in the dry
summer of 1740 and 1741, and some years after, they swarmed to
such a degree that parties of unreasonable sportsmen killed twenty
and sometimes thirty brace in a day.

But there was a nobler species of game in this forest, now extinct, which I have heard old people say abounded much before
shooting flying became so common, and that was the *heath-cock*,
black game, or *grouse*. When I was a little boy I recollect one
coming now and then to my father's table. The last pack remembered was killed about thirty-five years ago; and within these ten
years one solitary *grey hen* was sprung by some beagles in beating
for a hare. The sportsmen cried out, " A hen pheasant;" but a
gentleman present, who had often seen grouse in the north of England, assured me that it was a greyhen.

Nor does the loss of our black game prove the only gap in the
Fauna Selborniensis; for another beautiful link in the chain of beings

" earth from ascending from greater depths below them : for the snow lay where the drain
" had more than four feet depth of earth over it. It continued also to lie on thatch, tiles,
" and the tops of walls." See *Hales*'s Hæmastatics : p. 360. Quere, Might not such
observations be reduced to domestic use, by promoting the discovery of old obliterated
drains and wells about houses; and in Roman stations and camps lead to the finding of
pavements, baths and graves, and other hidden relics of curious antiquity ?

is wanting, I mean the *red deer*, which toward the beginning of this century amounted to about five hundred head, and made a stately appearance. There is an old keeper, now alive, named *Adams*, whose great grandfather (mentioned in a perambulation taken in 1635) grandfather, father and self, enjoyed the head keepership of *Wolmer forest* in succession for more than an hundred years. This person assures me, that his father has often told him, that Queen *Anne*, as she was journeying on the *Portsmouth* road, did not think the forest of *Wolmer* beneath her royal regard. For she came out of the great road at *Lippock*, which is just by, and, reposing herself on a bank smoothed for that purpose, lying about half a mile to the east of *Wolmer-pond*, and still called *Queen's-bank*, saw with great complacency and satisfaction the whole herd of red deer brought by the keepers along the vale before her, consisting then of about five hundred head. A sight this worthy the attention of the greatest sovereign! But he farther adds that, by means of the *Waltham blacks*, or, to use his own expression, as soon as they began *blacking*, they were reduced to about fifty head, and so continued decreasing till the time of the late Duke of *Cumberland*. It is now more than thirty years ago that his highness sent down an huntsman, and six yeomen-prickers, in scarlet jackets laced with gold, attended by the stag-hounds; ordering them to take every deer in this forest alive, and to convey them in carts to *Windsor*. In the course of the summer they caught every stag, some of which shewed extraordinary diversion: but, in the following winter, when the hinds were also carried off, such fine chases were exhibited as served the country people for matter of talk and wonder for years afterwards. I saw myself one of the yeomen-prickers single out a stag from the herd, and must confess that it was the most curious feat of activity I ever beheld, superior to any thing in Mr. *Astley's*

D   riding-

riding-school. The exertions made by the horse and deer much exceeded all my expectations; though the former greatly excelled the latter in speed. When the devoted deer was separated from his companions, they gave him, by their watches, law, as they called it, for twenty minutes; when, sounding their horns, the stop-dogs were permitted to pursue, and a most gallant scene ensued.

LETTER VII.

TO THE SAME.

THOUGH large herds of deer do much harm to the neighbourhood, yet the injury to the morals of the people is of more moment than the loss of their crops. The temptation is irresistible; for most men are sportsmen by constitution: and there is such an inherent spirit for hunting in human nature, as scarce any inhibitions can restrain. Hence, towards the beginning of this century all this country was wild about deer-stealing. Unless he was a *hunter*, as they affected to call themselves, no young person was allowed to be possessed of manhood or gallantry. The *Waltham blacks* at length committed such enormities, that government was forced to interfere with that severe and sanguinary act called the *black act*[m], which now comprehends more felonies than any law that ever was framed before. And, therefore, a late bishop of *Winchester*,

[m] Statute 9 *Geo.* I. c. 22.

when

when urged to re-ſtock *Waltham-chaſe*[n], refuſed, from a motive worthy of a prelate, replying that " it had done miſchief enough " already."

Our old race of deer-ſtealers are hardly extinct yet : it was but a little while ago that, over their ale, they uſed to recount the exploits of their youth ; ſuch as watching the pregnant hind to her lair, and, when the calf was dropped, paring it's feet with a penknife to the quick to prevent it's eſcape, till it was large and fat enough to be killed ; the ſhooting at one of their neighbours with a bullet in a turnip-field by moonſhine, miſtaking him for a deer ; and the loſing a dog in the following extraordinary manner :—Some fellows, ſuſpecting that a calf new-fallen was depoſited in a certain ſpot of thick fern, went, with a lurcher, to ſurpriſe it ; when the parent-hind ruſhed out of the brake, and, taking a vaſt ſpring with all her feet cloſe together, pitched upon the neck of the dog, and broke it ſhort in two.

Another temptation to idleneſs and ſporting was a number of rabbits, which poſſeſſed all the hillocks and dry places : but theſe being inconvenient to the huntſmen, on account of their burrows, when they came to take away the deer, they permitted the country people to deſtroy them all.

Such foreſts and waſtes, when their allurements to irregularities are removed, are of conſiderable ſervice to neighbourhoods that verge upon them, by furniſhing them with peat and turf for their firing ; with fuel for the burning their lime ; and with aſhes for their graſſes ; and by maintaining their geeſe and their ſtock of young cattle at little or no expenſe.

The manor-farm of the pariſh of *Greatham* has an admitted

---

[n] This chaſe remains un-ſtocked to this day ; the biſhop was Dr. *Hoadly.*

claim,

claim, I fee, (by an old record taken from the *Tower of London*) of turning all live ftock on the foreft, at proper feafons, *bidentibus exceptis*°. The reafon, I prefume, why fheep^P are excluded, is, becaufe, being fuch clofe grazers, they would pick out all the fineft graffes, and hinder the deer from thriving.

Though (by ftatute 4 and 5 *W.* and *Mary*) c. 23. " to burn on " any wafte, between *Candlemas* and *Midfummer*, any grig, ling, " heath and furze, gofs or fern, is punifhable with whipping " and confinement in the houfe of correction;" yet, in this foreft, about *March* or *April*, according to the drynefs of the feafon, fuch vaft heath-fires are lighted up, that they often get to a mafterlefs head, and, catching the hedges, have fometimes been communicated to the underwoods, woods, and coppices, where great damage has enfued. The plea for thefe burnings is, that, when the old coat of heath, &c. is confumed, young will fprout up, and afford much tender brouze for cattle; but, where there is large old furze, the fire, following the roots, confumes the very ground; fo that for hundreds of acres nothing is to be feen but fmother and defolation, the whole circuit round looking like the cinders of a volcano; and, the foil being quite exhaufted, no traces of vegetation are to be found for years. Thefe conflagrations, as they take place ufually with a north-eaft or eaft wind, much annoy this village with their fmoke, and often alarm the country; and, once in particular, I remember that a gentleman, who lives beyond *Andover*, coming to my houfe, when he got on the downs between that town and *Winchefter*, at twenty-five miles

---

° For this privilege the owner of that eftate ufed to pay to the king annually feven bufhels of oats.

P In *The Holt*, where a full ftock of fallow-deer has been kept up till lately, no fheep are admitted to this day.

distance, was surprised much with smoke and a hot smell of fire; and concluded that *Alresford* was in flames; but, when he came to that town, he then had apprehensions for the next village, and so on to the end of his journey.

On two of the most conspicuous eminences of this forest stand two *arbours* or *bowers*, made of the boughs of oaks; the one called *Waldon-lodge*, the other *Brimstone-lodge*: these the keepers renew annually on the feast of St. *Barnabas*, taking the old materials for a perquisite. The farm called *Blackmoor*, in this parish, is obliged to find the posts and brush-wood for the former; while the farms at *Greatham*, in rotation, furnish for the latter; and are all enjoined to cut and deliver the materials at the spot. This custom I mention, because I look upon it to be of very remote antiquity.

LETTER VIII.

TO THE SAME.

On the verge of the forest, as it is now circumscribed, are three considerable lakes, two in *Oakhanger*, of which I have nothing particular to say; and one called *Bin's*, or *Bean's pond*, which is worthy the attention of a naturalist or a sportsman. For, being crowded at the upper end with willows, and with the *carex cespitosa*[q], it affords such a safe and pleasing shelter to wild ducks,

[q] I mean that sort which, rising into tall hassocks, is called by the foresters *torrets*; a corruption, I suppose of turrets.

Note, In the beginning of the summer 1787 the royal forests of *Wolmer* and *Holt* were measured by persons sent down by government.

teals,

teals, fnipes, &c. that they breed there. In the winter this covert is alfo frequented by foxes, and fometimes by pheafants; and the bogs produce many curious plants. [For which confult letter XLII. to Mr. *Barrington.*]

By a *perambulation* of *Wolmer foreft* and *The Holt*, made in 1635, and in the eleventh year of *Charles* the Firft (which now lies before me), it appears that the limits of the former are much circum-fcribed. For, to fay nothing of the farther fide, with which I am not fo well acquainted, the bounds on this fide, in old times, came into *Binfwood*; and extended to the ditch of *Ward le ham-park*, in which ftands the curious mount called *King John's Hill*, and *Lodge Hill*; and to the verge of *Hartley Mauduit*, called *Mauduit-hatch*; comprehending alfo *Short-heath*, *Oakhanger*, and *Oakwoods*; a large diftrict, now private property, though once belonging to the royal domain.

It is remarkable that the term *purlieu* is never once mentioned in this long roll of parchment. It contains, befides the *perambulation*, a rough eftimate of the value of the timbers, which were confider-able, growing at that time in the diftrict of *The Holt*; and enu-merates the officers, fuperior and inferior, of thofe joint forefts, for the time being, and their oftenfible fees and perquifites. In thofe days, as at prefent, there were hardly any trees in *Wolmer foreft.*

Within the prefent limits of the foreft are three confiderable lakes, *Hogmer*, *Cranmer*, and *Wolmer*; all of which are ftocked with carp, tench, eels, and perch: but the fifh do not thrive well, becaufe the water is hungry, and the bottoms are a naked fand.

A circumftance refpecting thefe ponds, though by no means peculiar to them, I cannot pafs over in filence; and that is, that inftinct by which in fummer all the kine, whether oxen, cows,

calves

calves, or heifers, retire conftantly to the water during the hotter hours; where, being more exempt from flies, and inhaling the cool-nefs of that element, fome belly deep, and fome only to mid-leg, they ruminate and folace themfelves from about ten in the morning till four in the afternoon, and then return to their feeding. During this great proportion of the day they drop much dung, in which infects neftle; and fo fupply food for the fifh, which would be poorly fubfifted but from this contingency. Thus Nature, who is a great economift, converts the recreation of one animal to the fupport of another! *Thomfon*, who was a nice obferver of natural occur-rences, did not let this pleafing circumftance efcape him. He fays, in his *Summer*,

> " A various group the herds and flocks compofe :
> " ——————— on the graffy bank
> " Some ruminating lie ; while others ftand
> " Half in the flood, and, often bending, fip
> " The circling furface."

Wolmer-pond, fo called, I fuppofe, for eminence fake, is a vaft lake for this part of the world, containing, in it's whole circumfe-rence, 2646 yards, or very near a mile and an half. The length of the north-weft and oppofite fide is about 704 yards, and the breadth of the fouth-weft end about 456 yards. This meafure-ment, which I caufed to be made with good exactnefs, gives an area of about fixty-fix acres, exclufive of a large irregular arm at the north-eaft corner, which we did not take into the reckoning.

On the face of this expanfe of waters, and perfectly fecure from fowlers, lie all day long, in the winter feafon, vaft flocks of ducks, teals, and wigeons, of various denominations; where they preen and folace, and reft themfelves, till towards fun-fet, when they iffue forth in little parties (for in their natural ftate they are all

birds

birds of the night) to feed in the brooks and meadows; returning again with the dawn of the morning. Had this lake an arm or two more, and were it planted round with thick covert (for now it is perfectly naked), it might make a valuable decoy.

Yet neither it's extent, nor the clearnefs of it's water, nor the refort of various and curious fowls, nor it's picturefque groups of cattle, can render this *meer* fo remarkable as the great quantity of coins that were found in it's bed about forty years ago. But, as fuch difcoveries more properly belong to the *antiquities* of this place, I fhall fupprefs all particulars for the prefent, till I enter profeffedly on my feries of letters refpecting the more remote hiftory of this village and diftrict.

LETTER IX.

TO THE SAME.

By way of fupplement, I fhall trouble you once more on this fubject, to inform you that *Wolmer*, with her fifter foreft *Ayles Holt*, alias *Alice Holt* q, as it is called in old records, is held by grant from the crown for a term of years.

The grantees that the author remembers are Brigadier-General *Emanuel Scroope Howe*, and his lady, *Ruperta*, who was a natural daughter of Prince *Rupert* by *Margaret Hughs*; a Mr. *Mordaunt*, of

---

q " In Rot. Inquifit. de ftatu foreft. in Scaccar. 36. Ed. 3. it is called *Aifholt*."

In the fame, " Tit. *Woolmer* and *Aifholt* Hantifc. Dominus Rex habet unam capel-
" lam in *haia* fuâ de Kingefle." " *Haia, fepes, fepimentum, parcus :* a Gall. *haie* and
" *haye*." Spelman's Gloffary.

the

the *Peterborough* family, who married a dowager Lady *Pembroke*; *Henry Bilson Legge* and lady; and now Lord *Stawel*, their son.

The lady of General *Howe* lived to an advanced age, long surviving her husband; and, at her death, left behind her many curious pieces of mechanism of her father's constructing, who was a distinguished mechanic and artist [r], as well as warrior; and, among the rest, a very complicated clock, lately in possession of Mr. *Elmer*, the celebrated game-painter at *Farnham*, in the county of *Surrey*.

Though these two forests are only parted by a narrow range of enclosures, yet no two soils can be more different: for *The Holt* consists of a strong loam, of a miry nature, carrying a good turf, and abounding with oaks that grow to be large timber; while *Wolmer* is nothing but a hungry, sandy, barren waste.

The former, being all in the parish of *Binsted*, is about two miles in extent from north to south, and near as much from east to west; and contains within it many woodlands and lawns, and the *great lodge* where the grantees reside; and a smaller lodge called *Goose-green*; and is abutted on by the parishes of *Kingsley*, *Frinsham*, *Farnham*, and *Bentley*; all of which have right of common.

One thing is remarkable; that, though *The Holt* has been of old well stocked with fallow-deer, unrestrained by any pales or fences more than a common hedge, yet they were never seen within the limits of *Wolmer*; nor were the red deer of *Wolmer* ever known to haunt the thickets or glades of *The Holt*.

At present the deer of *The Holt* are much thinned and reduced by the night-hunters, who perpetually harass them in spite of the

---

[r] This prince was the inventor of *mezzotinto*.

E

efforts

efforts of numerous keepers, and the severe penalties that have been put in force againſt them as often as they have been deteĉted, and rendered liable to the laſh of the law. Neither fines nor imprifonments can deter them : ſo impoſſible is it to extinguiſh the ſpirit of ſporting, which ſeems to be inherent in human nature.

General *Howe* turned out ſome *German* wild boars and ſows in his foreſts, to the great terror of the neighbourhood ; and, at one time, a wild bull or buffalo : but the country roſe upon them and deſtroyed them.

A very large fall of timber, conſiſting of about one thouſand oaks, has been cut this ſpring (viz. 1784) in *The Holt foreſt* ; one fifth of which, it is ſaid, belongs to the grantee, Lord *Stawel.* He lays claim alſo to the lop and top : but the poor of the pariſhes of *Binſted* and *Frinſham, Bentley* and *Kingſley,* aſſert that it belongs to them ; and, aſſembling in a riotous manner, have actually taken it all away. One man, who keeps a team, has carried home, for his ſhare, forty ſtacks of wood. Forty-five of theſe people his Lordſhip has ſerved with actions. Theſe trees, which were very ſound, and in high perfection, were *winter-cut,* viz. in *February* and *March,* before the bark would run. In old times *The Holt* was eſtimated to be eighteen miles, computed meaſure, from water-carriage, *viz.* from the town of *Chertſey,* on the *Thames* ; but now it is not half that diſtance, ſince the *Wey* is made navigable up to the town of *Godalming* in the county of *Surrey.*

LETTER X.

TO THE SAME.

August 4, 1767.

IT has been my misfortune never to have had any neighbours whose studies have led them towards the pursuit of natural knowledge: so that, for want of a companion to quicken my industry and sharpen my attention, I have made but slender progress in a kind of information to which I have been attached from my childhood.

As to *swallows (hirundines rusticæ)* being found in a torpid state during the winter in the isle of *Wight*, or any part of this country, I never heard any such account worth attending to. But a clergyman, of an inquisitive turn, assures me, that, when he was a great boy, some workmen, in pulling down the battlements of a church tower early in the spring, found two or three *swifts (hirundines apodes)* among the rubbish, which were, at first appearance, dead; but, on being carried toward the fire, revived. He told me that, out of his great care to preserve them, he put them in a paper-bag, and hung them by the kitchen fire, where they were suffocated.

Another intelligent person has informed me that, while he was a schoolboy at *Brighthelmstone*, in *Sussex*, a great fragment of the chalk-cliff fell down one stormy winter on the beach; and that many people found *swallows* among the rubbish: but, on my questioning him whether he saw any of those birds himself; to my no small disappointment, he answered me in the negative; but that others assured him they did.

E 2 Young

Young broods of *swallows* began to appear this year on *July* the eleventh, and young *martins (hirundines urbicæ)* were then fledged in their nests. Both species will breed again once. For I see by my *fauna* of last year, that young broods came forth so late as *September* the eighteenth. Are not these late hatchings more in favour of hiding than migration? Nay, some young martins remained in their nests last year so late as *September* the twenty-ninth; and yet they totally disappeared with us by the fifth of *October*.

How strange is it that the *swift*, which seems to live exactly the same life with the *swallow* and *house-martin*, should leave us before the middle of *August* invariably! while the latter stay often till the middle of *October*; and once I saw numbers of house-martins on the seventh of *November*. The martins and *red-wing fieldfares* were flying in sight together; an uncommon assemblage of summer and winter-birds!

A little yellow bird (it is either a species of the *alauda trivialis*, or rather perhaps of the *motacilla trochilus)* still continues to make a sibilous shivering noise in the tops of tall woods. The *stoparola* of *Ray* (for which we have as yet no name in these parts) is called, in your *Zoology*, the *fly-catcher*. There is one circumstance characteristic of this bird, which seems to have escaped observation, and that is, it takes it's stand on the top of some stake or post, from whence it springs forth on it's prey, catching a fly in the air, and hardly ever touching the ground, but returning still to the same stand for many times together.

I perceive there are more than one species of the *motacilla trochilus*: Mr. *Derham* supposes, in *Ray's Philof. Letters*, that he has discovered three. In these there is again an instance of some very common birds that have as yet no *English* name.

Mr.

Mr. *Stillingfleet* makes a question whether the *black-cap (motacilla atricapilla)* be a bird of passage or not: I think there is no doubt of it: for, in *April*, in the first fine weather, they come trooping, all at once, into these parts, but are never seen in the winter. They are delicate songsters.

Numbers of *snipes* breed every summer in some moory ground on the verge of this parish. It is very amusing to see the cock bird on wing at that time, and to hear his piping and humming notes.

I have had no opportunity yet of procuring any of those mice which I mentioned to you in town. The person that brought me the last says they are plenty in harvest, at which time I will take care to get more; and will endeavour to put the matter out of doubt, whether it be a non-descript species or not.

I suspect much there may be two species of water-rats. *Ray* says, and *Linnæus* after him, that the water-rat is web-footed behind. Now I have discovered a rat on the banks of our little stream that is not web-footed, and yet is an exellent swimmer and diver: it answers exactly to the *mus amphibius* of *Linnæus* (See *Syst. Nat.*) which he says " *natat in fossis & urinatur.*" I should be glad to procure one " *plantis palmatis.*" *Linnæus* seems to be in a puzzle about his *mus amphibius,* and to doubt whether it differs from his *mus terrestris* ; which if it be, as he allows, the " *mus agrestis capite grandi brachyuros*" of *Ray,* is widely different from the water-rat, both in size, make, and manner of life.

As to the *falco,* which I mentioned in town, I shall take the liberty to send it down to you into *Wales*; presuming on your candour, that you will excuse me if it should appear as familiar to you as it is strange to me. Though mutilated " *qualem dices . . .* " *antehac fuisse, tales cum sint reliquiæ!*"

It

It haunted a marshy piece of ground in quest of wild-ducks and snipes : but, when it was shot, had just knocked down a rook, which it was tearing in pieces. I cannot make it answer to any of our *English* hawks; neither could I find any like it at the curious exhibition of stuffed birds in *Spring-Gardens*. I found it nailed up at the end of a barn, which is the countryman's museum.

The parish I live in is a very abrupt, uneven country, full of hills and woods, and therefore full of birds.

LETTER XI.

TO THE SAME.

SELBORNE, September 9, 1767.

IT will not be without impatience that I shall wait for your thoughts with regard to the *falco*; as to it's weight, breadth, &c. I wish I had set them down at the time : but, to the best of my remembrance, it weighed two pounds and eight ounces, and measured, from wing to wing, thirty-eight inches. It's *cere* and feet were yellow, and the circle of it's eyelids a bright yellow. As it had been killed some days, and the eyes were sunk, I could make no good observation on the colour of the pupils and the *irides*.

The most unusual birds I ever observed in these parts were a pair of *hoopoes* (*upupa*), which came several years ago in the summer, and frequented an ornamented piece of ground, which joins to my garden, for some weeks. They used to march about in a stately manner, feeding in the walks, many times in the day; and seemed

disposed

difpofed to breed in my outlet; but were frighted and perfecuted by idle boys, who would never let them be at reft.

Three *grofs-beaks (loxia coccothrauftes)* appeared fome years ago in my fields, in the winter; one of which I fhot: fince that, now and then one is occafionally feen in the fame dead feafon.

A *crofs-bill (loxia curviroftra)* was killed laft year in this neighbourhood.

Our ftreams, which are fmall, and rife only at the end of the village, yield nothing but the *bull's-head* or *miller's-thumb (gobius fluviatilis capitatus)*, the *trout (trutta fluviatilis)*, the *eel (anguilla)*, the *lampern (lampætra parva et fluviatilis)*, and the *ftickle-back (pifciculus aculeatus)*.

We are twenty miles from the fea, and almoft as many from a great river, and therefore fee but little of fea-birds. As to wild fowls, we have a few teems of *ducks* bred in the moors where the fnipes breed; and multitudes of *widgeons* and *teals* in hard weather frequent our lakes in the foreft.

Having fome acquaintance with a tame *brown owl*, I find that it cafts up the fur of mice, and the feathers of birds in pellets; after the manner of hawks: when full, like a dog, it hides what it cannot eat.

The young of the barn-owl are not eafily raifed, as they want a conftant fupply of frefh mice: whereas the young of the brown owl will eat indifcriminately all that is brought; fnails, rats, kittens, puppies, magpies, and any kind of carrion or offal.

The houfe-martins have eggs ftill, and fquab-young. The laft fwift I obferved was about the twenty-firft of *Auguft*; it was a ftraggler.

Red-ftarts, *fly-catchers*, *white-throats*, and *reguli non criftati*, ftill appear; but I have feen no *black-caps* lately.

I forgot

I forgot to mention that I once faw, in *Chrift Church* college quadrangle in *Oxford*, on a very funny warm morning, a *houfe martin* flying about, and fettling on the parapets, fo late as the twentieth of *November*.

At prefent I know only two fpecies of *bats*, the common *vefpertilio murinus* and the *vefpertilio auribus*.

I was much entertained laft fummer with a tame bat, which would take flies out of a perfon's hand. If you gave it any thing to eat, it brought it's wings round before the mouth, hovering and hiding it's head in the manner of birds of prey when they feed. The adroitnefs it fhewed in fhearing off the wings of the flies, which were always rejeƈted, was worthy of obfervation, and pleafed me much. Infeƈts feemed to be moft acceptable, though it did not refufe raw flefh when offered: fo that the notion, that bats go down chimnies and gnaw men's bacon, feems no improbable ftory. While I amufed myfelf with this wonderful quadruped, I faw it feveral times confute the vulgar opinion, that bats when down on a flat furface cannot get on the wing again, by rifing with great eafe from the floor. It ran, I obferved, with more difpatch than I was aware of; but in a moft ridiculous and grotefque manner.

Bats drink on the wing, like fwallows, by fipping the furface, as they play over pools and ftreams. They love to frequent waters, not only for the fake of drinking, but on account of infeƈts, which are found over them in the greateft plenty. As I was going, fome years ago, pretty late, in a boat from *Richmond* to *Sunbury*, on a warm fummer's evening, I think I faw myriads of bats between the two places: the air fwarmed with them all along the *Thames*, fo that hundreds were in fight at a time.

I am, &c.

LETTER

LETTER XII.

TO THE SAME.

SIR,   November 4, 1767.

IT gave me no small satisfaction to hear that the *falco*[s] turned out an uncommon one. I must confess I should have been better pleased to have heard that I had sent you a bird that you had never seen before; but that, I find, would be a difficult task.

I have procured some of the mice mentioned in my former letters, a young one and a female with young, both of which I have preserved in brandy. From the colour, shape, size, and manner of nesting, I make no doubt but that the species is nondescript. They are much smaller, and more slender, than the *mus domesticus medius* of *Ray*; and have more of the squirrel or dormouse colour: their belly is white; a straight line along their sides divides the shades of their back and belly. They never enter into houses; are carried into ricks and barns with the sheaves; abound in harvest; and build their nests amidst the straws of the corn above the ground, and sometimes in thistles. They breed as many as eight at a litter, in a little round nest composed of the blades of grass or wheat.

One of these nests I procured this autumn, most artificially platted, and composed of the blades of wheat; perfectly round, and about the size of a cricket-ball; with the aperture so ingeniously closed, that there was no discovering to what part it

[s] This hawk proved to be the *falco peregrinus*; a variety.

F   belonged.

belonged. It was so compact and well filled, that it would roll across the table without being discomposed, though it contained eight little mice that were naked and blind. As this nest was perfectly full, how could the dam come at her litter respectively so as to administer a teat to each? perhaps she opens different places for that purpose, adjusting them again when the business is over: but she could not possibly be contained herself in the ball with her young, which moreover would be daily increasing in bulk. This wonderful procreant cradle, an elegant instance of the efforts of instinct, was found in a wheat-field suspended in the head of a thistle.

A gentleman, curious in birds, wrote me word that his servant had shot one last *January*, in that severe weather, which he believed would puzzle me. I called to see it this summer, not knowing what to expect : but, the moment I took it in hand, I pronounced it the male *garrulus bohemicus* or *German* silk-tail, from the five peculiar crimson tags or points which it carries at the ends of five of the short remiges. It cannot, I suppose, with any propriety, be called an *English* bird : and yet I see, by *Ray's Philosoph. Letters,* that great flocks of them, feeding on haws, appeared in this kingdom in the winter of 1685.

The mention of haws puts me in mind that there is a total failure of that wild fruit, so conducive to the support of many of the winged nation. For the same severe weather, late in the spring, which cut off all the produce of the more tender and curious trees, destroyed also that of the more hardy and common.

Some birds, haunting with the missel-thrushes, and feeding on the berries of the yew-tree, which answered to the description of the *merula torquata* or *ring-ouzel*, were lately seen in this neighbourhood.

bourhood. I employed some people to procure me a specimen, but without success. See Letter VIII.

Query—Might not *Canary* birds be naturalized to this climate, provided their eggs were put, in the spring, into the nests of some of their congeners, as goldfinches, greenfinches, &c.? Before winter perhaps they might be hardened, and able to shift for themselves.

About ten years ago I used to spend some weeks yearly at *Sunbury*, which is one of those pleasant villages lying on the *Thames*, near *Hampton-court*. In the autumn, I could not help being much amused with those myriads of the swallow kind which assemble in those parts. But what struck me most was, that, from the time they began to congregate, forsaking the chimnies and houses, they roosted every night in the osier-beds of the aits of that river. Now this resorting towards that element, at that season of the year, seems to give some countenance to the northern opinion (strange as it is) of their retiring under water. A *Swedish* naturalist is so much persuaded of that fact, that he talks, in his calendar of *Flora*, as familiarly of the swallow's going under water in the beginning of *September*, as he would of his poultry going to roost a little before sunset.

An observing gentleman in *London* writes me word that he saw an house-martin, on the twenty-third of last *October*, flying in and out of it's nest in the *Borough*. And I myself, on the twenty-ninth of last *October* (as I was travelling through *Oxford*), saw four or five swallows hovering round and settling on the roof of the county-hospital.

Now is it likely that these poor little birds (which perhaps had not been hatched but a few weeks) should, at that late season of the

year,

year, and from so midland a county, attempt a voyage to *Goree* or *Senegal*, almost as far as the *equator* [t]?

I acquiesce entirely in your opinion—that, though most of the swallow kind may migrate, yet that some do stay behind and hide with us during the winter.

As to the short-winged soft-billed birds, which come trooping in such numbers in the spring, I am at a loss even what to suspect about them. I watched them narrowly this year, and saw them abound till about *Michaelmas*, when they appeared no longer. Subsist they cannot openly among us, and yet elude the eyes of the inquisitive: and, as to their hiding, no man pretends to have found any of them in a torpid state in the winter. But with regard to their migration, what difficulties attend that supposition! that such feeble bad fliers (who the summer long never flit but from hedge to hedge) should be able to traverse vast seas and continents in order to enjoy milder seasons amidst the regions of *Africa*!

LETTER XIII.

TO THE SAME.

SIR,

SELBORNE, Jan. 22, 1768.

As in one of your former letters you expressed the more satisfaction from my correspondence on account of my living in the most southerly county; so now I may return the compliment, and expect to have my curiosity gratified by your living much more to the North.

[t] See *Adanson*'s Voyage to *Senegal*.

For

For many years paft I have obferved that towards *Chriftmas* vaft flocks of chaffinches have appeared in the fields; many more, I ufed to think, than could be hatched in any one neighbourhood. But, when I came to obferve them more narrowly, I was amazed to find that they feemed to me to be almoft all hens. I communicated my fufpicions to fome intelligent neighbours, who, after taking pains about the matter, declared that they alfo thought them all moftly females; at leaft fifty to one. This extraordinary occurrence brought to my mind the remark of *Linnæus*; that " before winter all their hen chaffinches migrate through *Holland* " into *Italy*." Now I want to know, from fome curious perfon in the north, whether there are any large flocks of thefe finches with them in the winter, and of which fex they moftly confift? For, from fuch intelligence, one might be able to judge whether our female flocks migrate from the other end of the ifland, or whether they come over to us from the continent.

We have, in the winter, vaft flocks of the common linnets; more, I think, than can be bred in any one diftrict. Thefe, I obferve, when the fpring advances, affemble on fome tree in the funfhine, and join all in a gentle fort of chirping, as if they were about to break up their winter quarters and betake themfelves to their proper fummer homes. It is well known, at leaft, that the fwallows and the fieldfares do congregate with a gentle twittering before they make their refpective departure.

You may depend on it that the bunting, *emberiza miliaria*, does not leave this county in the winter. In *January* 1767 I faw feveral dozen of them, in the midft of a fevere froft, among the bufhes on the downs near *Andover:* in our woodland enclofed diftrict it is a rare bird.

Wagtails,

Wagtails, both white and yellow, are with us all the winter. Quails crowd to our fouthern coaft, and are often killed in numbers by people that go on purpofe.

Mr. *Stillingfleet*, in his Tracts, fays that "if the wheatear *(ænanthe)* "does not quit *England*, it certainly fhifts places; for about "harveft they are not to be found, where there was before great "plenty of them." This well accounts for the vaft quantities that are caught about that time on the fouth downs near *Lewes*, where they are efteemed a delicacy. There have been fhepherds, I have been credibly informed, that have made many pounds in a feafon by catching them in traps. And though fuch multitudes are taken, I never faw (and I am well acquainted with thofe parts) above two or three at a time: for they are never gregarious. They may perhaps migrate in general; and, for that purpofe, draw towards the coaft of *Suffex* in autumn: but that they do not all withdraw I am fure; becaufe I fee a few ftragglers in many counties, at all times of the year, efpecially about warrens and ftone quarries.

I have no acquaintance, at prefent, among the gentlemen of the navy: but have written to a friend, who was a fea-chaplain in the late war, defiring him to look into his minutes, with refpect to birds that fettled on their rigging during their voyage up or down the channel. What *Haffelquift* fays on that fubject is remarkable. there were little fhort-winged birds frequently coming on board his fhip all the way from our channel quite up to the *Levant*, efpecially before fqually weather.

What you fuggeft, with regard to *Spain*, is highly probable. The winters of *Andalufia* are fo mild, that, in all likelihood, the foft-billed birds that leave us at that feafon may find infects fufficient to fupport them there.

Some

Some young man, poffeffed of fortune, health, and leifure, fhould make an autumnal voyage into that kingdom; and fhould fpend a year there, inveftigating the natural hiftory of that vaft country. Mr. *Willughby*[u] paffed through that kingdom on fuch an errand; but he feems to have fkirted along in a fuperficial manner and an ill humour, being much difgufted at the rude diffolute manners of the people.

I have no friend left now at *Sunbury* to apply to about the fwallows roofting on the aits of the *Thames*: nor can I hear any more about thofe birds which I fufpected were *merulæ torquatæ*.

As to the fmall mice, I have farther to remark, that though they hang their nefts for breeding up amidft the ftraws of the ftanding corn, above the ground; yet I find that, in the winter, they burrow deep in the earth, and make warm beds of grafs: but their grand rendezvous feems to be in corn-ricks, into which they are carried at harveft. A neighbour houfed an oat-rick lately, under the thatch of which were affembled near an hundred, moft of which were taken; and fome I faw. I meafured them; and found that, from nofe to tail, they were juft two inches and a quarter, and their tails juft two inches long. Two of them, in a fcale, weighed down juft one copper halfpenny, which is about the third of an ounce avoirdupois: fo that I fuppofe they are the fmalleft quadrupeds in this ifland. A full-grown *mus medius domefticus* weighs, I find, one ounce lumping weight, which is more than fix times as much as the moufe above; and meafures from nofe to rump four inches and a quarter, and the fame in it's tail. We have had a very fevere froft and deep fnow this month. My thermometer was one day fourteen degrees and

[u] See *Ray's* Travels, p. 466.

an half below the freezing point, within doors. The tender ever-greens were injured pretty much. It was very providential that the air was ſtill, and the ground well covered with ſnow, elſe vegetation in general muſt have ſuffered prodigiouſly. There is reaſon to believe that ſome days were more ſevere than any ſince the year 1739-40.

I am, &c. &c.

LETTER XIV.

TO THE SAME.

DEAR SIR,   SELBORNE, March 12, 1768.

IF ſome curious gentleman would procure the head of a fallow-deer, and have it diſſected, he would find it furniſhed with two ſpiracula, or breathing-places, beſides the noſtrils; probably analogous to the *puncta lachrymalia* in the human head. When deer are thirſty they plunge their noſes, like ſome horſes, very deep under water, while in the act of drinking, and continue them in that ſituation for a conſiderable time: but, to obviate any incon-veniency, they can open two vents, one at the inner corner of each eye, having a communication with the noſe. Here ſeems to be an extraordinary proviſion of nature worthy our attention; and which has not, that I know of, been noticed by any naturaliſt. For it looks as if theſe creatures would not be ſuffocated, though both their mouths and noſtrils were ſtopped. This curious form-ation of the head may be of ſingular ſervice to beaſts of chaſe, by

affording

affording them free refpiration : and no doubt thefe additional noftrils are thrown open when they are hard run[x]. Mr. *Ray* obferved that, at *Malta*, the owners flit up the noftrils of fuch affes as were hard worked : for they, being naturally ftrait or fmall, did not admit air fufficient to ferve them when they travelled, or laboured, in that hot climate. And we know that grooms, and gentlemen of the turf, think large noftrils neceffary, and a perfec- tion, in hunters and running horfes.

Oppian, the *Greek* poet, by the following line, feems to have had fome notion that ftags have four fpiracula :

" Τετραδυμοι ʹρινες, πισυρες πνοιησι διαυλοι."
" Quadrifidæ nares, quadruplices ad refpirationem canales."

Opp. Cyn. Lib. ii. l. 181.

Writers, copying from one another, make *Ariftotle* fay that goats breathe at their ears ; whereas he afferts juft the con- trary :— " Αλκμαιων γαρ ουκ αληθη λεγει, Φαμενος αναπνειν τας αιγας " κατα τα ωτα." " *Alcmæon* does not advance what is true, when " he avers that goats breathe through their ears."—Hiftory of Animals. Book I. chap. xi.

[x] In anfwer to this account, Mr. *Pennant* fent me the following curious and pertinent reply. " I was much furprifed to find in the *antelope* fomething analogous to what you " mention as fo remarkable in deer. This animal alfo has a long flit beneath each eye, " which can be opened and fhut at pleafure. On holding an orange to one, the creature " made as much ufe of thofe orifices as of his noftrils, applying them to the fruit, and " feeming to fmell it through them."

G   LETTER

LETTER XV.

TO THE SAME.

DEAR SIR,   SELBORNE, March 30, 1768.

SOME intelligent country people have a notion that we have, in these parts, a species of the *genus muſtelinum*, besides the weasel, stoat, ferret, and polecat; a little reddish beast, not much bigger than a field mouse, but much longer, which they call a *cane*. This piece of intelligence can be little depended on; but farther inquiry may be made.

A gentleman in this neighbourhood had two milkwhite rooks in one nest. A booby of a carter, finding them before they were able to fly, threw them down and destroyed them, to the regret of the owner, who would have been glad to have preserved such a curi-oſity in his rookery. I saw the birds myself nailed againſt the end of a barn, and was surprised to find that their bills, legs, feet, and claws were milkwhite.

A shepherd saw, as he thought, some white larks on a down above my house this winter: were not these the *emberiza nivalis*, the snow-flake of the *Brit. Zool.?* No doubt they were.

A few years ago I saw a cock bullfinch in a cage, which had been caught in the fields after it was come to it's full colours. In about a year it began to look dingy; and, blackening every succeeding year, it became coal-black at the end of four. It's chief food was hempseed. Such influence has food on the colour of animals! The pied and mottled colours of domeſticated ani-mals are supposed to be owing to high, various, and unusual food.

I had

I had remarked, for years, that the root of the cuckoo-pint (*arum*) was frequently scratched out of the dry banks of hedges, and eaten in severe snowy weather. After observing, with some exactness, myself, and getting others to do the same, we found it was the thrush kind that searched it out. The root of the *arum* is remarkably warm and pungent.

Our flocks of female chaffinches have not yet forsaken us. The blackbirds and thrushes are very much thinned down by that fierce weather in *January*.

In the middle of *February* I discovered, in my tall hedges, a little bird that raised my curiosity: it was of that yellow-green colour that belongs to the *salicaria* kind, and, I think, was soft-billed. It was no *parus*; and was too long and too big for the golden-crowned wren, appearing most like the largest willow-wren. It hung sometimes with it's back downwards, but never continuing one moment in the same place. I shot at it, but it was so desultory that I missed my aim.

I wonder that the stone curlew, *charadrius oedicnemus*, should be mentioned by the writers as a rare bird: it abounds in all the campaign parts of *Hampshire* and *Sussex*, and breeds, I think, all the summer, having young ones, I know, very late in the autumn. Already they begin clamouring in the evening. They cannot, I think, with any propriety, be called, as they are by Mr. *Ray*, " *circa aquas versantes*;" for with us, by day at least, they haunt only the most dry, open, upland fields and sheep walks, far removed from water : what they may do in the night I cannot say. Worms are their usual food, but they also eat toads and frogs.

I can shew you some good specimens of my *new mice*. *Linnæus* perhaps would call the species *mus minimus*.

LETTER

LETTER XVI.

TO THE SAME.

DEAR SIR,
SELBORNE, April 18, 1768.

THE hiftory of the ftone curlew, *charadrius oedicnemus*, is as follows. It lays it's eggs, ufually two, never more than three, on the bare ground, without any neft, in the field; fo that the countryman, in ftirring his fallows, often deftroys them. The young run immediately from the egg like partridges, &c. and are withdrawn to fome flinty field by the dam, where they fculk among the ftones, which are their beft fecurity; for their feathers are fo exactly of the colour of our grey fpotted flints, that the moft exact obferver, unlefs he catches the eye of the young bird, may be eluded. The eggs are fhort and round; of a dirty white, fpotted with dark bloody blotches. Though I might not be able, juft when I pleafed, to procure you a bird, yet I could fhew you them almoft any day; and any evening you may hear them round the village, for they make a clamour which may be heard a mile. *Oedicnemus* is a moft apt and expreffive name for them, fince their legs feem fwoln like thofe of a gouty man. After harveft I have fhot them before the pointers in turnip-fields.

I make no doubt but there are three fpecies of the *willow-wrens*: two I know perfectly; but have not been able yet to procure the third. No two birds can differ more in their notes, and that conftantly, than thofe two that I am acquainted with; for the one has a joyous, eafy, laughing note; the other a harfh loud chirp.

The

The former is every way larger, and three quarters of an inch longer, and weighs two drams and an half; while the latter weighs but two: so the fongfter is one fifth heavier than the chirper. The chirper (being the firft fummer-bird of paffage that is heard, the wryneck fometimes excepted) begins his two notes in the middle of *March*, and continues them through the fpring and fummer till the end of *Auguft*, as appears by my journals. The legs of the larger of thefe two are flefh-coloured; of the lefs, black.

The *grafshopper-lark* began his fibilous note in my fields laft *Saturday*. Nothing can be more amufing than the whifper of this little bird, which feems to be clofe by though at an hundred yards diftance; and, when clofe at your ear, is fcarce any louder than when a great way off. Had I not been a little acquainted with infects, and known that the grafshopper kind is not yet hatched, I fhould have hardly believed but that it had been a *locufta* whifpering in the bufhes. The country people laugh when you tell them that it is the note of a bird. It is a moft artful creature, fculking in the thickeft part of a bufh; and will fing at a yard diftance, provided it be concealed. I was obliged to get a perfon to go on the other fide of the hedge where it haunted; and then it would run, creeping like a moufe, before us for an hundred yards together, through the bottom of the thorns; yet it would not come into fair fight: but in a morning early, and when undifturbed, it fings on the top of a twig, gaping and fhivering with it's wings. Mr. *Ray* himfelf had no knowledge of this bird, but received his account from Mr. *Johnfon*, who apparently confounds it with the *reguli non criftati*, from which it is very diftinct. See *Ray's Philof. Letters*, p. 108.

The

The fly-catcher (*ſtoparola*) has not yet appeared: it uſually breeds in my vine. The *redſtart* begins to ſing: it's note is ſhort and imperfect, but is continued till about the middle of *June*. The *willow-wrens* (the ſmaller ſort) are horrid peſts in a garden, deſtroying the peaſe, cherries, currants, &c.; and are ſo tame that a gun will not ſcare them.

A LIST *of the* SUMMER BIRDS *of* PASSAGE *diſcovered in this neighbourhood, ranged ſomewhat in the Order in which they appear:*

	Linnæi Nomina.
Smalleſt willow-wren,	*Motacilla trochilus:*
Wryneck,	*Jynx torquilla:*
Houſe-ſwallow,	*Hirundo ruſtica:*
Martin,	*Hirundo urbica:*
Sand-martin,	*Hirundo riparia:*
Cuckoo,	*Cuculus canorus:*
Nightingale,	*Motacilla luſcinia:*
Blackcap,	*Motacilla atricapilla:*
Whitethroat,	*Motacilla ſylvia:*
Middle willow-wren,	*Motacilla trochilus:*
Swift,	*Hirundo apus:*
Stone curlew, ?	*Charadrius oedicnemus?*
Turtle-dove, ?	*Turtur aldrovandi?*
Graſshopper-lark,	*Alauda trivialis:*
Landrail,	*Rallus crex:*
Largeſt willow-wren,	*Motacilla trochilus:*
Redſtart,	*Motacilla phænicurus:*
Goatſucker, or fern-owl,	*Caprimulgus europæus:*
Fly-catcher,	*Muſcicapa griſola.*

My

My countrymen talk much of a bird that makes a clatter with it's bill againſt a dead bough, or ſome old pales, calling it a jar-bird. I procured one to be ſhot in the very fact; it proved to be the *ſitta europæa* (*the nuthatch.*) Mr. *Ray* ſays that the leſs ſpotted *woodpecker* does the ſame. This noiſe may be heard a furlong or more.

Now is the only time to aſcertain the ſhort-winged ſummer birds; for, when the leaf is out, there is no making any remarks on ſuch a reſtleſs tribe; and, when once the young begin to appear, it is all confuſion: there is no diſtinction of genus, ſpecies, or ſex.

In breeding-time ſnipes play over the moors, piping and hum-ming: they always hum as they are deſcending. Is not their hum ventriloquous like that of the turkey? Some ſuſpect it is made by their wings.

This morning I ſaw the golden-crowned wren, whoſe crown glitters like burniſhed gold. It often hangs like a titmouſe, with it's back downwards.

Yours, &c. &c.

LETTER XVII.

TO THE SAME.

DEAR SIR, SELBORNE, June 18, 1768.

ON *Wedneſday* laſt arrived your agreeable letter of *June* the 10th. It gives me great ſatisfaction to find that you purſue theſe ſtudies ſtill with ſuch vigour, and are in ſuch forwardneſs with regard to reptiles and fiſhes.

The

The reptiles, few as they are, I am not acquainted with, so well as I could wish, with regard to their natural history. There is a degree of dubiousness and obscurity attending the propagation of this class of animals, something analagous to that of the *cryptogamia* in the sexual system of plants: and the case is the same with regard to some of the fishes; as the eel, &c.

The method in which toads procreate and bring forth seems to be very much in the dark. Some authors say that they are viviparous: and yet *Ray* classes them among his oviparous animals; and is silent with regard to the manner of their bringing forth. Perhaps they may be ἔσω μὲν ὠοτόκοι, ἔξω δὲ ζωοτόκοι, as is known to be the case with the viper.

The copulation of frogs (or at least the appearance of it; for *Swammerdam* proves that the male has no *penis intrans*) is notorious to every body: because we see them sticking upon each others backs for a month together in the spring: and yet I never saw, or read, of toads being observed in the same situation. It is strange that the matter with regard to the venom of toads has not been yet settled. That they are not noxious to some animals is plain: for ducks, buzzards, owls, stone curlews, and snakes, eat them, to my knowledge, with impunity. And I well remember the time, but was not eye-witness to the fact (though numbers of persons were) when a quack, at this village, ate a toad to make the country-people stare; afterwards he drank oil.

I have been informed also, from undoubted authority, that some ladies (ladies you will say of peculiar taste) took a fancy to a toad, which they nourished summer after summer, for many years, till he grew to a monstrous size, with the maggots which turn to flesh flies. The reptile used to come forth every evening from an hole under the garden-steps; and was taken up, after supper, on the

table

table to be fed. But at laſt a tame raven, kenning him as he put forth his head, gave him ſuch a ſevere ſtroke with his horny beak as put out one eye. After this accident the creature languiſhed for ſome time and died.

I need not remind a gentleman of your extenſive reading of the excellent account there is from Mr. *Derham*, in *Ray*'s Wiſdom of God in the Creation (p. 365), concerning the migration of frogs from their breeding ponds. In this account he at once ſubverts that fooliſh opinion of their dropping from the clouds in rain; ſhewing that it is from the grateful coolneſs and moiſture of thoſe ſhowers that they are tempted to ſet out on their travels, which they defer till thoſe fall. Frogs are as yet in their tadpole ſtate; but, in a few weeks, our lanes, paths, fields, will ſwarm for a few days with myriads of thoſe emigrants, no larger than my little finger nail. *Swammerdam* gives a moſt accurate account of the method and ſituation in which the male impregnates the ſpawn of the female. How wonderful is the œconomy of Providence with regard to the limbs of ſo vile a reptile! While it is an *aquatic* it has a fiſh-like tail, and no legs: as ſoon as the legs ſprout, the tail drops off as uſeleſs, and the animal betakes itſelf to the land!

Merret, I truſt, is widely miſtaken when he advances that the *rana arborea* is an *Engliſh* reptile; it abounds in *Germany* and *Switzerland*.

It is to be remembered that the *ſalamandra aquatica* of *Ray* (the water-newt or eft) will frequently bite at the angler's bait, and is often caught on his hook. I uſed to take it for granted that the *ſalamandra aquatica* was hatched, lived, and died, in the water. But *John Ellis*, Eſq. F. R. S. (the coralline *Ellis*) aſſerts, in a letter to the Royal Society, dated *June* the 5th, 1766, in his ac-

H   count

count of the *mud inguana*, an amphibious *bipes* from *South Carolina*, that the water-eft, or newt, is only the *larva* of the land-eft, as tadpoles are of frogs. Left I fhould be fufpected to mifunder-ftand his meaning, I fhall give it in his own words. Speaking of the *opercula* or coverings to the gills of the *mud inguana*, he pro-ceeds to fay that " The form of thefe pennated coverings " approach very near to what I have fome time ago obferved in " the *larva* or *aquatic* ftate of our *Englifh lacerta*, known by the " name of eft, or newt; which ferve them for coverings to their " gills, and for fins to fwim with while in this ftate; and which " they lofe, as well as the fins of their tails, when they *change* their " ftate and *become land animals*, as I have obferved, by keeping " them alive for fome time myfelf."

Linnæus, in his *Syftema Naturæ*, hints at what Mr. *Ellis* advances more than once.

Providence has been fo indulgent to us as to allow of but one venomous reptile of the ferpent kind in thefe kingdoms, and that is the viper. As you propofe the good of mankind to be an object of your publications, you will not omit to mention common fallad-oil as a fovereign remedy againft the bite of the viper. As to the blind worm *(anguis fragilis*, fo called becaufe it fnaps in funder with a fmall blow), I have found, on examination, that it is per-fectly innocuous. A neighbouring yeoman (to whom I am indebted for fome good hints) killed and opened a female viper about the twenty-feventh of *May* : he found her filled with a chain of eleven eggs, about the fize of thofe of a blackbird; but none of them were advanced fo far towards a ftate of maturity as to contain any rudiments of young. Though they are oviparous, yet they are viviparous alfo, hatching their young within their bellies, and then bringing them forth. Whereas fnakes lay
chains

chains of eggs every summer in my melon beds, in spite of all that my people can do to prevent them; which eggs do not hatch till the spring following as I have often experienced. Several intelligent folks affure me that they have seen the viper open her mouth and admit her helplefs young down her throat on sudden surprifes, juft as the female opoffum does her brood into the pouch under her belly, upon the like emergencies; and yet the *London* viper-catchers infift on it, to Mr. *Barrington*, that no fuch thing ever happens. The ferpent kind eat, I believe, but once in a year; or, rather, but only juft at one feafon of the year. Country people talk much of a water-fnake, but, I am pretty fure, without any reafon; for the common fnake (*coluber natrix*) delights much to fport in the water, perhaps with a view to procure frogs and other food.

I cannot well guefs how you are to make out your twelve fpecies of reptiles, unlefs it be by the various fpecies, or rather varieties, of our *lacerti*, of which *Ray* enumerates five. I have not had opportunity of afcertaining thefe; but remember well to have feen, formerly, feveral beautiful green *lacerti* on the funny fandbanks near *Farnham*, in *Surrey*; and *Ray* admits there are fuch in *Ireland*.

LETTER

LETTER XVIII.

TO THE SAME.

DEAR SIR,   SELBORNE, July 27, 1768.

I RECEIVED your obliging and communicative letter of *June* the
28th, while I was on a visit at a gentleman's house, where I had
neither books to turn to, nor leisure to sit down, to return you an
answer to many queries, which I wanted to resolve in the best
manner that I am able.

A person, by my order, has searched our brooks, but could
find no such fish as the *gasterosteus pungitius*: he found the
gasterosteus aculeatus in plenty. This morning, in a basket, I
packed a little earthen pot full of wet moss, and in it some stickle-
backs, male and female; the females big with spawn: some
lamperns; some bulls heads; but I could procure no minnows.
This basket will be in *Fleet-street* by eight this evening; so I
hope *Mazel* will have them fresh and fair to-morrow morning. I
gave some directions, in a letter, to what particulars the engraver
should be attentive.

Finding, while I was on a visit, that I was within a reasonable
distance of *Ambresbury*, I sent a servant over to that town, and
procured several living specimens of loaches, which he brought,
safe and brisk, in a glass decanter. They were taken in the gul-
lies that were cut for watering the meadows. From these fishes
(which measured from two to four inches in length) I took the
following description: " The loach, in it's general aspect, has
" a pellucid appearance: it's back is mottled with irregular
" collections of small black dots, not reaching much below the

" *linea*

" *linea lateralis*, as are the back and tail fins: a black line runs
" from each eye down to the nofe; it's belly is of a filvery
" white; the upper jaw projects beyond the lower, and is fur-
" rounded with fix feelers, three on each fide: it's pectoral fins
" are large, it's ventral much fmaller; the fin behind it's anus
" fmall; it's dorfal-fin large, containing eight fpines; it's tail,
" where it joins to the tail-fin, *remarkably broad*, without any
" tapernefs, fo as to be characteriftic of this genus: the tail-fin
" is broad, and fquare at the end. From the breadth and muf-
" cular ftrength of the tail it appears to be an active nimble
" fish."

In my vifit I was not very far from *Hungerford*, and did not
forget to make fome inquiries concerning the wonderful method
of curing cancers by means of toads. Several intelligent perfons, both gentry and clergy, do, I find, give a great deal of
credit to what was afferted in the papers: and I myfelf dined
with a clergyman who feemed to be perfuaded that what is related
is matter of fact; but, when I came to attend to his account, I
thought I difcerned circumftances which did not a little invalidate
the woman's ftory of the manner in which fhe came by her fkill.
She fays of herfelf " that, labouring under a virulent cancer, fhe
" went to fome church where there was a vaft crowd: on going
" into a pew, fhe was accofted by a ftrange clergyman; who,
" after expreffing compaffion for her fituation, told her that if fhe
" would make fuch an application of living toads as is mentioned
" fhe would be well." Now is it likely that this unknown gentleman fhould exprefs fo much tendernefs for this fingle fufferer,
and not feel any for the many thoufands that daily languifh under
this terrible diforder? Would he not have made ufe of this
invaluable noftrum for his own emolument; or, at leaft, by

fome

some means of publication or other, have found a method of making it public for the good of mankind ? In short, this woman (as it appears to me) having set up for a cancer-doctress, finds it expedient to amuse the country with this dark and mysterious relation.

The water-eft has not, that I can discern, the least appearance of any gills; for want of which it is continually rising to the surface of the water to take in fresh air. I opened a big-bellied one indeed, and found it full of spawn. Not that this circumstance at all invalidates the assertion that they are *larvæ* : for the *larvæ* of insects are full of eggs, which they exclude the instant they enter their last state. The water-eft is continually climbing over the brims of the vessel, within which we keep it in water, and wandering away: and people every summer see numbers crawling, out of the pools where they are hatched, up the dry banks. There are varieties of them, differing in colour; and some have fins up their tail and back, and some have not.

LETTER XIX.

TO THE SAME.

DEAR SIR,
SELBORNE, Aug. 17, 1768.

I HAVE now, past dispute, made out three distinct species of the willow-wrens *(motacillæ trochili)* which *constantly* and *invariably* use distinct notes. But, at the same time, I am obliged to confess that I know nothing of your willow-lark [y]. In my letter

[y] *Brit. Zool.* edit. 1776, octavo, p. 381.

of

of *April* the 18th, I had told you peremptorily that I knew your willow-lark, but had not feen it then : but, when I came to procure it, it proved, in all refpects, a very *motacilla trochilus* ; only that it is a fize larger than the two other, and the yellow-green of the whole upper part of the body is more vivid, and the belly of a clearer white. I have fpecimens of the three forts now lying before me; and can difcern that there are three gradations of fizes, and that the leaft has black legs, and the other two flefh-coloured ones. The yelloweft bird is confiderably the largeft, and has it's quill-feathers and fecondary feathers tipped with white, which the others have not. This laft haunts only the tops of trees in high beechen woods, and makes a fibilous grafshopper-like noife, now and then, at fhort intervals, fhivering a little with it's wings when it fings ; and is, I make no doubt now, the *regulus non criftatus* of *Ray*; which he fays " *cantat voce ftridulâ locuftæ.*" Yet this great ornithologift never fufpected that there were three fpecies.

LETTER XX.

TO THE SAME.

SELBORNE, October 8, 1768.

IT is, I find, in *zoology* as it is in *botany:* all nature is fo full, that that diftrict produces the greateft variety which is the moft examined. Several birds, which are faid to belong to the north only, are, it feems, often in the fouth. I have difcovered this

fummer

fummer three fpecies of birds with us, which writers mention as only to be feen in the northern counties. The firft that was brought me (on the 14th of *May*), was the fandpiper, *tringa hypoleucus*: it was a cock bird, and haunted the banks of fome ponds near the village; and, as it had a companion, doubtlefs intended to have bred near that water. Befides, the owner has told me fince, that, on recollection, he has feen fome of the fame birds round his ponds in former fummers.

The next bird that I procured (on the 21ft of *May*) was a male red-backed butcher bird, *lanius collurio*. My neighbour, who fhot it, fays that it might eafily have efcaped his notice, had not the outcries and chattering of the white-throats and other fmall birds drawn his attention to the bufh where it was: it's craw was filled with the legs and wings of beetles.

The next rare birds (which were procured for me laft week) were fome ring-oufels, *turdi torquati*.

This week twelve months a gentleman from *London*, being with us, was amufing himfelf with a gun, and found, he told us, on an old yew hedge where there were berries, fome birds like black-birds, with rings of white round their necks: a neighbouring farmer alfo at the fame time obferved the fame; but, as no fpeci-mens were procured, little notice was taken. I mentioned this circumftance to you in my letter of *November* the 4th, 1767: (you however paid but fmall regard to what I faid, as I had not feen thefe birds myfelf): but laft week the aforefaid farmer, feeing a large flock, twenty or thirty of thefe birds, fhot two cocks and two hens: and fays, on recollection, that he remembers to have obferved thefe birds again laft fpring, about *Lady-day*, as it were, on their return to the north. Now perhaps thefe oufels are not the oufels of the north of *England*, but belong to the more northern parts of

Europe;

Europe; and may retire before the exceffive rigor of the frofts in thofe parts; and return to breed in the fpring, when the cold abates. If this be the cafe, here is difcovered a new bird of winter paffage, concerning whofe migrations the writers are filent: but if thefe birds fhould prove the oufels of the north of *England*, then here is a migration difclofed within our own kingdom never before remarked. It does not yet appear whether they retire beyond the bounds of our ifland to the fouth; but it is moft probable that they ufually do, or elfe one cannot fuppofe that they would have continued fo long unnoticed in the fouthern counties. The oufel is larger than a blackbird, and feeds on haws; but laft autumn (when there were no haws) it fed on yew-berries: in the fpring it feeds on ivy-berries, which ripen only at that feafon, in *March* and *April*.

I muft not omit to tell you (as you have been fo lately on the ftudy of reptiles) that my people, every now and then of late, draw up with a bucket of water from my well, which is 63 feet deep, a large black warty lizard with a fin-tail and yellow belly. How they firft came down at that depth, and how they were ever to have got out thence without help, is more than I am able to fay.

My thanks are due to you for your trouble and care in the examination of a buck's head. As far as your difcoveries reach at prefent, they feem much to corroborate my fufpicions; and I hope Mr. —— may find reafon to give his decifion in my favour; and then, I think, we may advance this extraordinary provifion of nature as a new inftance of the wifdom of God in the creation.

As yet I have not quite done with my hiftory of the *oedicnemus*, or ftone-curlew; for I fhall defire a gentleman in *Suffex* (near whofe houfe thefe birds congregate in vaft flocks in the autumn)

I

to obferve nicely when they leave him, (if they do leave him) and when they return again in the fpring: I was with this gentleman lately, and faw feveral fingle birds.

LETTER XXI.

TO THE SAME.

DEAR SIR,   SELBORNE, Nov. 28, 1768.

WITH regard to the *oedicnemus*, or ftone-curlew, I intend to write very foon to my friend near *Chicheſter*, in whofe neighbourhood thefe birds feem moft to abound; and ſhall urge him to take particular notice when they begin to congregate, and afterwards to watch them moft narrowly whether they do not withdraw themfelves during the dead of the winter. When I have obtained information with refpect to this circumftance, I ſhall have finiſhed my hiftory of the *ftone-curlew*; which I hope will prove to your fatiffaction, as it will be, I truft, very near the truth. This gentleman, as he occupies a large farm of his own, and is abroad early and late, will be a very proper fpy upon the motions of thefe birds: and befides, as I have prevailed on him to buy the Naturalift's Journal (with which he is much delighted), I ſhall expect that he will be very exact in his dates. It is very extraordinary, as you obferve, that a bird fo common with us ſhould never ſtraggle to you.

And here will be the propereft place to mention, while I think of it, an anecdote which the above-mentioned gentleman told me

when

when I was laſt at his houſe; which was that, in a warren joining to his outlet, many daws *(corvi monedulæ)* build every year in the rabbit-burrows under ground. The way he and his brothers uſed to take their neſts, while they were boys, was by liſtening at the mouths of the holes; and, if they heard the young ones cry, they twiſted the neſt out with a forked ſtick. Some water-fowls (*viz.* the puffins) breed, I know, in that manner; but I ſhould never have ſuſpected the daws of building in holes on the flat ground.

Another very unlikely ſpot is made uſe of by daws as a place to breed in, and that is *Stonehenge.* Theſe birds depoſit their neſts in the interſtices between the upright and the impoſt ſtones of that amazing work of antiquity: which circumſtance alone ſpeaks the prodigious height of the upright ſtones, that they ſhould be tall enough to ſecure thoſe neſts from the annoyance of ſhepherd-boys, who are always idling round that place.

One of my neighbours laſt *Saturday, November* the 26th, ſaw a martin in a ſheltered bottom: the ſun ſhone warm, and the bird was hawking briſkly after flies. I am now perfectly ſatisfied that they do not all leave this iſland in the winter.

You judge very right, I think, in ſpeaking with reſerve and caution concerning the cures done by toads: for, let people advance what they will on ſuch ſubjects, yet there is ſuch a propenſity in mankind towards deceiving and being deceived, that one cannot ſafely relate any thing from common report, eſpecially in print, without expreſſing ſome degree of doubt and ſuſpicion.

Your approbation, with regard to my new diſcovery of the migration of the ring-ouſel, gives me ſatisfaction; and I find you concur with me in ſuſpecting that they are foreign birds which viſit us. You will be ſure, I hope, not to omit to make inquiry whether your ring-ouſels leave your rocks in the autumn. What

puzzles

puzzles me moſt, is the very ſhort ſtay they make with us; for in about three weeks they are all gone. I ſhall be very curious to remark whether they will call on us at their return in the ſpring, as they did laſt year.

I want to be better informed with regard to icthyology. If fortune had ſettled me near the ſea-ſide, or near ſome great river, my natural propenſity would ſoon have urged me to have made myſelf acquainted with their productions: but as I have lived moſtly in inland parts, and in an upland diſtrict, my knowledge of fiſhes extends little farther than to thoſe common ſorts which our brooks and lakes produce.

I am, &c.

LETTER XXII.

TO THE SAME.

DEAR SIR,   SELBORNE, Jan. 2, 1769.

As to the peculiarity of jackdaws building with us under the ground in rabbit-burrows, you have, in part, hit upon the reaſon; for, in reality, there are hardly any towers or ſteeples in all this country. And perhaps, *Norfolk* excepted, *Hampſhire* and *Suſſex* are as meanly furniſhed with churches as almoſt any counties in the kingdom. We have many livings of two or three hundred pounds a year, whoſe houſes of worſhip make little better appearance than dovecots. When I firſt ſaw *Northamptonſhire, Cambridgeſhire*

and

and *Huntingdonshire*, and the fens of *Lincolnshire*, I was amazed at the number of spires which presented themselves in every point of view. As an admirer of prospects, I have reason to lament this want in my own country; for such objects are very necessary ingredients in an elegant landscape.

What you mention with respect to reclaimed toads raises my curiosity. An ancient author, though no naturalist, has well remarked that " *Every kind of beasts, and of birds, and of serpents, and* " *things in the sea, is tamed, and hath been tamed, of mankind*[z]."

It is a satisfaction to me to find that a green lizard has actually been procured for you in *Devonshire*; because it corroborates my discovery, which I made many years ago, of the same sort, on a sunny sandbank near *Farnham*, in *Surrey*. I am well acquainted with the south hams of *Devonshire*; and can suppose that district, from it's southerly situation, to be a proper habitation for such animals in their best colours.

Since the ring-ousels of your vast mountains do certainly not forsake them against winter, our suspicions that those which visit this neighbourhood about *Michaelmas* are not *English* birds, but driven from the more northern parts of *Europe* by the frosts, are still more reasonable; and it will be worth your pains to endeavour to trace from whence they come, and to inquire why they make so very short a stay.

In your account of your error with regard to the two species of herons, you incidentally gave me great entertainment in your description of the heronry at *Cressi-hall*; which is a curiosity I never could manage to see. Fourscore nests of such a bird on one tree is a rarity which I would ride half as many miles to have a

---

* *James*, chap. iii. 7.

fight

fight of. Pray be fure to tell me in your next whofe feat *Creffi-hall* is, and near what town it lies[a]. I have often thought that thofe vaft extents of fens have never been fufficiently explored. If half a dozen gentlemen, furnifhed with a good ftrength of water-fpaniels, were to beat them over for a week, they would certainly find more fpecies.

There is no bird, I believe, whofe manners I have ftudied more than that of the *caprimulgus* (the goat-fucker), as it is a wonderful and curious creature: but I have always found that though fometimes it may chatter as it flies, as I know it does, yet in general it utters it's jarring note fitting on a bough; and I have for many an half hour watched it as it fat with it's under mandible quivering, and particularly this fummer. It perches ufually on a bare twig, with it's head lower than it's tail, in an attitude well expreffed by your draughtfman in the folio *Britifh Zoology*. This bird is moft punctual in beginning it's fong exactly at the clofe of day; fo exactly that I have known it ftrike up more than once or twice juft at the report of the *Portfmouth* evening gun, which we can hear when the weather is ftill. It appears to me paft all doubt that it's notes are formed by organic impulfe, by the powers of the parts of it's windpipe, formed for found, juft as cats pur. You will credit me, I hope, when I affure you that, as my neighbours were affembled in an[b] hermitage on the fide of a fteep hill where we drink tea, one of thefe churn-owls came and fettled on the crofs of that little ftraw edifice and began to chatter, and continued his note for many minutes: and we were all ftruck with wonder to find that the organs of that little animal, when put in motion, gave a fenfible vibration to the whole building! This

[a] *Creffi-hall* is near *Spalding*, in *Lincolnfhire*.
[b] See the vignette in this book.

bird

bird alfo fometimes makes a fmall fqueak, repeated four or five times; and I have obferved that to happen when the cock has been purfuing the hen in a toying way through the boughs of a tree.

It would not be at all ftrange if your bat, which you have procured, fhould prove a new one, fince five fpecies have been found in a neighbouring kingdom. The great fort that I mentioned is certainly a non-defcript: I faw but one this fummer, and that I had no opportunity of taking.

Your account of the *Indian-grafs* was entertaining. I am no angler myfelf; but inquiring of thofe that are, what they fuppofed that part of their tackle to be made of? they replied " of the in-
" teftines of a filkworm."

Though I muft not pretend to great fkill in entomology, yet I cannot fay that I am ignorant of that kind of knowledge: I may now and then perhaps be able to furnifh you with a little information.

The vaft rains ceafed with us much about the fame time as with you, and fince we have had delicate weather. Mr. *Barker*, who has meafured the rain for more than thirty years, fays, in a late letter, that more has fallen this year than in any he ever attended to; though, from *July* 1763 to *January* 1764, more fell than in any feven months of this year.

LETTER

LETTER XXIII.

TO THE SAME.

DEAR SIR,   SELBORNE, February 28, 1769.

It is not improbable that the *Guernsey* lizard and our green lizards may be specifically the same; all that I know is, that, when some years ago many *Guernsey* lizards were turned loose in *Pembroke* college garden, in the university of *Oxford*, they lived a great while, and seemed to enjoy themselves very well, but never bred. Whether this circumstance will prove any thing either way I shall not pretend to say.

I return you thanks for your account of *Cressi-hall*; but recollect, not without regret, that in *June* 1746 I was visiting for a week together at *Spalding*, without ever being told that such a curiosity was just at hand. Pray send me word in your next what sort of tree it is that contains such a quantity of herons' nests; and whether the heronry consists of a whole grove or wood, or only of a few trees.

It gave me satisfaction to find we accorded so well about the *caprimulgus*: all I contended for was to prove that it often chatters sitting as well as flying; and therefore the noise was voluntary, and from organic impulse, and not from the resistance of the air against the hollow of it's mouth and throat.

If ever I saw any thing like actual migration, it was last *Michaelmas-day*. I was travelling, and out early in the morning: at first there was a vast fog; but, by the time that I was got seven or eight miles from home towards the coast, the sun broke out into a

delicate

delicate warm day. We were then on a large heath or common, and I could difcern, as the mift began to break away, great numbers of fwallows (*hirundines ruſticæ*) cluſtering on the ſtunted ſhrubs and buſhes, as if they had rooſted there all night. As ſoon as the air became clear and pleaſant they all were on the wing at once; and, by a placid and eaſy flight, proceeded on ſouthward towards the ſea: after this I did not ſee any more flocks, only now and then a ſtraggler.

I cannot agree with thoſe perſons that aſſert that the ſwallow kind diſappear ſome and ſome gradually, as they come, for the bulk of them ſeem to withdraw at once: only ſome ſtragglers ſtay behind a long while, and do never, there is the greateſt reaſon to believe, leave this iſland. Swallows ſeem to lay themſelves up, and to come forth in a warm day, as bats do continually of a warm evening, after they have diſappeared for weeks. For a very reſpectable gentleman aſſured me that, as he was walking with ſome friends under *Merton-wall* on a remarkably hot noon, either in the laſt week in *December* or the firſt week in *January*, he eſpied three or four ſwallows huddled together on the moulding of one of the windows of that college. I have frequently remarked that ſwallows are ſeen later at *Oxford* than elſewhere: is it owing to the vaſt maſſy buildings of that place, to the many waters round it, or to what elſe?

When I uſed to riſe in a morning laſt autumn, and ſee the ſwallows and martins cluſtering on the chimnies and thatch of the neighbouring cottages, I could not help being touched with a ſecret delight, mixed with ſome degree of mortification: with delight, to obſerve with how much ardour and punctuality thoſe poor little birds obeyed the ſtrong impulſe towards migration, or hiding, imprinted on their minds by their great Creator; and with ſome degree of mortification, when I reflected that, after all our

K

pains

pains and inquiries, we are yet not quite certain to what regions they do migrate; and are still farther embarassed to find that some do not actually migrate at all.

These reflections made so strong an impression on my imagination, that they became productive of a composition that may perhaps amuse you for a quarter of an hour when next I have the honour of writing to you.

LETTER XXIV.

TO THE SAME.

DEAR SIR,   SELBORNE, May 29, 1769.

THE *scarabæus fullo* I know very well, having seen it in collections; but have never been able to discover one wild in it's natural state. Mr. *Banks* told me he thought it might be found on the seacoast.

On the thirteenth of *April* I went to the sheep-down, where the *ring-ousels* have been observed to make their appearance at spring and fall, in their way perhaps to the north or south; and was much pleased to see three birds about the usual spot. We shot a cock and a hen; they were plump and in high condition. The hen had but very small rudiments of eggs within her, which proves they are late breeders; whereas those species of the thrush kind that remain with us the whole year have fledged young before that time. In their crops was nothing very distinguishable, but somewhat that seemed like blades of vegetables nearly digested. In

autumn

autumn they feed on haws and yew-berries, and in the spring on ivy-berries. I dreſſed one of theſe birds, and found it juicy and well flavoured. It is remarkable that they make but a few days ſtay in their ſpring viſit, but reſt near a fortnight at *Michaelmas*. Theſe birds, from the obſervations of three ſprings and two autumns, are moſt punctual in their return; and exhibit a new migration unnoticed by the writers, who ſuppoſed they never were to be ſeen in any of the ſouthern counties.

One of my neighbours lately brought me a new *ſalicaria*, which at firſt I ſuſpected might have proved your willow-lark[c], but, on a nicer examination, it anſwered much better to the deſcription of that ſpecies which you ſhot at *Reveſby*, in *Lincolnſhire*. My bird I deſcribe thus: " It is a ſize leſs than the graſshopper-" lark; the head, back, and coverts of the wings, of a duſky " brown, without thoſe dark ſpots of the graſshopper-lark; over " each eye is a milkwhite ſtroke; the chin and throat are white, " and the under parts of a yellowiſh white; the rump is tawny, " and the feathers of the tail ſharp-pointed; the bill is duſky and " ſharp, and the legs are duſky; the hinder claw long and " crooked." The perſon that ſhot it ſays that it ſung ſo like a reed-ſparrow that he took it for one; and that it ſings all night: but this account merits farther inquiry. For my part, I ſuſpect it is a ſecond ſort of *locuſtella*, hinted at by Dr. *Derham* in *Ray's Letters:* ſee p. 108. He alſo procured me a graſshopper-lark.

The queſtion that you put with regard to thoſe genera of animals that are peculiar to *America*, viz. how they came there, and whence? is too puzzling for me to anſwer; and yet ſo obvious as often to have ſtruck me with wonder. If one looks into the writers

For this *ſalicaria* ſee letter *Auguſt* 30, 1769.

on

on that fubject little fatisfaction is to be found. Ingenious men will readily advance plaufible arguments to fupport whatever theory they fhall chufe to maintain; but then the misfortune is, every one's hypothefis is each as good as another's, fince they are all founded on conjecture. The late writers of this fort, in whom may be feen all the arguments of thofe that have gone before, as I remember, ftock *America* from the weftern coaft of *Africa* and the fouth of *Europe*; and then break down the Ifthmus that bridged over the *Atlantic*. But this is making ufe of a violent piece of machinery: it is a difficulty worthy of the interpofition of a god! " *Incredulus odi*".

———————

TO THOMAS PENNANT, ESQUIRE.

THE NATURALIST's SUMMER-EVENING WALK.

——— equidem credo, quia fit divinitus illis
Ingenium.   VIRG. GEORG.

WHEN day declining fheds a milder gleam,
What time the may-fly[d] haunts the pool or ftream;
When the ftill owl fkims round the graffy mead,
What time the timorous hare limps forth to feed;

[d] The angler's may-fly, the *ephemera vulgata Linn.* comes forth from it's aurelia ftate, and emerges out of the water about fix in the evening, and dies about eleven at night, determining the date of it's fly ftate in about five or fix hours. They ufually begin to appear about the 4th of *June*, and continue in fucceffion for near a fortnight. See *Swammerdam, Derham, Scopoli, &c.*

                                                        Then

Then be the time to steal adown the vale,
And listen to the vagrant^e cuckoo's tale;
To hear the clamorous^f curlew call his mate,
Or the soft quail his tender pain relate;
To see the swallow sweep the dark'ning plain
Belated, to support her infant train;
To mark the swift in rapid giddy ring
Dash round the steeple, unsubdu'd of wing:
Amusive birds!—say where your hid retreat
When the frost rages and the tempests beat;
Whence your return, by such nice instinct led,
When spring, soft season, lifts her bloomy head?
Such baffled searches mock man's prying pride,
The GOD of NATURE is your secret guide!

   While deep'ning shades obscure the face of day
To yonder bench leaf-shelter'd let us stray,
'Till blended objects fail the swimming sight,
And all the fading landscape sinks in night;
To hear the drowsy dor come brushing by
With buzzing wing, or the shrill^g cricket cry;
To see the feeding bat glance through the wood;
To catch the distant falling of the flood;
While o'er the cliff th' awaken'd churn-owl hung
Through the still gloom protracts his chattering song;
While high in air, and pois'd upon his wings,
Unseen, the soft enamour'd^h woodlark sings:

---

   e Vagrant cuckoo; so called because, being tied down by no incubation or attend-
ance about the nutrition of it's young, it wanders without control.

   f *Charadrius oedicnemus.*   g *Gryllus campestris.*

   h In hot summer nights woodlarks soar to a prodigious height and hang singing in
the air.

These,

Thefe, NATURE's works, the curious mind employ,
Infpire a foothing melancholy joy:
As fancy warms, a pleafing kind of pain
Steals o'er the cheek, and thrills the creeping vein!

   Each rural fight, each found, each fmell, combine;
The tinkling fheep-bell, or the breath of kine;
The new-mown hay that fcents the fwelling breeze,
Or cottage-chimney fmoking through the trees.

   The chilling night-dews fall:—away, retire;
For fee, the glow-worm lights her amorous fire[i]!
Thus, e'er night's veil had half obfcur'd the fky,
Th' impatient damfel hung her lamp on high:
True to the fignal, by love's meteor led,
Leander haften'd to his Hero's bed[k].

I am, &c.

LETTER XXV.

TO THE SAME.

DEAR SIR,   SELBORNE, Aug. 30, 1769.

IT gives me fatisfaction to find that my account of the *oufel migration* pleafes you. You put a very fhrewd queftion when you afk me how I know that their autumnal migration is fouthward?

---

[i] The light of the female glow-worm (as fhe often crawls up the ftalk of a grafs to make herfelf more confpicuous) is a fignal to the male, which is a flender dufky *fcarabæus*.

[k] See the ftory of *Hero* and *Leander*.

Was

Was not candour and openness the very life of natural history, I should pass over this query just as a sly commentator does over a crabbed passage in a classic; but common ingenuousness obliges me to confess, not without some degree of shame, that I only reasoned in that case from analogy. For as all other autumnal birds migrate from the northward to us, to partake of our milder winters, and return to the northward again when the rigorous cold abates, so I concluded that the ring-ousels did the same, as well as their congeners the fieldfares; and especially as ring-ousels are known to haunt cold mountainous countries: but I have good reason to suspect since that they may come to us from the westward; because I hear, from very good authority, that they breed on *Dartmore*; and that they forsake that wild district about the time that our visitors appear, and do not return till late in the spring.

I have taken a great deal of pains about your *falicaria* and mine, with a white stroke over it's eye and a tawny rump. I have surveyed it alive and dead, and have procured several specimens; and am perfectly persuaded myself (and trust you will soon be convinced of the same) that it is no more nor less than the *passer arundinaceus minor* of *Ray*. This bird, by some means or other, seems to be entirely omitted in the *British Zoology*; and one reason probably was because it is so strangely classed in *Ray*, who ranges it among his *picis affines*. It ought no doubt to have gone among his *aviculæ caudâ unicolore*, and among your slender-billed small birds of the same division. *Linnæus* might with great propriety have put it into his genus of *motacilla*; and the *motacilla falicaria* of his *fauna suecica* seems to come the nearest to it. It is no uncommon bird, haunting the sides of ponds and rivers where there is covert, and the reeds and sedges of moors. The country people in some places call it the *sedge-bird*. It sings incessantly night and

day

day during the breeding-time, imitating the note of a sparrow, a swallow, a sky-lark; and has a strange hurrying manner in it's song. My specimens correspond most minutely to the description of your *fen salicaria* shot near *Revesby*. Mr. *Ray* has given an excellent characteristic of it when he says, " *Rostrum & pedes in* " *hâc aviculâ multò majores sunt quâm pro corporis ratione.*" See letter *May 29,* 1769.

I have got you the egg of an *oedicnemus,* or stone-curlew, which was picked up in a fallow on the naked ground: there were two; but the finder inadvertently crushed one with his foot before he saw them.

When I wrote to you last year on reptiles, I wish I had not forgot to mention the faculty that snakes have of stinking *se defendendo.* I knew a gentleman who kept a tame snake, which was in it's person as sweet as any animal while in good humour and unalarmed; but as soon as a stranger, or a dog or cat, came in, it fell to hissing, and filled the room with such nauseous effluvia as rendered it hardly supportable. Thus the squnck, or stonck, of *Ray's Synop. Quadr.* is an innocuous and sweet animal; but, when pressed hard by dogs and men, it can eject such a most pestilent and fetid smell and excrement, that nothing can be more horrible.

A gentleman sent me lately a fine specimen of the *lanius minor cinerascens cum maculâ in scapulis albâ, Raii*; which is a bird that, at the time of your publishing your two first volumes of *British Zoology,* I find you had not seen. You have described it well from *Edwards's* drawing.

LETTER

LETTER XXVI.

TO THE SAME.

DEAR SIR,   SELBORNE, December 8, 1769.

I was much gratified by your communicative letter on your return from *Scotland*, where you spent, I find, some considerable time, and gave yourself good room to examine the natural curiosities of that extensive kingdom, both those of the islands, as well as those of the highlands. The usual bane of such expeditions is hurry; because men seldom allot themselves half the time they should do : but, fixing on a day for their return, post from place to place, rather as if they were on a journey that required dispatch, than as philosophers investigating the works of nature. You must have made, no doubt, many discoveries, and laid up a good fund of materials for a future edition of the *British Zoology*; and will have no reason to repent that you have bestowed so much pains on a part of *Great-Britian* that perhaps was never so well examined before.

It has always been matter of wonder to me that fieldfares, which are so congenerous to thrushes and blackbirds, should never chuse to breed in *England*: but that they should not think even the highlands cold and northerly, and sequestered enough, is a circumstance still more strange and wonderful. The ring-ousel, you find, stays in *Scotland* the whole year round; so that we have reason to conclude that those migrators that visit us for a short space every autumn do not come from thence.

L   And

And here, I think, will be the proper place to mention that those birds were moſt punctual again in their migration this autumn, appearing, as before, about the 30th of *September:* but their flocks were larger than common, and their ſtay protracted ſomewhat beyond the uſual time. If they came to ſpend the whole winter with us, as ſome of their congeners do, and then left us, as they do, in ſpring, I ſhould not be ſo much ſtruck with the occurrence, ſince it would be ſimilar to that of the other winter birds of paſſage; but when I ſee them for a fortnight at *Michaelmas,* and again for about a week in the middle of *April,* I am ſeized with wonder, and long to be informed whence theſe travellers come, and whither they go, ſince they ſeem to uſe our hills merely as an inn or baiting place.

Your account of the greater brambling, or ſnow-fleck, is very amuſing; and ſtrange it is that ſuch a ſhort-winged bird ſhould delight in ſuch perilous voyages over the northern ocean! Some country people in the winter time have every now and then told me that they have ſeen two or three white larks on our downs; but, on conſidering the matter, I begin to ſuſpect that theſe are ſome ſtragglers of the birds we are talking of, which ſometimes perhaps may rove ſo far to the ſouthward.

It pleaſes me to find that white hares are ſo frequent on the *Scottiſh* mountains, and eſpecially as you inform me that it is a diſtinct ſpecies; for the quadrupeds of *Britain* are ſo few, that every new ſpecies is a great acquiſition.

The eagle-owl, could it be proved to belong to us, is ſo majeſtic a bird, that it would grace our *fauna* much. I never was informed before where wild-geeſe are known to breed.

You admit, I find, that I have proved your *fen-ſalicaria* to be the leſſer reed-ſparrow of *Ray:* and I think you may be ſecure

that

that I am right; for I took very particular pains to clear up that matter, and had some fair specimens; but, as they were not well preserved, they are decayed already. You will, no doubt, insert it in it's proper place in your next edition. Your additional plates will much improve your work.

De Buffon, I know, has described the water shrew-mouse: but still I am pleased to find you have discovered it in *Lincolnshire*, for the reason I have given in the article of the white hare.

As a neighbour was lately plowing in a dry chalky field, far removed from any water, he turned out a water-rat, that was curiously laid up in an *hybernaculum* artificially formed of grass and leaves. At one end of the *burrow* lay above a gallon of potatoes regularly stowed, on which it was to have supported itself for the winter. But the difficulty with me is how this *amphibius mus* came to fix it's winter station at such a distance from the water. Was it determined in it's choice of that place by the mere accident of finding the potatoes which were planted there; or is it the constant practice of the aquatic-rat to forsake the neighbourhood of the water in the colder months?

Though I delight very little in analogous reasoning, knowing how fallacious it is with respect to natural history; yet, in the following instance, I cannot help being inclined to think it may conduce towards the explanation of a difficulty that I have mentioned before, with respect to the invariable early retreat of the *hirundo apus*, or swift, so many weeks before it's congeners; and that not only with us, but also in *Andalusia*, where they also begin to retire about the beginning of *August*.

The great large bat[1] (which by the by is at present a non-

---

[1] The little bat appears almost every month in the year; but I have never seen the large ones till the end of *April*, nor after *July*. They are most common in *June*, but never in any plenty: are a rare species with us.

descript

descript in *England,* and what I have never been able yet to procure) retires or migrates very early in the summer: it also ranges very high for it's food, feeding in a different region of the air; and that is the reason I never could procure one. Now this is exactly the case with the swifts; for they take their food in a more exalted region than the other species, and are very seldom seen hawking for flies near the ground, or over the surface of the water. From hence I would conclude that these *hirundines,* and the larger bats, are supported by some sorts of high-flying gnats, scarabs, or *phalænæ,* that are of short continuance; and that the short stay of these strangers is regulated by the defect of their food.

By my journal it appears that curlews clamoured on to *October* the thirty-first; since which I have not seen or heard any. Swallows were observed on to *November* the third.

LETTER XXVII.

TO THE SAME.

DEAR SIR,   SELBORNE, Feb. 22, 1770.

Hedge-hogs abound in my gardens and fields. The manner in which they eat their roots of the plantain in my grass-walks is very curious: with their upper mandible, which is much longer than their lower, they bore under the plant, and so eat the root off upwards, leaving the tuft of leaves untouched. In this respect they

are

are ferviceable, as they deſtroy a very troubleſome weed; but they deface the walks in ſome meaſure by digging little round holes. It appears, by the dung that they drop upon the turf, that beetles are no inconſiderable part of their food. In *June* laſt I procured a litter of four or five young hedge-hogs, which appeared to be about five or ſix days old: they, I find, like puppies, are born blind, and could not ſee when they came to my hands. No doubt their ſpines are ſoft and flexible at the time of their birth, or elſe the poor dam would have but a bad time of it in the critical moment of parturition: but it is plain that they ſoon harden; for theſe little pigs had ſuch ſtiff prickles on their backs and ſides as would eaſily have fetched blood, had they not been handled with caution. Their ſpines are quite white at this age; and they have little hanging ears, which I do not remember to be diſcernible in the old ones. They can, in part, at this age draw their ſkin down over their faces; but are not able to contract themſelves into a ball, as they do, for the ſake of defence, when full grown. The reaſon, I ſuppoſe, is, becauſe the curious muſcle that enables the creature to roll itſelf up in a ball was not then arrived at it's full tone and firmneſs. Hedge-hogs make a deep and warm *hybernaculum* with leaves and moſs, in which they conceal themſelves for the winter: but I never could find that they ſtored in any winter proviſion, as ſome quadrupeds certainly do.

I have diſcovered an anecdote with reſpect to the fieldfare *(turdus pilaris)*, which I think is particular enough: this bird, though it ſits on trees in the day-time, and procures the greateſt part of it's food from white-thorn hedges; yea, moreover, builds on very high trees; as may be ſeen by the *fauna ſuecica*; yet always appears with us to rooſt on the ground. They are ſeen to come in flocks juſt before it is dark, and to ſettle and neſtle

among

among the heath on our foreft. And befides, the larkers, in dragging their nets by night, frequently catch them in the wheat-ftubbles; while the bat-fowlers, who take many red-wings in the hedges, never entangle any of this fpecies. Why thefe birds, in the matter of roofting, fhould differ from all their congeners, and from themfelves alfo with refpect to their proceedings by day, is a fact for which I am by no means able to account.

I have fomewhat to inform you of concerning the *moofe-deer*; but in general foreign animals fall feldom in my way: my little intelligence is confined to the narrow fphere of my own obfervations at home.

LETTER XXVIII.

TO THE SAME.

SELBORNE, March 1770.

ON *Michaelmas-day* 1768 I managed to get a fight of the female moofe belonging to the duke of *Richmond*, at *Goodwood*; but was greatly difappointed, when I arrived at the fpot, to find that it died, after having appeared in a languifhing way for fome time, on the morning before. However, underftanding that it was not ftripped, I proceeded to examine this rare quadruped: I found it in an old green-houfe, flung under the belly and chin by ropes, and in a ftanding pofture; but, though it had been dead for fo fhort a time, it was in fo putrid a ftate that the ftench was hardly

2   fupportable.

supportable. The grand diftinction between this deer, and any other fpecies that I have ever met with, confifted in the ftrange length of it's legs; on which it was tilted up much in the manner of the birds of the *gralla* order. I meafured it, as they do an horfe, and found that, from the ground to the wither, it was juft five feet four inches; which height anfwers exactly to fixteen hands, a growth that few horfes arrive at : but then, with this length of legs, it's neck was remarkably fhort, no more than twelve inches; fo that, by ftraddling with one foot forward and the other backward, it grazed on the plain ground, with the greateft difficulty, between it's legs : the ears were vaft and lopping, and as long as the neck; the head was about twenty inches long, and afs-like; and had fuch a redundancy of upper lip as I never faw before, with huge noftrils. This lip, travellers fay, is efteemed a dainty difh in *North America.* It is very reafonable to fuppofe that this creature fupports itfelf chiefly by browfing of trees, and by wading after water plants; towards which way of livelihood the length of legs and great lip muft contribute much. I have read fomewhere that it delights in eating the *nymphæa,* or water-lily. From the fore-feet to the belly behind the fhoulder it meafured three feet and eight inches : the length of the legs before and behind confifted a great deal in the *tibia,* which was ftrangely long; but, in my hafte to get out of the ftench, I forgot to meafure that joint exactly. It's fcut feemed to be about an inch long; the colour was a grizzly black; the mane about four inches long; the fore-hoofs were upright and fhapely, the hind flat and fplayed. The fpring before it was only two years old, fo that moft probably it was not then come to it's growth. What a vaft tall beaft muft a full grown ftag be ! I have been told fome arrive at ten feet and an half! This poor creature had at firft a female companion of the

same

same species, which died the spring before. In the same garden was a young stag, or red deer, between whom and this moose it was hoped that there might have been a breed; but their inequality of height must have always been a bar to any commerce of the amorous kind. I should have been glad to have examined the teeth, tongue, lips, hoofs, &c. minutely; but the putrefaction precluded all farther curiosity. This animal, the keeper told me, seemed to enjoy itself best in the extreme frost of the former winter. In the house they shewed me the horn of a male moose, which had no front-antlers, but only a broad palm with some snags on the edge. The noble owner of the dead moose proposed to make a skeleton of her bones.

Please to let me hear if my female moose corresponds with that you saw; and whether you think still that the *American* moose and *European* elk are the same creature.   I am,

With the greatest esteem, &c.

LETTER XXIX.

TO THE SAME.

DEAR SIR,   SELBORNE, May 12, 1770.

LAST month we had such a series of cold turbulent weather, such a constant succession of frost, and snow, and hail, and tempest, that the regular migration or appearance of the summer birds was much interrupted. Some did not shew themselves (at
least

leaſt were not heard) till weeks after their uſual time; as the *black-cap* and *white-throat*; and ſome have not been heard yet, as the *graſshopper-lark* and largeſt *willow-wren*. As to the *fly-catcher*, I have not ſeen it; it is indeed one of the lateſt, but ſhould appear about this time: and yet, amidſt all this meteorous ſtrife and war of the elements, two ſwallows diſcovered themſelves as long ago as the eleventh of *April*, in froſt and ſnow; but they withdrew quickly, and were not viſible again for many days. Houſe-martins, which are always more backward than ſwallows, were not obſerved till *May* came in.

Among the *monogamous* birds ſeveral are to be found, after pairing-time, ſingle, and of each ſex: but whether this ſtate of celibacy is matter of choice or neceſſity, is not ſo eaſily diſcover-able. When the houſe-ſparrows deprive my martins of their neſts, as ſoon as I cauſe one to be ſhot, the other, be it cock or hen, preſently procures a mate, and ſo for ſeveral times following.

I have known a dove-houſe infeſted by a pair of white owls, which made great havock among the young pigeons: one of the owls was ſhot as ſoon as poſſible; but the ſurvivor readily found a mate, and the miſchief went on. After ſome time the new pair were both deſtroyed, and the annoyance ceaſed.

Another inſtance I remember of a ſportſman, whoſe zeal for the increaſe of his game being greater than his humanity, after pairing-time he always ſhot the cock-bird of every couple of partridges upon his grounds; ſuppoſing that the rivalry of many males interrupted the breed: he uſed to ſay, that, though he had widowed the ſame hen ſeveral times, yet he found ſhe was ſtill provided with a freſh paramour, that did not take her away from her uſual haunt.

Again; I knew a lover of ſetting, an old ſportſman, who has

M often

often told me that foon after harveft he has frequently taken fmall coveys of partridges, confifting of cock-birds alone; thefe he pleafantly ufed to call old bachelors.

There is a propenfity belonging to common houfe-cats that is very remarkable; I mean their violent fondnefs for fifh, which appears to be their moft favourite food: and yet nature in this inftance feems to have planted in them an appetite that, unaffifted, they know not how to gratify: for of all quadrupeds cats are the leaft difpofed towards water; and will not, when they can avoid it, deign to wet a foot, much lefs to plunge into that element.

Quadrupeds that prey on fifh are amphibious: fuch is the otter, which by nature is fo well formed for diving, that it makes great havock among the inhabitants of the waters. Not fuppofing that we had any of thofe beafts in our fhallow brooks, I was much pleafed to fee a male otter brought to me, weighing twenty-one pounds, that had been fhot on the bank of our ftream below the *Priory*, where the rivulet divides the parifh of *Selborne* from *Harteley-wood.*

LETTER XXX.

TO THE SAME.

DEAR SIR,   SELBORNE, Aug. 1, 1770.

THE *French*, I think, in general are ftrangely prolix in their natural hiftory. What *Linnæus* fays with refpect to infects holds good in every other branch : " *Verbofitas præfentis fæculi, calamitas* " *artis.*"

Pray

Pray how do you approve of *Scopoli*'s new work? as I admire his *Entomologia*, I long to see it.

I forgot to mention in my last letter (and had not room to insert in the former) that the male moose, in rutting time, swims from island to island, in the lakes and rivers of *North-America*, in pursuit of the females. My friend, the chaplain, saw one killed in the water as it was on that errand in the river St. *Lawrence*: it was a monstrous beast, he told me; but he did not take the dimensions.

When I was last in town our friend Mr. *Barrington* most obligingly carried me to see many curious sights. As you were then writing to him about horns, he carried me to see many strange and wonderful specimens. There is, I remember, at Lord *Pembroke*'s, at *Wilton*, an horn room furnished with more than thirty different pairs; but I have not seen that house lately.

Mr. *Barrington* shewed me many astonishing collections of stuffed and living birds from all quarters of the world. After I had studied over the latter for a time, I remarked that every species almost that came from distant regions, such as *South America*, the coast of *Guinea*, &c. were thick-billed birds of the *loxia* and *fringilla* genera; and no *motacillæ*, or *muscicapæ*, were to be met with. When I came to consider, the reason was obvious enough; for the hard-billed birds subsist on seeds which are easily carried on board; while the soft-billed birds, which are supported by worms and insects, or, what is a *succedaneum* for them, fresh raw meat, can meet with neither in long and tedious voyages. It is from this defect of food that our collections (curious as they are) are defective, and we are deprived of some of the most delicate and lively genera.   I am, &c.

LETTER XXXI.

DEAR SIR,　　　　　　　　　　SELBORNE, Sept. 14, 1770.

You saw, I find, the ring-ousels again among their native crags; and are farther assured that they continue resident in those cold regions the whole year. From whence then do our ring-ousels migrate so regularly every *September*, and make their appearance again, as if in their return, every *April?* They are more early this year than common, for some were seen at the usual hill on the fourth of this month.

An observing *Devonshire* gentleman tells me that they frequent some parts of *Dartmoor*, and breed there; but leave those haunts about the end of *September* or beginning of *October*, and return again about the end of *March*.

Another intelligent person assures me that they breed in great abundance all over the *Peak* of *Derby*, and are called there *Tor-ousels*; withdraw in *October* and *November*, and return in spring. This information seems to throw some light on my new migration.

Scopoli's[m] new work (which I have just procured) has it's merit in ascertaining many of the birds of the *Tirol* and *Carniola*. Monographers, come from whence they may, have, I think, fair pretence to challenge some regard and approbation from the

---

[m] *Annus Primus Historico-Naturalis.*

lovers

lovers of natural hiftory; for, as no man can alone inveftigate all the works of nature, thefe partial writers may, each in their department, be more accurate in their difcoveries, and freer from errors, than more general writers; and fo by degrees may pave the way to an univerfal correct natural hiftory. Not that *Scopoli* is fo circumftantial and attentive to the life and converfation of his birds as I could wifh: he advances fome falfe facts; as when he fays of the *hirundo urbica* that " *pullos extra nidum non* " *nutrit.*" This affertion I know to be wrong from repeated obfervation this fummer; for houfe-martins do feed their young flying, though it muft be acknowledged not fo commonly as the houfe-fwallow; and the feat is done in fo quick a manner as not to be perceptible to indifferent obfervers. He alfo advances fome (I was going to fay) improbable facts; as when he fays of the woodcock that " *pullos roftro portat fugiens ab hofte.*" But candour forbids me to fay abfolutely that any fact is falfe, becaufe I have never been witnefs to fuch a fact. I have only to remark that the long unweildy bill of the woodcock is perhaps the worft adapted of any among the winged creation for fuch a feat of natural affection. I am, &c.

LETTER

LETTER XXXII.

TO THE SAME.

DEAR SIR, SELBORNE, *October* 29, 1770.

AFTER an ineffectual search in *Linnæus*, *Brisson*, &c. I begin to suspect that I discern my brother's *hirundo hyberna* in *Scopoli*'s new discovered *hirundo rupestris*, p. 167. His description of " *Supra murina, subtus albida; rectrices maculâ ovali albâ in latere* " *interno; pedes nudi, nigri; rostrum nigrum; remiges obscuriores quam* " *plumæ dorsales; rectrices remigibus concolores; caudâ emarginatâ, nec* " *forcipatâ;*" agrees very well with the bird in question: but when he comes to advance that it is " *statura hirundinis urbicæ,*" and that " *definitio hirundinis ripariæ Linnæi huic quoque convenit,*" he in some measure invalidates all he has said; at least he shews at once that he compares them to these species merely from memory: for I have compared the birds themselves, and find they differ widely in every circumstance of shape, size, and colour. However, as you will have a specimen, I shall be glad to hear what your judgment is in the matter.

Whether my brother is forestalled in his non-descript or not, he will have the credit of first discovering that they spend their winters under the warm and sheltery shores of *Gibraltar* and *Barbary*.

Scopoli's characters of his ordines and genera are clean, just, and expressive, and much in the spirit of *Linnæus*. These few remarks are the result of my first perusal of *Scopoli*'s *Annus Primus*.

The

The bane of our ſcience is the comparing one animal to the other by memory: for want of caution in this particular *Scopoli* falls into errors: he is not ſo full with regard to the manners of his indigenous birds as might be wiſhed, as you juſtly obſerve: his Latin is eaſy, elegant, and expreſſive, and very ſuperior to *Kramer's*[n].

I am pleaſed to ſee that my deſcription of the *moose* correſponds ſo well with yours.   I am, &c.

———————

LETTER XXXIII.

TO THE SAME.

DEAR SIR,   SELBORNE, Nov. 26, 1770.

I WAS much pleaſed to ſee, among the collection of birds from *Gibraltar*, ſome of thoſe ſhort-winged *Engliſh* ſummer-birds of paſſage, concerning whoſe departure we have made ſo much inquiry. Now if theſe birds are found in *Andaluſia* to migrate to and from *Barbary*, it may eaſily be ſuppoſed that thoſe that come to us may migrate back to the continent, and ſpend their winters in ſome of the warmer parts of *Europe*. This is certain, that many ſoft-billed birds that come to *Gibraltar* appear there only in ſpring and autumn, ſeeming to advance in pairs towards the northward, for the ſake of breeding during the ſummer months;

[n] See his *Elenchus vegetabilium et animalium per Auſtriam inferiorem, &c.*

and

and retiring in parties and broods towards the fouth at the decline of the year: fo that the rock of *Gibraltar* is the great rendezvous, and plac e of obfervation, from whence they take their departure each way towards *Europe* or *Africa*. It is therefore no mean difcovery, I think, to find that our fmall fhort-winged fummer birds of paffage are to be feen fpring and autumn on the very fkirts of *Europe*; it is a prefumptive proof of their emigrations.

Scopoli feems to me to have found the *hirundo melba*, the great *Gibraltar* fwift, in *Tirol*, without knowing it. For what is his *hirundo alpina* but the afore-mentioned bird in other words? Says he " *Omnia prioris*" (meaning the fwift); " *fed pectus album*; " *paulo major priore*." I do not fuppofe this to be a new fpecies. It is true alfo of the *melba*, that " *nidificat in excelfis Alpium rupibus*." *Vid. Annum Primum*.

My *Suffex* friend, a man of obfervation and good fenfe, but no naturalift, to whom I applied on account of the *ftone-curlew*, *oedicnemus*, fends me the following account: " In looking over " my Naturalift's Journal for the month of *April*, I find the *ftone-* " *curlews* are firft mentioned on the feventeenth and eighteenth, " which date feems to me rather late. They live with us all the " fpring and fummer, and at the beginning of autumn prepare " to take leave by getting together in flocks. They feem to me " a bird of paffage that may travel into fome dry hilly country " fouth of us, probably Spain, becaufe of the abundance of " fheep-walks in that country; for they fpend their fummers " with us in fuch diftricts. This conjecture I hazard, as I have " never met with any one that has feen them in *England* in the " winter. I believe they are not fond of going near the water, " but feed on earth-worms, that are common on fheep-walks and " downs. They breed on fallows and lay-fields abounding with

" grey

" grey moffy flints, which much refemble their young in colour;
" among which they fkulk and conceal themfelves. They
" make no neft, but lay their eggs on the bare ground,
" producing in common but two at a time. There is reafon to
" think their young run foon after they are hatched; and that
" the old ones do not feed them, but only lead them about at
" the time of feeding, which, for the moft part, is in the night."
Thus far my friend.

In the manners of this bird you fee there is fomething very
analogous to the buftard, whom it alfo fomewhat refembles in
afpect and make, and in the ftructure of it's feet.

For a long time I have defired my relation to look out for
thefe birds in *Andalufia*; and now he writes me word that, for the
firft time, he faw one dead in the market on the third of *September*.

When the *oedicnemus* flies it ftretches out it's legs ftraight behind,
like an heron. I am &c.

LETTER XXXIV.

TO THE SAME.

DEAR SIR, SELBORNE, March 30, 1771.

THERE is an infect with us, efpecially on chalky diftricts, which
is very troublefome and teafing all the latter end of the fummer,
getting into people's fkins, efpecially thofe of women and children,

N and

and raifing tumours which itch intolerably. This animal (which we call an harveft bug) is very minute, fcarce difcernible to the naked eye; of a bright fcarlet colour, and of the genus of *Acarus*. They are to be met with in gardens on kidneybeans, or any legumens; but prevail only in the hot months of fummer. Warreners, as fome have affured me, are much infefted by them on chalky downs; where thefe infects fwarm fometimes to fo infinite a degree as to difcolour their nets, and to give them a reddifh caft, while the men are fo bitten as to be thrown into fevers.

There is a fmall long fhining fly in thefe parts very troublefome to the houfewife, by getting into the chimnies, and laying it's eggs in the bacon while it is drying: thefe eggs produce maggots called *jumpers*, which, harbouring in the gammons and beft parts of the hogs, eat down to the bone, and make great wafte. This fly I fufpect to be a variety of the *mufca putris* of *Linnæus*: it is to be feen in the fummer in farm-kitchens on the bacon-racks and about the mantle-pieces, and on the ceilings.

The infect that infefts turnips and many crops in the garden (deftroying often whole fields while in their feedling leaves) is an animal that wants to be better known. The country people here call it the *turnip-fly* and *black-dolphin*; but I know it to be one of the *coleoptera*; the " *chryfomela oleracea, faltatoria, femoribus* " *pofticis craffiffimis.*" In very hot fummers they abound to an amazing degree, and, as you walk in a field or in a garden, make a pattering like rain, by jumping on the leaves of the turnips or cabbages.

There is an *Oeftrus*, known in thefe parts to every ploughboy; which, becaufe it is omitted by *Linnæus*, is alfo paffed over by late writers; and that is the *curvicauda* of old *Moufet*, mentioned

by

by *Derham* in his Physico-theology, p. 250: an insect worthy of remark for depositing it's eggs as it flies in so dextrous a manner on the single hairs of the legs and flanks of grass-horses. But then *Derham* is mistaken when he advances that this *Oestrus* is the parent of that wonderful star-tailed maggot which he mentions afterwards; for more modern entomologists have discovered that singular production to be derived from the egg, of the *musca chamæleon:* see *Geoffroy*, t. 17, f. 4.

A full history of noxious insects hurtful in the field, garden, and house, suggesting all the known and likely means of destroying them, would be allowed by the public to be a most useful and important work. What knowledge there is of this sort lies scattered, and wants to be collected; great improvements would soon follow of course. A knowledge of the properties, œconomy, propagation, and in short of the life and conversation of these animals, is a necessary step to lead us to some method of preventing their depredations.

As far as I am a judge, nothing would recommend entomology more than some neat plates that should well express the *generic distinctions* of insects according to *Linnæus*; for I am well assured that many people would study insects, could they set out with a more adequate notion of those distinctions than can be conveyed at first by words alone.

LETTER

LETTER XXXV.

TO THE SAME.

DEAR SIR,   SELBORNE, 1771.

HAPPENING to make a vifit to my neighbour's peacocks, I could not help obferving that the trains of thofe magnificent birds appear by no means to be their tails; thofe long feathers growing not from their *uropygium*, but all up their backs. A range of fhort brown ftiff feathers, about fix inches long, fixed in the *uropygium*, is the real tail, and ferves as the *fulcrum* to prop the train, which is long and top-heavy, when fet an end. When the train is up, nothing appears of the bird before but it's head and neck; but this would not be the cafe were thofe long feathers fixed only in the rump, as may be feen by the turkey-cock when in a ftrutting attitude. By a ftrong mufcular vibration thefe birds can make the fhafts of their long feathers clatter like the fwords of a fword-dancer; they then trample very quick with their feet, and run backwards towards the females.

I fhould tell you that I have got an uncommon *calculus ægogropila*, taken out of the ftomach of a fat ox; it is perfectly round, and about the fize of a large *Seville* orange; fuch are, I think, ufually flat.

LETTER

LETTER XXXVI.

TO THE SAME.

DEAR SIR, Sept. 1771.

THE summer through I have seen but two of that large species of bat which I call *vespertilio altivolans*, from it's manner of feeding high in the air: I procured one of them, and found it to be a male; and made no doubt, as they accompanied together, that the other was a female: but, happening in an evening or two to procure the other likewise, I was somewhat disappointed, when it appeared to be also of the same sex. This circumstance, and the great scarcity of this sort, at least in these parts, occasions some suspicions in my mind whether it is really a species, or whether it may not be the male part of the more known species, one of which may supply many females; as is known to be the case in sheep, and some other quadrupeds. But this doubt can only be cleared by a farther examination, and some attention to the sex, of more specimens: all that I know at present is, that my two were amply furnished with the parts of generation much resembling those of a boar.

In the extent of their wings they measured fourteen inches and an half; and four inches and an half from the nose to the tip of the tail: their heads were large, their nostrils bilobated, their shoulders broad and muscular; and their whole bodies fleshy and plump. Nothing could be more sleek and soft than their fur, which was of a bright chesnut colour; their maws were full of

food,

food, but so macerated that the quality could not be distinguished; their livers, kidnies, and hearts, were large, and their bowels covered with fat. They weighed each, when entire, full one ounce and one drachm. Within the ear there was somewhat of a peculiar structure that I did not understand perfectly; but refer it to the observation of the curious anatomist. These creatures sent forth a very rancid and offensive smell.

LETTER XXXVII.

TO THE SAME.

DEAR SIR,   SELBORNE, 1771.

ON the twelfth of *July* I had a fair opportunity of contemplating the motions of the *caprimulgus,* or fern-owl, as it was playing round a large oak that swarmed with *scarabæi solstitiales,* or fern-chafers. The powers of it's wing were wonderful, exceeding, if possible, the various evolutions and quick turns of the swallow genus. But the circumstance that pleased me most was, that I saw it distinctly, more than once, put out it's short leg while on the wing, and, by a bend of the head, deliver somewhat into it's mouth. If it takes any part of it's prey with it's foot, as I have now the greatest reason to suppose it does these chafers, I no longer wonder at the use of it's middle toe, which is curiously furnished with a serrated claw.

Swallows

Swallows and martins, the bulk of them I mean, have forfaken us fooner this year than ufual; for, on *September* the twenty-fecond, they rendezvoufed in a neighbour's walnut-tree, where it feemed probable they had taken up their lodging for the night. At the dawn of the day, which was foggy, they arofe all together in infinite numbers, occafioning fuch a rufhing from the ftrokes of their wings againft the hazy air, as might be heard to a confiderable diftance : fince that no flock has appeared, only a few ftragglers.

Some fwifts ftaid late, till the twenty-fecond of *Auguft*—a rare inftance ! for they ufually withdraw within the firft week°.

On *September* the twenty-fourth three or four ring-oufels appeared in my fields for the firft time this feafon : how punctual are thefe vifitors in their autumnal and fpring migrations !

LETTER XXXVIII.

TO THE SAME.

DEAR SIR,   SELBORNE, March 15, 1773.

By my journal for laft autumn it appears that the houfe-martins bred very late, and ftaid very late in thefe parts; for, on the firft of *October*, I faw young martins in their neft nearly fledged; and again, on the twenty-firft of *October*, we had at the next houfe a neft full of young martins juft ready to fly ; and the old ones were hawking for infects with great alertnefs. The next

° See letter liii. to Mr. *Barrington.*

morning

morning the brood forſook their neſt, and were flying round the village. From this day I never ſaw one of the ſwallow kind till *November* the third; when twenty, or perhaps thirty, houſe-martins were playing all day long by the ſide of the hanging wood, and over my fields. Did theſe ſmall weak birds, ſome of which were neſtlings twelve days ago, ſhift their quarters at this late ſeaſon of the year to the other ſide of the northern tropic? Or rather, is it not more probable that the next church, ruin, chalk-cliff, ſteep covert, or perhaps ſandbank, lake or pool (as a more northern naturaliſt would ſay), may become their *hybernaculum*, and afford them a ready and obvious retreat?

We now begin to expeƈt our vernal migration of ring-ouſels every week. Perſons worthy of credit aſſure me that ring-ouſels were ſeen at *Chriſtmas* 1770 in the foreſt of *Bere*, on the ſouthern verge of this county. Hence we may conclude that their migrations are only internal, and not extended to the continent ſouthward, if they do at firſt come at all from the northern parts of this iſland only, and not from the north of *Europe*. Come from whence they will, it is plain, from the fearleſs diſregard that they ſhew for men or guns, that they have been little accuſtomed to places of much reſort. Navigators mention that in the *Iſle of Aſcenſion*, and other ſuch deſolate diſtriƈts, birds are ſo little acquainted with the human form that they ſettle on men's ſhoulders; and have no more dread of a ſailor than they would have of a goat that was grazing. A young man at *Lewes*, in *Suſſex*, aſſured me that about ſeven years ago ring-ouſels abounded ſo about that town in the autumn that he killed ſixteen himſelf in one afternoon: he added further, that ſome had appeared ſince in every autumn; but he could not find that any had been obſerved before the ſeaſon in which he ſhot ſo many. I myſelf have found theſe birds in

little

little parties in the autumn cantoned all along the *Suffex* downs, wherever there were fhrubs and bufhes, from *Chichefter* to *Lewes*; particularly in the autumn of 1770. I am, &c.

LETTER XXXIX.

TO THE SAME.

DEAR SIR,   SELBORNE, Nov. 9, 1773.

As you defire me to fend you fuch obfervations as may occur, I take the liberty of making the following remarks, that you may, according as you think me right or wrong, admit or rejeƈt what I here advance, in your intended new edition of the *Britifh Zoology*.

The ofprey [p] was fhot about a year ago at *Frinfham-pond*, a great lake, at about fix miles from hence, while it was fitting on the handle of a plough and devouring a fifh: it ufed to precipitate itfelf into the water, and fo take it's prey by furprife.

A great afh-coloured [q] butcher-bird was fhot laft winter in *Tifted-park*, and a red backed butcher-bird at *Selborne*: they are *raræ aves* in this county.

Crows [r] go in pairs the whole year round.

Cornifh choughs [s] abound, and breed on *Beachy-head* and on all the cliffs of the *Suffex* coaft.

[p] Britifh Zoology, vol. 1, p. 128.   [q] p. 161.   [r] p. 167.   [s] p. 198.

O   The

The common wild-pigeon,[t] or ſtock-dove, is a bird of paſſage in the ſouth of *England*, ſeldom appearing till towards the end of *November*; is uſually the lateſt winter-bird of paſſage. Before our beechen woods were ſo much deſtroyed we had myriads of them, reaching in ſtrings for a mile together as they went out in a morning to feed. They leave us early in ſpring; where do they breed?

The people of *Hampſhire* and *Suſſex* call the miſſel-bird[u] the ſtorm-cock, becauſe it ſings early in the ſpring in blowing ſhowery weather; it's ſong often commences with the year: with us it builds much in orchards.

A gentleman aſſures me he has taken the neſts of ring-ouſels[x] on *Dartmoor*: they build in banks on the ſides of ſtreams.

Titlarks[y] not only ſing ſweetly as they ſit on trees, but alſo as they play and toy about on the wing; and particularly while they are deſcending, and ſometimes as they ſtand on the ground.

Adanſon's[z] teſtimony ſeems to me to be a very poor evidence that *European* ſwallows migrate during our winter to *Senegal*: he does not talk at all like an ornithologiſt; and probably ſaw only the ſwallows of that country, which I know build within Governor *O'Hara*'s hall againſt the roof. Had he known *European* ſwallows, would he not have mentioned the ſpecies?

The *houſe-ſwallow* waſhes by dropping into the water as it flies: this *ſpecies* appears commonly about a week before the *houſe-martin*, and about ten or twelve days before the *ſwift*.

In 1772 there were young houſe-martins[a] in their neſt till *October* the twenty-third.

---

[t] p. 216.   [u] p. 224.   [x] p. 229.   [y] Vol. 2, p 237.
[z] p. 242.   [a] 244.

The

The *swift*[b] appears about *ten* or *twelve* days later than the *house-swallow* : viz. about the twenty-fourth or twenty-sixth of *April*.

Whin-chats and *stone-chatters*[c] stay with us the whole year.

Some wheat-ears[d] continue with us the winter through.

Wagtails, all sorts, remain with us all the winter.

Bulfinches,[e] when fed on hempseed, often become wholly black.

We have vast flocks of *female* chaffinches[f] all the winter, with hardly any males among them.

When you say that in breeding-time the cock-snipes[g] make a bleating noise, and I a drumming (perhaps I should have rather said an humming), I suspect we mean the same thing. However, while they are playing about on the wing they certainly make a loud piping with their mouths : but whether that bleating or humming is ventriloquous, or proceeds from the motion of their wings, I cannot say; but this I know, that when this noise happens the bird is always descending, and his wings are violently agitated.

Soon after the lapwings[h] have done breeding they congregate, and, leaving the moors and marshes, betake themselves to downs and sheep-walks.

Two years ago[i] last spring the little auk was found alive and unhurt, but fluttering and unable to rise, in a lane a few miles from *Alresford*, where there is a great lake : it was kept awhile, but died.

I saw young teals[k] taken alive in the ponds of *Wolmer-forest* in the beginning of *July* last, along with flappers, or young wild-ducks.

b 245.   c 270. 271.   d 269.   e 300.   f 306.   g 358.
h 360.   i 409.   k 475.

O 2   Speaking

Speaking of the *swift*,[1] that page says " *it's drink the dew*;" whereas it fhould be " it drinks on the wing;" for all the fwallow kind fip their water as they fweep over the face of pools or rivers: like *Virgil*'s bees, they drink flying; " *flumina fumma libant*." In this method of drinking perhaps this genus may be peculiar.

Of the fedge-bird[m] be pleafed to fay it fings moft part of the night; it's notes are hurrying, but not unpleafing, and imitative of feveral birds; as the fparrow, fwallow, fky-lark. When it happens to be filent in the night, by throwing a ftone or clod into the bufhes where it fits you immediately fet it a finging; or in other words, though it flumbers fometimes, yet as foon as it is awakened it reaffumes it's fong.

LETTER XL.

TO THE SAME.

DEAR SIR,   SELBORNE, Sept. 2, 1774.

BEFORE your letter arrived, and of my own accord, I had been remarking and comparing the tails of the male and female fwallow, and this ere any young broods appeared; fo that there was no danger of confounding the dams with their *pulli* : and befides, as they were then always in pairs, and bufied in the employ of nidification, there could be no room for miftaking the fexes, nor the individuals of different chimnies the one for the other. From all my obfervations, it conftantly appeared that each fex

[1] p. 15.   [m] p. 16.

has

has the long feathers in it's tail that give it that forked fhape; with this difference, that they are longer in the tail of the male than in that of the female.

Nightingales, when their young firft come abroad, and are helplefs, make a plaintive and a jarring noife; and alfo a fnapping or cracking, purfuing people along the hedges as they walk: thefe laft founds feem intended for menace and defiance.

The grafshopper-lark chirps all night in the height of fummer.

Swans turn white the fecond year, and breed the third.

Weafels prey on moles, as appears by their being fometimes caught in mole-traps.

Sparrow-hawks fometimes breed in old crows' nefts, and the keftril in churches and ruins.

There are fuppofed to be two forts of eels in the ifland of *Ely*. The threads fometimes difcovered in eels are perhaps their young: the generation of eels is very dark and myfterious.

Hen-harriers breed on the ground, and feem never to fettle on trees.

When redftarts fhake their tails they move them horizontally, as dogs do when they fawn: the tail of a wagtail, when in motion, bobs up and down like that of a jaded horfe.

Hedge-fparrows have a remarkable flirt with their wings in breeding-time; as foon as frofty mornings come they make a very piping plaintive noife.

Many birds which become filent about *Midfummer* reaffume their notes again in *September*; as the thrufh, blackbird, wood-lark, willow-wren, &c.; hence *Auguft* is by much the moft mute month, the fpring, fummer, and autumn through. Are birds induced to fing again becaufe the temperament of autumn refembles that of fpring?

Linnæus

Linnæus ranges plants geographically; palms inhabit the tropics, graffes the temperate zones, and moffes and lichens the polar circles; no doubt animals may be claffed in the fame manner with propriety.

Houfe-fparrows build under eaves in the fpring; as the weather becomes hotter they get out for coolnefs, and neft in plum-trees and apple-trees. Thefe birds have been known fometimes to build in rooks' nefts, and fometimes in the forks of boughs under rooks' nefts.

As my neighbour was houfing a rick he obferved that his dogs devoured all the little red mice that they could catch, but rejected the common mice; and that his cats ate the common mice, refufing the red.

Red-breafts fing all through the fpring, fummer, and autumn. The reafon that they are called autumn fongfters is, becaufe in the two firft feafons their voices are drowned and loft in the general chorus; in the latter their fong becomes diftinguifhable. Many fongfters of the autumn feem to be the young cock red-breafts of that year: notwithftanding the prejudices in their favour, they do much mifchief in gardens to the fummer-fruits [n].

The titmoufe, which early in *February* begins to make two quaint notes, like the whetting of a faw, is the marfh titmoufe: the great titmoufe fings with three cheerful joyous notes, and begins about the fame time.

Wrens fing all the winter through, froft excepted.

Houfe-martins came remarkably late this year both in *Hampfhire* and *Devonfhire*: is this circumftance for or againft either hiding or migration?

---

[n] They eat alfo the berries of the ivy, the honey-fuckle, and the *euonymus europæus*, or fpindle-tree.

Moft

Most birds drink sipping at intervals; but pigeons take a long continued draught, like quadrupeds.

Notwithstanding what I have said in a former letter, no grey crows were ever known to breed on *Dartmoor*; it was my mistake.

The appearance and flying of the *scarabæus solstitialis*, or fern-chafer, commence with the month of *July*, and cease about the end of it. These scarabs are the constant food of *caprimulgi*, or fern owls, through that period. They abound on the chalky downs and in some sandy districts, but not in the clays.

In the garden of the *Black-bear* inn in the town of *Reading* is a stream or canal running under the stables and out into the fields on the other side of the road: in this water are many carps, which lie rolling about in sight, being fed by travellers, who amuse themselves by tossing them bread: but as soon as the weather grows at all severe these fishes are no longer seen, because they retire under the stables, where they remain till the return of spring. Do they lie in a torpid state? if they do not, how are they supported?

The note of the white-throat, which is continually repeated, and often attended with odd gesticulations on the wing, is harsh and displeasing. These birds seem of a pugnacious disposition; for they sing with an erected crest and attitudes of rivalry and defiance; are shy and wild in breeding-time, avoiding neighbourhoods, and haunting lonely lanes and commons; nay even the very tops of the *Sussex-downs*, where there are bushes and covert; but in *July* and *August* they bring their broods into gardens and orchards, and make great havock among the summer-fruits.

The black-cap has in common a full, sweet, deep, loud, and wild pipe; yet that strain is of short continuance, and his motions are desultory; but when that bird sits calmly and engages in song

in

in earneſt, he pours forth very ſweet, but inward melody, and expreſſes great variety of ſoft and gentle modulations, ſuperior perhaps to thoſe of any of our warblers, the nightingale excepted.

Black-caps moſtly haunt orchards and gardens; while they warble their throats are wonderfully diſtended.

The ſong of the redſtart is ſuperior, though ſomewhat like that of the white-throat: ſome birds have a few more notes than others. Sitting very placidly on the top of a tall tree in a village, the cock ſings from morning to night: he affects neighbourhoods, and avoids ſolitude, and loves to build in orchards and about houſes; with us he perches on the vane of a tall maypole.

The fly-catcher is of all our ſummer birds the moſt mute and the moſt familiar; it alſo appears the laſt of any. It builds in a vine, or a ſweetbriar, againſt the wall of an houſe, or in the hole of a wall, or on the end of a beam or plate, and often cloſe to the poſt of a door where people are going in and out all day long. This bird does not make the leaſt pretenſion to ſong, but uſes a little inward wailing note when it thinks it's young in danger from cats or other annoyances: it breeds but once, and retires early.

Selborne pariſh alone can and has exhibited at times more than half the birds that are ever ſeen in all *Sweden*; the former has produced more than one hundred and twenty ſpecies, the latter only two hundred and twenty-one. Let me add alſo that it has ſhewn near half the ſpecies that were ever known in *Great-Britian* [P].

On a retroſpect, I obſerve that my long letter carries with it a quaint and magiſterial air, and is very ſententious; but, when I recollect that you requeſted ſtricture and anecdote, I hope you will pardon the didactic manner for the ſake of the information it may happen to contain.

[P] *Sweden* 221, *Great-Britian* 252 ſpecies.

LETTER

LETTER XLI.

TO THE SAME.

It is matter of curious inquiry to trace out how thofe fpecies of foft-billed birds, that continue with us the winter through, fubfift during the dead months. The imbecility of birds feems not to be the only reafon why they fhun the rigour of our winters; for the robuft *wry-neck* (fo much refembling the hardy race of *wood-peckers*) migrates, while the feeble little *golden-crowned wren*, that fhadow of a bird, braves our fevereft frofts without availing himfelf of houfes or villages, to which moft of our winter-birds crowd in diftrefsful feafons, while this keeps aloof in fields and woods; but perhaps this may be the reafon why they may often perifh, and why they are almoft as rare as any bird we know.

I have no reafon to doubt but that the foft-billed birds, which winter with us, fubfift chiefly on infects in their *aurelia* ftate. All the fpecies of *wagtails* in fevere weather haunt fhallow ftreams near their fpring-heads, where they never freeze; and, by wading, pick out the aurelias of the genus of *Phryganeæ*, &c.

Hedge-fparrows frequent finks and gutters in hard weather, where they pick up crumbs and other fweepings: and in mild weather they procure worms, which are ftirring every month in the year, as any one may fee that will only be at the trouble of taking a candle to a grafs-plot on any mild winter's night. Red-breafts and wrens in the winter haunt out-houfes, ftables, and barns,

q See *Derham*'s Phyfico-theology, p. 235.

P   where

where they find fpiders and flies that have laid themfelves up during the cold feafon. But the grand fupport of the foft-billed birds in winter is that infinite profufion of *aureliæ* of the *lepidoptera ordo*, which is faftened to the twigs of trees and their trunks; to the pales and walls of gardens and buildings; and is found in every cranny and cleft of rock or rubbifh, and even in the ground itfelf.

Every fpecies of titmoufe winters with us; they have what I call a kind of intermediate bill between the hard and the foft, between the *Linnæan* genera of *fringilla* and *motacilla*. One fpecies alone fpends it's whole time in the woods and fields, never retreating for fuccour in the fevereft feafons to houfes and neighbourhoods; and that is the delicate long-tailed titmoufe, which is almoft as minute as the golden-crowned wren: but the blue titmoufe, or nun *(parus cæruleus)*, the cole-moufe *(parus ater)*, the great black-headed titmoufe *(fringillago)*, and the marfh titmoufe *(parus paluftris)*, all refort, at times, to buildings; and in hard weather particularly. The great titmoufe, driven by ftrefs of weather, much frequents houfes; and, in deep fnows, I have feen this bird, while it hung with it's back downwards (to my no fmall delight and admiration), draw ftraws lengthwife from out the eaves of thatched houfes, in order to pull out the flies that were concealed between them, and that in fuch numbers that they quite defaced the thatch, and gave it a ragged appearance.

The blue *titmoufe*, or *nun*, is a great frequenter of houfes, and a general devourer. Befides infects, it is very fond of flefh; for it frequently picks bones on dunghills: it is a vaft admirer of fuer, and haunts butchers' fhops. When a boy, I have known twenty in a morning caught with fnap moufe-traps, baited with tallow or fuet. It will alfo pick holes in apples left on the ground, and

be

be well entertained with the feeds on the head of a fun-flower. The blue, marfh, and great titmice will, in very fevere weather, carry away barley and oat ftraws from the fides of ricks.

How the *wheat-ear* and *whin-chat* fupport themfelves in winter cannot be fo eafily afcertained, fince they fpend their time on wild heaths and warrens; the former efpecially, where there are ftone quarries: moft probably it is that their maintenance arifes from the *aureliæ* of the *lepidoptera ordo*, which furnifh them with a plentiful table in the wildernefs. **I am, &c.**

LETTER XLII.

TO THE SAME.

DEAR SIR, SELBORNE, March 9, 1775.

SOME future *faunift*, a man of fortune, will, I hope, extend his vifits to the kingdom of *Ireland*; a new field, and a country little known to the naturalift. He will not, it is to be wifhed, under-take that tour unaccompanied by a botanift, becaufe the mountains have fcarcely been fufficiently examined; and the foutherly coun-ties of fo mild an ifland may poffibly afford fome plants little to be expected within the *Britifh* dominions. A perfon of a thinking turn of mind will draw many juft remarks from the modern im-provements of that country, both in arts and agriculture, where premiums obtained long before they were heard of with us. The manners of the wild natives, their fuperftitions, their prejudices,

P 2 their

their fordid way of life, will extort from him many ufeful reflections. He fhould alfo take with him an able draughtfman; for he muft by no means pafs over the noble caftles and feats, the extenfive and picturefque lakes and waterfalls, and the lofty ftupendous mountains, fo little known, and fo engaging to the imagination when defcribed and exhibited in a lively manner: fuch a work would be well received.

As I have feen no modern map of *Scotland*, I cannot pretend to fay how accurate or particular any fuch may be; but this I know, that the beft old maps of that kingdom are very defective.

The great obvious defect that I have remarked in all maps of *Scotland* that have fallen in my way is, a want of a *coloured line*, or *ftroke*, that fhall exactly define the juft limits of that diftrict called *The Highlands*. Moreover, all the great avenues to that mountainous and romantic country want to be well diftinguifhed. The military roads formed by general *Wade* are fo great and Roman-like an undertaking that they well merit attention. My old map, *Moll*'s Map, takes notice of *Fort William*; but could not mention the other forts that have been erected long fince: therefore a good reprefentation of the chain of forts fhould not be omitted.

The celebrated zigzag up the *Coryarich* muft not be paffed over. *Moll* takes notice of *Hamilton* and *Drumlanrig*, and fuch capital houfes; but a new furvey, no doubt, fhould reprefent every feat and caftle remarkable for any great event, or celebrated for it's paintings, &c. Lord *Breadalbane*'s feat and beautiful *policy* are too curious and extraordinary to be omitted.

The feat of the Earl of *Eglintoun*, near *Glafgow*, is worthy of notice. The pine-plantations of that nobleman are very grand and extenfive indeed. I am, &c.

LETTER

LETTER XLIII.

TO THE SAME.

A PAIR of *honey-buzzards, buteo apivorus, five vespivorus Raii,* built them a large shallow nest, composed of twigs and lined with dead beechen leaves, upon a tall slender beech near the middle of *Selborne-hanger,* in the summer of 1780. In the middle of the month of *June* a bold boy climbed this tree, though standing on so steep and dizzy a situation, and brought down an egg, the only one in the nest, which had been sat on for some time, and contained the embrio of a young bird. The egg was smaller, and not so round as those of the common buzzard; was dotted at each end with small red spots, and surrounded in the middle with a broad bloody zone.

The hen-bird was shot, and answered exactly to Mr. *Ray*'s description of that species; had a black *cere,* short thick legs, and a long tail. When on the wing this species may be easily distinguished from the *common buzzard* by it's hawk-like appearance, small head, wings not so blunt, and longer tail. This specimen contained in it's craw some limbs of frogs and many grey snails without shells. The *irides* of the eyes of this bird were of a beautiful bright yellow colour.

About the tenth of *July* in the same summer a pair of *sparrow-hawks* bred in an old crow's nest on a low beech in the same hanger; and as their brood, which was numerous, began to grow up,

up, became fo daring and ravenous, that they were a terror to all
the dames in the village that had chickens or ducklings under
their care. A boy climbed the tree, and found the young fo
fledged that they all efcaped from him; but difcovered that a
good houfe had been kept: the larder was well-ftored with
provifions; for he brought down a young blackbird, jay, and
houfe-martin, all clean picked, and fome half devoured. The
old birds had been obferved to make fad havock for fome days
among the new-flown fwallows and martins, which, being but
lately out of their nefts, had not acquired thofe powers and command of wing that enable them, when more mature, to fet fuch
enemies at defiance.

LETTER XLIV.

TO THE SAME.

DEAR SIR,   SELBORNE, Nov. 30, 1780.

EVERY incident that occafions a renewal of our correfpondence
will ever be pleafing and agreeable to me.

As to the wild *wood-pigeon*, the *oenas*, or *vinago*, of *Ray*, I am
much of your mind; and fee no reafon for making it the origin of
the common *houfe-dove*: but fuppofe thofe that have advanced that
opinion may have been mifled by another appellation, often given
to the *oenas*, which is that of *ftock-dove*,

                                                        Unlefs

Unless the stock-dove in the winter varies greatly in manners from itself in summer, no species seems more unlikely to be domesticated, and to make an *house-dove*. We very rarely see the latter settle on trees at all, nor does it ever haunt the woods; but the former, as long as it stays with us, from *November* perhaps to *February*, lives the same wild life with the *ring-dove, palumbus torquatus*; frequents coppices and groves, supports itself chiefly by mast, and delights to roost in the tallest beeches. Could it be known in what manner stock-doves build, the doubt would be settled with me at once, provided they construct their nests on trees, like the *ring-dove*, as I much suspect they do.

You received, you say, last spring a *stock-dove* from *Sussex*; and are informed that they sometimes breed in that country. But why did not your correspondent determine the place of it's nidification, whether on rocks, cliffs, or trees? If he was not an adroit ornithologist I should doubt the fact, because people with us perpetually confound the *stock-dove* with the *ring-dove*.

For my own part, I readily concur with you in supposing that house-doves are derived from the *small blue rock-pigeon*, for many reasons. In the first place the wild stock-dove is manifestly larger than the common house-dove, against the usual rule of domestication, which generally enlarges the breed. Again, those two remarkable *black spots* on the remiges of each wing of the stock-dove, which are so characteristic of the species, would not, one should think, be totally lost by it's being reclaimed; but would often break out among its descendants. But what is worth an hundred arguments is, the instance you give in Sir *Roger Mostyn*'s house-doves in *Caernarvonshire*; which, though tempted by plenty of food and gentle treatment, can never be prevailed on to

inhabit

inhabit their cote for any time; but, as foon as they begin to breed, betake themfelves to the faftneffes of *Ormfhead*, and depofit their young in fafety amidft the inacceffible caverns, and precipices of that ftupendous promontory.

" Naturam expellas furcâ . . . tamen ufque recurret."

I have confulted a fportfman, now in his feventy-eighth year, who tells me that fifty or fixty years back, when the beechen woods were much more extenfive than at prefent, the number of wood-pigeons was aftonifhing; that he has often killed near twenty in a day; and that with a long wild-fowl piece he has fhot feven or eight at a time on the wing as they came wheeling over his head: he moreover adds, which I was not aware of, that often there were among them little parties of fmall *blue doves*, which he calls *rockiers*. The food of thefe numberlefs emigrants was beech-maft and fome acorns; and particularly barley, which they collected in the ftubbles. But of late years, fince the vaft increafe of turnips, that vegetable has furnifhed a great part of their fupport in hard weather; and the holes they pick in thefe roots greatly damage the crop. From this food their flefh has contracted a rancidnefs which occafions them to be rejected by nicer judges of eating, who thought them before a delicate difh. They were fhot not only as they were feeding in the fields, and efpecially in fnowy weather, but alfo at the clofe of the evening, by men who lay in ambufh among the woods and groves to kill them as they came in to rooft[r]. Thefe are the principal circumftances relating to this

---

[r] Some old fportfmen fay that the main part of thefe flocks ufed to withdraw as foon as the heavy *Chriftmas* frofts were over.

wonderful

wonderful *internal* migration, which with us takes place towards the end of *November*, and ceafes early in the fpring. Laft winter we had in *Selborne* high wood about an hundred of thefe doves; but in former times the flocks were fo vaft, not only with us but all the diftrict round, that on mornings and evenings they traverfed the air, like rooks, in ftrings, reaching for a mile together. When they thus rendezvoufed here by thoufands, if they happened to be fuddenly roufed from their rooft-trees on an evening,

" Their rifing all at once was like the found
" Of thunder heard remote."——

It will by no means be foreign to the prefent purpofe to add, that I had a relation in this neighbourhood who made it a practice, for a time, whenever he could procure the eggs of a *ring-dove*, to place them under a pair of doves that were fitting in his own pigeon-houfe; hoping thereby, if he could bring about a coalition, to enlarge his breed, and teach his own doves to beat out into the woods and to fupport themfelves by maft : the plan was plaufible, but fomething always interrupted the fuccefs; for though the birds were ufually hatched, and fometimes grew to half their fize, yet none ever arrived at maturity. I myfelf have feen thefe foundlings in their neft difplaying a ftrange ferocity of nature, fo as fcarcely to bear to be looked at, and fnapping with their bills by way of menace. In fhort, they always died, perhaps for want of proper fuftenance : but the owner thought that by their fierce and wild demeanour they frighted their fofter-mothers, and fo were ftarved.

Virgil, as a familiar occurrence, by way of fimile, defcribes a dove haunting the cavern of a rock in fuch engaging numbers, that I cannot refrain from quoting the paffage : and *John Dryden*

Q   has

has rendered it so happily in our language, that without farther excuse I shall add his translation also.

" Qualis speluncâ subitò commota Columba,
" Cui domus, et dulces latebroso in pumice nidi,
" Fertur in arva volans, plausumque exterrita pennis
" Dat tecto ingentem—mox aere lapsa quieto,
" Radit iter liquidum, celeres neque commovet alas."

" As when a dove her rocky hold forsakes,
" Rous'd, in a fright her sounding wings she shakes;
" The cavern rings with clattering:—out she flies,
" And leaves her callow care, and cleaves the skies:
" At first she flutters:—but at length she springs
" To smoother flight, and shoots upon her wings."

I am, &c.

LETTER

LETTER I.

TO THE HONOURABLE DAINES BARRINGTON.

DEAR SIR,   SELBORNE, June 30, 1769.

WHEN I was in town laſt month I partly engaged that I would ſometime do myſelf the honour to write to you on the ſubject of natural hiſtory: and I am the more ready to fulfil my promiſe, becauſe I ſee you are a gentleman of great candour, and one that will make allowances; eſpecially where the writer profeſſes to be an *out-door naturaliſt*, one that takes his obſervations from the ſubject itſelf, and not from the writings of others.

The following is a LIST *of the* SUMMER BIRDS *of* PASSAGE *which I have diſcovered in this neighbourhood, ranged ſomewhat in the order which they appear:*

	RAII NOMINA.	USUALLY APPEARS ABOUT
1. Wryneck,	*Jynx, ſive torquilla:*	The middle of *March:* harſh note.
2. Smalleſt willow-wren,	*Regulus non criſtatus:*	*March* 23: chirps till *September.*
3. Swallow,	*Hirundo domeſtica:*	*April* 13.
4. Martin,	*Hirundo ruſtica:*	Ditto.
5. Sand-martin,	*Hirundo riparia:*	Ditto.
6. Black-cap,	*Atricapilla:*	Ditto: a ſweet wild note.
7. Nightingale,	*Luſcinia:*	Beginning of *April.*
8. Cuckoo,	*Cuculus:*	Middle of *April.*
9. Middle willow-wren,	*Regulus non criſtatus:*	Ditto: a ſweet plantive note.
10. White-throat,	*Ficedulæ affinis:*	Ditto: mean note; ſings on till *September.*

Q 2   11. Red-ſtart,

11. Red-start,	*Ruticilla:*	Ditto: more agreeable song.
12. Stone-curlew,	*Oedicnemus:*	{ End of *March*: loud nocturnal whistle.
13. Turtle-dove,	*Turtur.*	
14. Grasshopper-lark,	*Alauda minima locustæ voce:*	{ Middle *April*: a small sibilous note, till the end of *July*.
15. Swift,	*Hirundo apus:*	About *April* 27.
16. Less reed-sparrow,	*Passer arundinaceus minor:*	{ A sweet polyglot, but hurrying: it has the notes of many birds.
17. Land-rail,	*Ortyigometra:*	A loud harsh note, crex, crex.
18. Largest willow-wren,	*Regulus non cristatus:*	{ Cantat voce stridulâ locustæ; end of *April*, on the tops of high beeches.
19. Goatsucker, or fern-owl,	*Caprimulgus:*	{ Beginning of *May*; chatters by night with a singular noise.
20. Fly-catcher,	*Stoparola:*	{ May 12. A very mute bird: This is the latest summer bird of passage.

This assemblage of curious and amusing birds belongs to ten several genera of the *Linnæan* system; and are all of the *ordo* of *passeres* save the *jynx* and *cuculus*, which are *picæ*, and the *charadrius* (*oedicnemus*) and *rallus* (*ortygometra*), which are *grallæ*.

These birds, as they stand numerically, belong to the following *Linnæan* genera:

1,	*Jynx:*	13.	*Columba:*
2, 6, 7, 9, 10, 11, 16, 18.	*Motacilla:*	17.	*Rallus:*
3, 4, 5, 15.	*Hirundo:*	19.	*Caprimulgus:*
8.	*Cuculus:*	14.	*Alauda:*
12.	*Charadrius:*	20.	*Muscicapa.*

Most

Moft foft-billed birds live on infects, and not on grain and feeds; and therefore at the end of fummer they retire: but the following foft-billed birds, though infect-eaters, ftay with us the year round:

RAII NOMINA.

Redbreaft,	*Rubecula:*	These frequent houfes; and haunt out-buildings in the winter: eat fpiders.
Wren,	*Paffer troglodytes:*	
Hedge-fparrow,	*Curruca:*	Haunt finks for crumbs and other fweepings.
White-wagtail,	*Motacilla alba:*	Thefe frequent fhallow rivulets near the fpring heads, where they never freeze: eat the aureliæ of Phryganea. The fmalleft birds that walk.
Yellow wagtail,	*Motacilla flava:*	
Grey wagtail,	*Motacilla cinerea:*	
Wheat-ear,	*Oenanthe:*	Some of thefe are to be feen with us the winter through.
Whin-chat,	*Oenanthe fecunda.*	
Stone-chatter,	*Oenanthe tertia.*	
Golden-crowned wren,	*Regulus criftatus.*	This is the fmalleft Britifh bird: haunts the tops of tall trees; ftays the winter through.

A LIST *of the* WINTER BIRDS *of* PASSAGE *round this neighbourhood, ranged fomewhat in the order in which they appear:*

RAII NOMINA.

1. Ring-oufel,	*Merula torquata:*	This is a new migration, which I have lately difcovered about *Michaelmas* week, and again about the fourteenth of *March.*
2. Redwing,	*Turdus iliacus:*	About old *Michaelmas.*
3. Fieldfare,	*Turdus pilaris:*	Though a percher by day, roofts on the ground.
4. Royfton-crow,	*Cornix cinerea:*	Moft frequent on downs,
5. Woodcock,	*Scolopax:*	Appears about old *Michaelmas.*
6. Snipe,	*Gallinago minor:*	Some fnipes conftantly breed with us.
7. Jack-fnipe,	*Gallinago minima.*	
8. Wood-pigeon	*Oenas:*	Seldom appears till late: not in fuch plenty as formerly.

9. Wild-

9. Wild-swan,	*Cygnus ferus:*	On some large waters.
10. Wild-goose,	*Anser ferus:*	
11. Wild-duck,	*Anas torquata minor:*	
12. Pochard,	*Anas fera fusca:*	
13. Wigeon,	*Penelope:*	
14. Teal, breeds with us in *Wolmer-forest*	*Querquedula:*	On our lakes and streams.
15. Cross-beak,	*Coccothraustes:*	These are only wanderers that appear
16. Gross-b'll,	*Loxia:*	occasionally, and are not observant of
17. Silk-tail,	*Garrulus bohemicus.*	any regular migration.

These birds, as they stand numerically, belong to the following *Linnæan* genera:

1, 2, 3,	*Turdus:*	9, 10, 11, 12, 13, 14,		
4,	*Corvus:*			*Anas:*
5, 6, 7,	*Scolopax:*	15, 16,		*Loxia:*
8,	*Columba:*	17.		*Ampelis.*

Birds that sing in the night are but few.

Nightingale,	*Luscinia:*	" In shadiest covert hid." MILTON.
Woodlark,	*Alauda arborea:*	Suspended in mid air.
Less reed-sparrow,	*Passer arundinaceus minor:*	Among reeds and willows.

I should now proceed to such birds as continue to sing after *Midsummer*, but, as they are rather numerous, they would exceed the bounds of this paper: besides, as this is now the season for remarking on that subject, I am willing to repeat my observations on some birds concerning the continuation of whose song I seem at present to have some doubt.   I am, &c.

————

LETTER

LETTER II.

TO THE SAME.

DEAR SIR, SELBORNE, Nov. 2, 1769.

WHEN I did myself the honour to write to you about the end of last *June* on the subject of natural history, I sent you a list of the summer-birds of passage which I have observed in this neighbourhood; and also a list of the winter-birds of passage: I mentioned besides those soft-billed birds that stay with us the winter through in the south of *England*, and those that are remarkable for singing in the night.

According to my proposal, I shall now proceed to such birds (singing birds strictly so called) as continue in full song till after *Midsummer*; and shall range them somewhat in the order in which they first begin to open as the spring advances.

RAII NOMINA.

1. Wood-lark,	*Alauda arborea:*	In *January*, and continues to sing through all the summer and autumn.
2. Song-thrush,	*Turdus simpliciter dictus:*	In *February* and on to *August*, re-assume their song in autumn.
3. Wren,	*Passer troglodytes:*	All the year, hard frost excepted.
4. Redbreast,	*Rubecula:*	Ditto.
5. Hedge-sparrow,	*Curruca:*	Early in *February* to *July* the 10th.
6. Yellowhammer,	*Emberiza flava:*	Early in *February*, and on through *July* to *August* the 21st.
7. Skylark,	*Alauda vulgaris:*	In *February*, and on to *October*.
8. Swallow,	*Hirundo domestica:*	From *April* to *September*.
9. Black-cap,	*Atricapilla:*	Beginning of *April* to *July* 13th.

10. Titlark,

10. Titlark,	*Alauda pratorum :*	From middle of *April* to *July* the 16th.
11. Blackbird,	*Merula vulgaris :*	Sometimes in *February* and *March*, and so on to *July* the twenty-third; reassumes in autumn.
12. White-throat,	*Ficedulæ affinis :*	In *April*, and on to *July* 23.
13. Goldfinch,	*Carduelis :*	*April*, and through to *September* 16.
14. Greenfinch,	*Chloris :*	On to *July* and *August* 2.
15. Less reed-sparrow.	*Passer arundinaceus minor :*	*May*, on to beginning of *July*.
16. Common linnet,	*Linaria vulgaris :*	Breeds and whistles on till *August*; reassumes it's note when they begin to congregate in *October*, and again early before the flocks separate.

Birds that cease to be in full song, and are usually silent at or before *Midsummer* :

17. Middle willow-wren,	*Regulus non cristatus :*	Middle of *June* : begins in *April*.
18. Redstart,	*Ruticilla :*	Ditto : begins in *May*.
19. Chaffinch,	*Fringilla :*	Beginning of *June* : sings first in *February*.
20. Nightingale,	*Luscinia :*	Middle of *June* : sings first in *April*.

Birds that sing for a short time, and very early in the spring :

| 21. Missel-bird, | *Turdus viscivorus :* | *January* the 2d, 1770, in *February*. Is called in *Hampshire* and *Sussex* the storm cock, because it's song is supposed to forebode windy wet weather : is the largest singing bird we have. |
| 22. Great titmouse, or ox-eye, | *Fringillago :* | In *February*, *March*, *April* : reassumes for a short time in *September*. |

Birds

Birds that have somewhat of a note or song, and yet are hardly to be called singing birds:

	RAII NOMINA.	
23. Golden-crowned wren,	} *Regulus cristatus:*	{ It's note as minute as it's person; frequents the tops of high oaks and firs : the smallest *British* bird.
24. Marsh-titmouse,	*Parus palustris:*	{ Haunts great woods: two harsh sharp notes.
25. Small willow-wren,	} *Regulus non cristatus:*	Sings in *March,* and on to *September.*
26. Largest ditto,	*Ditto:*	{ *Cantat voce stridulâ locustæ;* from end of *April* to *August.*
27. Grasshopper-lark,	{ *Alauda minima voce locustæ:*	{ Chirps all night, from the middle of *April* to the end of *July.*
28. Martin,	*Hirundo agrestis:*	{ All the breeding time; from *May* to *September.*
29. Bullfinch,	*Pyrrhula.*	
30. Bunting,	*Emberiza alba:*	From the end of *January* to *July.*

All singing birds, and those that have any pretensions to song, not only in *Britain,* but perhaps the world through, come under the *Linnæan ordo* of *passeres.*

The above-mentioned birds, as they stand numerically, belong to the following *Linnæan* genera.

1, 7, 10, 27.	*Alauda:*	8, 28.	*Hirundo.*
2, 11, 21.	*Turdus:*	13, 16, 19.	*Fringilla.*
3, 4, 5, 9, 12, 15, 17, 18, 20, 23, 25, 26.	} *Motacilla:*	22, 24.	*Parus.*
6, 30.	*Emberiza:*	14, 29.	*Loxia.*

Birds

Birds that sing as they fly are but few.

	RAII NOMINA.	
Skylark,	*Alauda vulgaris :*	Rising, suspended, and falling.
Titlark,	*Alauda pratorum :*	In it's descent; also sitting on trees, and walking on the ground.
Woodlark,	*Alauda arborea :*	Suspended; in hot summer nights all night long.
Blackbird,	*Merula :*	Sometimes from bush to bush.
White-throat,	*Ficedulæ affinis :*	Uses when singing on the wing odd jerks and gesticulations.
Swallow,	*Hirundo domestica :*	In soft sunny weather.
Wren,	*Passer troglodytes :*	Sometimes from bush to bush.

Birds that breed most early in these parts :

Raven,	*Corvus :*	Hatches in *February* and *March.*
Song thrush,	*Turdus :*	In *March.*
Blackbird,	*Merula :*	In *March.*
Rook,	*Cornix frugilega :*	Builds the beginning of *March.*
Woodlark,	*Alauda arborea :*	Hatches in *April.*
Ring dove,	*Palumbus torquatus :*	Lays the beginning of April.

All birds that continue in full song till after *Midsummer* appear to me to breed more than once.

Most kinds of birds seem to me to be wild and shy somewhat in proportion to their bulk; I mean in this island, where they are much pursued and annoyed: but in *Ascension Island*, and many other desolate places, mariners have found fowls so unacquainted with an human figure, that they would stand still to be taken; as is the case with boobies, &c. As an example of what is advanced, I remark that the *golden-crested wren* (the smallest *British* bird) will stand unconcerned till you come within three or four yards of it, while the *bustard (otis)*, the largest *British* land fowl, does not care to admit a person within so many furlongs.   I am, &c.

LETTER

LETTER III.

TO THE SAME.

DEAR SIR,   SELBORNE, Jan. 15, 1770.

IT was no small matter of satisfaction to me to find that you were not displeased with my little *methodus* of birds. If there was any merit in the sketch, it must be owing to it's punctuality. For many months I carried a list in my pocket of the birds that were to be remarked, and, as I rode or walked about my business, I noted each day the continuance or omission of each bird's song; so that I am as sure of the certainty of my facts as a man can be of any transaction whatsoever.

I shall now proceed to answer the several queries which you put in your two obliging letters, in the best manner that I am able. Perhaps *Eastwick*, and it's environs, where you heard so very few birds, is not a woodland country, and therefore not stocked with such songsters. If you will cast your eye on my last letter, you will find that many species continued to warble after the beginning of *July*.

The titlark and yellowhammer breed late, the latter very late; and therefore it is no wonder that they protract their song: for I lay it down as a maxim in ornithology, that as long as there is any incubation going on there is music. As to the redbreast and wren, it is well known to the most incurious observer that they whistle the year round, hard frost excepted; especially the latter.

R 2   It

It was not in my power to procure you a black-cap, or a lefs reed-fparrow, or fedge-bird, alive. As the firft is undoubtedly, and the laft, as far as I can yet fee, a fummer bird of paffage, they would require more nice and curious management in a cage than I fhould be able to give them: they are both diftinguifhed fongfters. The note of the former has fuch a wild fweetnefs that it always brings to my mind thofe lines in a fong in " *As You Like It.*"

> " And tune his merry note
> " Unto the *wild* bird's throat."   Shakespeare.

The latter has a furprifing variety of notes refembling the fong of feveral other birds; but then it has alfo an hurrying manner, not at all to it's advantage: it is notwithftanding a delicate polyglot.

It is new to me that titlarks in cages fing in the night; perhaps only caged birds do fo. I once knew a tame redbreaft in a cage that always fang as long as candles were in the room; but in their wild ftate no one fuppofes they fing in the night.

I fhould be almoft ready to doubt the fact, that there are to be feen much fewer birds in *July* than in any former month, notwith-ftanding fo many young are hatched daily. Sure I am that it is far otherwife with refpect to the *fwallow tribe*, which increafes prodigioufly as the fummer advances: and I faw, at the time men-tioned, many hundreds of young wagtails on the banks of the *Cherwell*, which almoft covered the meadows. If the matter ap-pears as you fay in the other fpecies, may it not be owing to the dams being engaged in incubation, while the young are concealed by the leaves?

Many

Many times have I had the curiofity to open the ftomachs of *woodcocks* and *fnipes*; but nothing ever occurred that helped to explain to me what their fubfiftence might be: all that I could ever find was a foft mucus, among which lay many pellucid fmall gravels. I am, &c.

LETTER IV.

TO THE SAME.

DEAR SIR,   SELBORNE, Feb. 19, 1770.

YOUR obfervation that " the *cuckoo* does not depofit it's egg in- " difcriminately in the neft of the firft bird that comes in it's way, " but probably looks out a nurfe in fome degree congenerous, " with whom to intruft it's young," is perfectly new to me; and ftruck me fo forcibly, that I naturally fell into a train of thought that led me to confider whether the fact was fo, and what reafon there was for it. When I came to recollect and inquire, I could not find that any cuckoo had ever been feen in thefe parts, except in the neft of the *wagtail*, the *hedge-fparrow*, the *titlark*, the *white-throat*, and the *redbreaft*, all foft-billed infectivorous birds. The excellent Mr. *Willughby* mentions the neft of the *palumbus* (*ring-dove*), and of the *fringilla* (*chaffinch*), birds that fubfift on acorns and grains, and fuch hard food: but then he does not mention them as of his own knowledge; but fays after-

wards

wards that he faw himfelf a *wagtail* feeding a cuckoo. It appears hardly poffible that a foft-billed bird fhould fubfift on the fame food with the hard-billed: for the former have thin membranaceous ftomachs fuited to their foft food; while the latter, the granivorous tribe, have ftrong mufcular gizzards, which, like mills, grind, by the help of fmall gravels and pebbles, what is fwallowed. This proceeding of the cuckoo, of dropping it's eggs as it were by chance, is fuch a monftrous outrage on maternal affection, one of the firft great dictates of nature; and fuch a violence on inftinct; that, had it only been related of a bird in the *Brafils*, or *Peru*, it would never have merited our belief. But yet, fhould it farther appear that this fimple bird, when divefted of that natural ςτοργη that feems to raife the kind in general above themfelves, and infpire them with extraordinary degrees of cunning and addrefs, may be ftill endued with a more enlarged faculty of difcerning what fpecies are fuitable and congenerous nurfing-mothers for it's difregarded eggs and young, and may depofit them only under *their* care, this would be adding wonder to wonder, and inftancing, in a frefh manner, that the methods of Providence are not fubjected to any mode or rule, but aftonifh us in new lights, and in various and changeable appearances.

What was faid by a very ancient and fublime writer concerning the defect of natural affection in the oftrich, may be well applied to the bird we are talking of:

" *She is hardened againft her young ones, as though they were not*
" *her's:*
" *Becaufe God hath deprived her of wifdom, neither hath he imparted*
" *to her underftanding* [s]."

[s] Job xxxix. 16, 17.

Query.

Query. Does each female cuckoo lay but one egg in a feafon, or does fhe drop feveral in different nefts according as opportunity offers? I am, &c.

LETTER V.

TO THE SAME.

DEAR SIR, SELBORNE, April 12, 1770.

I HEARD many birds of feveral fpecies fing laft year after *Mid-fummer*; enough to prove that the fummer folftice is not the period that puts a ftop to the mufic of the woods. The yellowhammer no doubt perfifts with more fteadinefs than any other; but the woodlark, the wren, the redbreaft, the fwallow, the white-throat, the goldfinch, the common linnet, are all undoubted inftances of the truth of what I advanced.

If this fevere feafon does not interrupt the regularity of the fummer migrations, the blackcap will be here in two or three days. I wifh it was in my power to procure you one of thofe fongfters; but I am no birdcatcher; and fo little ufed to birds in a cage, that I fear if I had one it would foon die for want of fkill in feeding.

Was

Was your reed-sparrow, which you kept in a cage, the thick-billed reed-sparrow of the Zoology, p. 320; or was it the less reed-sparrow of *Ray*, the *sedge-bird* of Mr. *Pennant*'s last publication, p. 16?

As to the matter of long-billed birds growing fatter in moderate frosts, I have no doubt within myself what should be the reason. The thriving at those times appears to me to arise altogether from the gentle check which the cold throws upon insensible perspiration. The case is just the same with blackbirds, &c.; and farmers and warreners observe, the first, that their hogs fat more kindly at such times, and the latter that their rabbits are never in such good case as in a gentle frost. But when frosts are severe, and of long continuance, the case is soon altered; for then a want of food soon overbalances the repletion occasioned by a checked perspiration. I have observed, moreover, that some human constitutions are more inclined to plumpness in winter than in summer.

When birds come to suffer by severe frost, I find that the first that fail and die are the redwing-fieldfares, and then the song-thrushes.

You wonder, with good reason, that the hedge-sparrows, &c. can be induced at all to sit on the egg of the cuckoo without being scandalized at the vast disproportioned size of the supposititious egg; but the brute creation, I suppose, have very little idea of size, colour, or number. For the common hen, I know, when the fury of incubation is on her, will sit on a single shapeless stone instead of a nest full of eggs that have been withdrawn: and, moreover, a hen-turkey, in the same circumstances, would sit on in the empty nest till she perished with hunger.

I think the matter might easily be determined whether a cuckoo lays one or two eggs, or more, in a season, by opening a female

during

during the laying-time. If more than one was come down out of the ovary, and advanced to a good fize, doubtlefs then fhe would that fpring lay more than one.

I will endeavour to get a hen, and to examine.

Your fuppofition that there may be fome natural obftruction in finging birds while they are mute, and that when this is removed the fong recommences, is new and bold: I wifh you could difcover fome good grounds for this fufpicion.

I was glad you were pleafed with my fpecimen of the *caprimulgus*, or fern-owl; you were, I find, acquainted with the bird before.

When we meet I fhall be glad to have fome converfation with you concerning the propofal you make of my drawing up an account of the animals in this neighbourhood. Your partiality towards my fmall abilities perfuades you, I fear, that I am able to do more than is in my power: for it is no fmall undertaking for a man unfupported and alone to begin a natural hiftory from his own autopfia! Though there is endlefs room for obfervation in the field of nature, which is boundlefs, yet inveftigation (where a man endeavours to be fure of his facts) can make but flow progrefs; and all that one could collect in many years would go into a very narrow compafs.

Some extracts from your ingenious " Inveftigations of the difference between the prefent temperature of the air in Italy," &c. have fallen in my way; and gave me great fatisfaction: they have removed the objections that always arofe in my mind whenever I came to the paffages which you quote. Surely the judicious *Virgil*, when writing a didactic poem for the region of *Italy*, could never think of defcribing freezing rivers, unlefs fuch feverity of weather pretty frequently occurred!

P. S. Swallows appear amidft fnows and froft.

S

LETTER

LETTER VI.

TO THE SAME.

DEAR SIR,

THE severity and turbulence of last month so interrupted the regular process of summer migration, that some of the birds do but just begin to shew themselves, and others are apparently thinner than usual; as the white-throat, the black-cap, the redstart, the fly-catcher. I well remember that after the very severe spring in the year 1739-40 summer birds of passage were very scarce. They come probably hither with a south-east wind, or when it blows between those points; but in that unfavourable year the winds blowed the whole spring and summer through from the opposite quarters. And yet amidst all these disadvantages two swallows, as I mentioned in my last, appeared this year as early as the eleventh of *April* amidst frost and snow; but they withdrew again for a time.

I am not pleased to find that some people seem so little satisfied with *Scopoli*'s new publication [t]; there is room to expect great things from the hands of that man, who is a good naturalist: and one would think that an history of the birds of so distant and southern a region as *Carniola* would be new and interesting. I could wish to see that work, and hope to get it sent down. Dr. *Scopoli* is physician to the wretches that work in the quicksilver mines of that district.

When you talked of keeping a reed-sparrow, and giving it feeds, I could not help wondering; because the reed-sparrow which I mentioned to you *(passer arundinaceus minor Raii)* is a soft-billed bird; and most probably migrates hence before winter;

[t] This work he calls his *Annus Primus Historico Naturalis.*

whereas

whereas the bird you kept *(paſſer torquatus Raii)* abides all the year, and is a thick-billed bird. I queſtion whether the latter be much of a ſongſter; but in this matter I want to be better informed. The former has a variety of hurrying notes, and ſings all night. Some part of the ſong of the former, I ſuſpect, is attributed to the latter. We have plenty of the ſoft-billed ſort; which Mr. *Pennant* had entirely left out of his *Britiſh Zoology*, till I reminded him of his omiſſion. See *Britiſh Zoology* laſt publiſhed, p. 16 [x].

I have ſomewhat to advance on the different manners in which different birds fly and walk; but as this is a ſubject that I have not enough conſidered, and is of ſuch a nature as not to be contained in a ſmall ſpace, I ſhall ſay nothing further about it at preſent [y].

No doubt the reaſon why the ſex of birds in their firſt plumage is ſo difficult to be diſtinguiſhed is, as you ſay, " becauſe they " are not to pair and diſcharge their parental functions till the " enſuing ſpring." As colours ſeem to be the chief external ſexual diſtinction in many birds, theſe colours do not take place till ſexual attachments begin to obtain. And the caſe is the ſame in quadrupeds; among whom, in their younger days, the ſexes differ but little: but, as they advance to maturity, horns and ſhaggy manes, beards and brawny necks, &c. &c. ſtrongly diſcriminate the male from the female. We may inſtance ſtill farther in our own ſpecies, where a beard and ſtronger features are uſually characteriſtic of the male ſex: but this ſexual diverſity does not take place in earlier life; for a beautiful youth ſhall be ſo like a beautiful girl that the difference ſhall not be diſcernible;

> " Quem ſi puellarum inſereres choro,
> " Mirè ſagaces falleret hoſpites
> " Diſcrimen obſcurum, ſolutis
> " Crinibus, ambiguoque vultu."   Hor.

[x] See letter xxv. to Mr. *Pennant*.   [y] See letter xlii. to Mr. *Barrington*.

LETTER

LETTER VII.

TO THE SAME.

DEAR SIR, RINGMER, near LEWES, Oct. 8, 1770.

I AM glad to hear that *Kuckalm* is to furnish you with the birds of *Jamaica*; a sight of the *hirundines* of that hot and distant island would be a great entertainment to me.

The *Anni of Scopoli* are now in my possession; and I have read the *Annus Primus* with satisfaction: for though some parts of this work are exceptionable, and he may advance some mistaken observations; yet the ornithology of so distant a country as *Carniola* is very curious. Men that undertake only one district are much more likely to advance natural knowledge than those that grasp at more than they can possibly be acquainted with: every kingdom, every province, should have it's own *monographer*.

The reason perhaps why he mentions nothing of *Ray*'s Ornithology may be the extreme poverty and distance of his country, into which the works of our great naturalist may have never yet found their way. You have doubts, I know, whether this Ornithology is genuine, and really the work of *Scopoli*: as to myself, I think I discover strong tokens of authenticity; the style corresponds with that of his *Entomology*; and his characters of his Ordines and Genera are many of them new, expressive, and masterly. He has ventured to alter some of the *Linnæan genera* with sufficient shew of reason.

It might perhaps be mere accident that you saw so many swifts and no swallows at *Staines*; because, in my long observation of

those

thofe birds, I never could difcover the leaft degree of rivalry or hoftility between the fpecies.

Ray remarks that birds of the *gallinæ order*, as cocks and hens, partridges, and pheafants, &c. are *pulveratrices*, fuch as duft themfelves, ufing that method of cleanfing their feathers, and ridding themfelves of their vermin. As far as I can obferve, many birds that duft themfelves never wafh : and I once thought that thofe birds that wafh themfelves would never duft; but here I find myfelf miftaken; for common houfe-fparrows are great *pulveratrices*, being frequently feen grovelling and wallowing in dufty roads; and yet they are great wafhers. Does not the fkylark duft?

Query. Might not *Mahomet* and his followers take one method of purification from thefe *pulveratrices*? becaufe I find from travellers of credit, that if a ftrict muffulman is journeying in a fandy defert where no water is to be found, at ftated hours he ftrips off his clothes, and moft fcrupuloufly rubs his body over with fand or duft.

A countryman told me he had found a young *fern-owl* in the neft of a fmall bird on the ground; and that it was fed by the little bird. I went to fee this extraordinary phenomenon, and found that it was a young cuckoo hatched in the neft of a titlark: it was become vaftly too big for it's neft, appearing

— — — — — in tenui re

Majores pennas nido extendiffe — —

and was very fierce and pugnacious, purfuing my finger, as I teazed it, for many feet from the neft, and fparring and buffetting with it's wings like a game-cock. The dupe of a dam appeared at a diftance, hovering about with meat in it's mouth, and expreffing the greateft folicitude.

In

In *July* I saw several cuckoos skimming over a large pond; and found, after some observation, that they were feeding on the *libellulæ*, or *dragon-flies*; some of which they caught as they settled on the weeds, and some as they were on the wing. Notwithstanding what *Linnæus* says, I cannot be induced to believe that they are birds of prey.

This district affords some birds that are hardly ever heard of at *Selborne*. In the first place considerable flocks of *cross-beaks (loxiæ curviroftræ)* have appeared this summer in the pine-groves belonging to this house; the *water-ousel* is said to haunt the mouth of the *Lewes* river, near *Newhaven*; and the *Cornish chough* builds, I know, all along the chalky cliffs of the *Sussex* shore.

I was greatly pleased to see little parties of *ring-ousels* (my newly discovered migraters) scattered, at intervals, all along the *Sussex* downs from *Chichester* to *Lewes*. Let them come from whence they will, it looks very suspicious that they are cantoned along the coast in order to pass the channel when severe weather advances. They visit us again in *April*, as it should seem, in their return; and are not to be found in the dead of winter. It is remarkable that they are very tame, and seem to have no manner of apprehensions of danger from a person with a gun. There are bustards on the wide downs near *Brighthelmstone*. No doubt you are acquainted with the *Sussex* downs: the prospects and rides round *Lewes* are most lovely!

As I rode along near the coast I kept a very sharp look out in the lanes and woods, hoping I might, at this time of the year, have discovered some of the summer short-winged birds of passage crowding towards the coast in order for their departure: but it was very extraordinary that I never saw a redstart, white-throat, black-cap, uncrested wren, fly-catcher, &c. And I remember to have made the same remark in former years, as I usually come to

this

this place annually about this time. The birds moſt common along the coaſt at preſent are the ſtone-chatters, whinchats, buntings, linnets, ſome few wheat-ears, titlarks, &c. Swallows and houſe-martins abound yet, induced to prolong their ſtay by this ſoft, ſtill, dry ſeaſon.

A land tortoiſe, which has been kept for thirty years in a little walled court belonging to the houſe where I now am viſiting, retires under ground about the middle of *November*, and comes forth again about the middle of *April*. When it firſt appears in the ſpring it diſcovers very little inclination towards food ; but in the height of ſummer grows voracious : and then as the ſummer de-clines it's appetite declines ; ſo that for the laſt ſix weeks in autumn it hardly eats at all. Milky plants, ſuch as lettuces, dandelions, ſowthiſtles, are it's favourite diſh. In a neighbouring village one was kept till by tradition it was ſuppoſed to be an hundred years old. An inſtance of vaſt longevity in ſuch a poor reptile !

LETTER VIII.

TO THE SAME.

DEAR SIR, SELBORNE, Dec. 20, 1770.

THE birds that I took for *aberdavines* were reed-ſparrows *(paſſeres torquati.)*

There are doubtleſs many home internal migrations within this kingdom that want to be better underſtood : witneſs thoſe vaſt flocks of hen chaffinches that appear with us in the winter without hardly any cocks among them. Now was there a due proportion of

4

of each fex, it fhould feem very improbable that any one diftrict fhould produce fuch numbers of thefe little birds ; and much more when only one half of the fpecies appears : therefore we may conclude that the *fringillæ cælebes*, for fome good purpofes, have a peculiar migration of their own in which the fexes part. Nor fhould it feem fo wonderful that the intercourfe of fexes in this fpecies of birds fhould be interrupted in winter ; fince in many animals, and particularly in bucks and does, the fexes herd feparately, except at the feafon when commerce is neceffary for the continuance of the breed. For this matter of the chaffinches fee *Fauna Suecica*, p. 85, and *Syftema Naturæ*, p. 318. I fee every winter vaft flights of hen chaffinches, but none of cocks.

Your method of accounting for the periodical motions of the *Britifh* finging birds, or birds of flight, is a very probable one ; fince the matter of food is a great regulator of the actions and proceedings of the brute creation : there is but one that can be fet in competition with it, and that is love. But I cannot quite acquiefce with you in one circumftance when you advance that, " when they have " thus feafted, they again feparate into fmall parties of five or fix, " and get the beft fare they can within a certain diftrict, having " no inducement to go in queft of frefh-turned earth." Now if you mean that the bufinefs of congregating is quite at an end from the conclufion of wheat-fowing to the feafon of barley and oats, it is not the cafe with us ; for larks and chaffinches, and particularly linnets, flock and congregate as much in the very dead of winter as when the hufbandman is bufy with his ploughs and harrows.

Sure there can be no doubt but that woodcocks and fieldfares leave us in the fpring, in order to crofs the feas, and to retire to fome diftricts more fuitable to the purpofe of breeding. That the

former

former pair before they retire, and that the hens are forward with egg, I myself, when I was a fportfman, have often experienced. It cannot indeed be denied but that now and then we hear of a woodcock's neft, or young birds, difcovered in fome part or other of this ifland : but then they are always mentioned as rarities, and fomewhat out of the common courfe of things : but as to redwings and fieldfares, no fportfman or naturalift has ever yet, that I could hear, pretended to have found the neft or young of thofe fpecies in any part of thefe kingdoms. And I the more admire at this inftance as extraordinary, fince, to all appearance, the fame food in fummer as well as in winter might fupport them here which maintains their congeners, the blackbirds and thrufhes, did they chufe to ftay the fummer through. From hence it appears that it is not food alone which determines fome fpecies of birds with regard to their ftay or departure. Fieldfares and redwings difappear fooner or later according as the warm weather comes on earlier or later. For I well remember, after that dreadful winter 1739-40, that cold north-eaft winds continued to blow on through *April* and *May*, and that thefe kinds of birds (what few remained of them) did not depart as ufual, but were feen lingering about till the beginning of *June.*

The beft authority that we can have for the nidification of the birds above mentioned in any diftrict, is the teftimony of faunifts that have written profeffedly the natural hiftory of particular countries. Now, as to the fieldfare, *Linnæus*, in his Fauna Suecica, fays of it that " *maximis in arboribus nidificat:*" and of the redwing he fays, in the fame place, that " *nidificat in mediis arbufculis, five fepibus: ova fex cæruleo-viridia maculis nigris variis.*" Hence we may be affured that fieldfares and redwings breed in *Sweden. Scopoli* fays, in his *Annus Primus*, of the woodcock, that " *nupta ad nos venit circa æquinoc-* " *tium vernale :*" meaning in *Tirol*, of which he is a native. And

T   afterwards

afterwards he adds " *nidificat in paludibus alpinis : ova ponit* 3 - - - 5." It does not appear from *Kramer* that woodcocks breed at all in *Auſtria:* but he ſays " *Avis hæc ſeptentrionalium provinciarum æſtivo* " *tempore incola eſt; ubi plerumque nidificat. Appropinquante hyeme* " *auſtraliores provincias petit : hinc circa plenilunium menſis Octobris plerum-* " *que Auſtriam tranſmigrat. Tunc rurſus circa plenilunium potiſſimum menſis* " *Martii per Auſtriam matrimonio juncta ad ſeptentrionales provincias* " *redit.*" For the whole paſſage (which I have abridged) ſee *Elenchus,* &c. p. 351. This ſeems to be a full proof of the migration of woodcocks ; though little is proved concerning the place of breeding.

P. S. There fell in the county of *Rutland,* in three weeks of this preſent very wet weather, ſeven inches and an half of rain, which is more than has fallen in any three weeks for theſe thirty years paſt in that part of the world. A mean quantity in that county for one year is twenty inches and an half.

LETTER IX.

TO THE SAME.

DEAR SIR,   FYFIELD, near ANDOVER, Feb. 12, 1771.

You are, I know, no great friend to migration ; and the well atteſted accounts from various parts of the kingdom ſeem to juſtify you in your ſuſpicions, that at leaſt many of the ſwallow kind do

                                                                not

not leave us in the winter, but lay themselves up like insects and bats, in a torpid state, and slumber away the more uncomfortable months till the return of the sun and fine weather awakens them.

But then we must not, I think, deny migration in general; because migration certainly does subsist in some places, as my brother in *Andalusia* has fully informed me. Of the motions of these birds he has ocular demonstration, for many weeks together, both spring and fall: during which periods myriads of the swallow kind traverse the Straits from north to south, and from south to north, according to the season. And these vast migrations consist not only of *hirundines* but of *bee-birds, hoopoes, oro pendolos, or golden thrushes,* &c. &c. and also of many of our *soft-billed summer birds of passage*; and moreover of birds which never leave us, such as all the various sorts of hawks and kites. Old *Belon*, two hundred years ago, gives a curious account of the incredible armies of hawks and kites which he saw in the spring-time traversing the *Thracian Bosphorus* from *Asia* to *Europe*. Besides the above mentioned, he remarks that the procession is swelled by whole troops of eagles and vultures.

Now it is no wonder that birds residing in *Africa* should retreat before the sun as it advances, and retire to milder regions, and especially birds of prey, whose blood being heated with hot animal food, are more impatient of a sultry climate: but then I cannot help wondering why kites and hawks, and such hardy birds as are known to defy all the severity of *England,* and even of *Sweden* and all north *Europe*, should want to migrate from the south of *Europe,* and be dissatisfied with the winters of *Andalusia.*

It does not appear to me that much stress may be laid on the difficulty and hazard that birds must run in their migrations, by reason of vast oceans, cross winds, &c.; because, if we reflect, a bird may travel from *England* to the equator without launching out and

T 2   exposing

exposing itself to boundless seas, and that by crossing the water at *Dover*, and again at *Gibraltar*. And I with the more confidence advance this obvious remark, because my brother has always found that some of his birds, and particularly the swallow kind, are very sparing of their pains in crossing the *Mediterranean* : for when arrived at *Gibraltar* they do not

— — — " Rang'd in figure wedge their way,
— — — — — " And set forth
" Their airy caravan high over seas
" Flying, and over lands with mutual wing
" Easing their flight:" — — — — MILTON.

but scout and hurry along in little detached parties of six or seven in a company; and sweeping low, just over the surface of the land and water, direct their course to the opposite continent at the narrowest passage they can find. They usually slope across the bay to the south-west, and so pass over opposite to *Tangier*, which, it seems, is the narrowest space.

In former letters we have considered whether it was probable that woodcocks in moon-shiny nights cross the *German* ocean from *Scandinavia*. As a proof that birds of less speed may pass that sea, considerable as it is, I shall relate the following incident, which, though mentioned to have happened so many years ago, was strictly matter of fact :—As some people were shooting in the parish of *Trotton*, in the county of *Sussex*, they killed a duck in that dreadful winter 1708-9, with a silver collar about it's neck [z], on which were engraven the arms of the king of *Denmark*. This anecdote the rector of *Trotton* at that time has often told to a near relation of mine; and, to the best of my remembrance, the collar was in the possession of the rector.

[z] I have read a like anecdote of a swan.

At

At prefent I do not know any body near the fea-fide that will take the trouble to remark at what time of the moon woodcocks firft come: if I lived near the fea myfelf I would foon tell you more of the matter. One thing I ufed to obferve when I was a fportfman, that there were times in which woodcocks were fo fluggifh and fleepy that they would drop again when flufhed juft before the fpaniels, nay juft at the muzzle of a gun that had been fired at them: whether this ftrange lazinefs was the effect of a recent fatiguing journey I fhall not prefume to fay.

Nightingales not only never reach *Northumberland* and *Scotland*, but alfo, as I have been always told, *Devonfhire* and *Cornwall*. In thofe two laft counties we cannot attribute the failure of them to the want of warmth: the defect in the weft is rather a prefumptive argument that thefe birds come over to us from the continent at the narroweft paffage, and do not ftroll fo far weftward.

Let me hear from your own obfervation whether fkylarks do not duft. I think they do: and if they do, whether they wafh alfo.

The *alauda pratenfis of Ray* was the poor dupe that was educating the booby of a cuckoo mentioned in my letter of *October* laft.

Your letter came too late for me to procure a ring-oufel for Mr. *Tunftal* during their autumnal vifit; but I will endeavour to get him one when they call on us again in *April*. I am glad that you and that gentleman faw my *Andalufian* birds; I hope they anfwered your expectation. *Royfton*, or grey crows, are winter birds that come much about the fame time with the woodcock: they, like the fieldfare and redwing, have no apparent reafon for migration; for as they fare in the winter like their congeners, fo might they in all appearance in the fummer. Was not *Tenant*, when a boy, miftaken? did he not find a miffel-thrufh's neft, and take it for the neft of a fieldfare?

The

The ftock-dove, or wood-pigeon, *œnas Raii*, is the laft winter bird of paffage which appears with us; and is not feen till towards the end of *November*: about twenty years ago they abounded in the diftrict of *Selborne*; and ftrings of them were feen morning and evening that reached a mile or more: but fince the beechen woods have been greatly thinned they are much decreafed in number. The ring-dove, *palumbus Raii*, ftays with us the whole year, and breeds feveral times through the fummer.

Before I received your letter of *October* laft I had juft remarked in my journal that the trees were unufually green. This uncommon verdure lafted on late into *November*; and may be accounted for from a late fpring, a cool and moift fummer; but more particularly from vaft armies of chafers, or tree-beetles, which, in many places, reduced whole woods to a leaflefs naked ftate. Thefe trees fhot again at *Midfummer*, and then retained their foliage till very late in the year.

My mufical friend, at whofe houfe I am now vifiting, has tried all the owls that are his near neighbours with a pitch-pipe fet at concert-pitch, and finds they all hoot in B flat. He will examine the nightingales next fpring.　　　　I am, &c. &c.

LETTER

LETTER X.

TO THE SAME.

DEAR SIR, SELBORNE, Aug. 1, 1771.

FROM what follows, it will appear that neither owls nor cuckoos keep to one note. A friend remarks that many (moſt) of his owls hoot in B flat; but that one went almoſt half a note below A. The pipe he tried their notes by was a common half-crown pitch-pipe, ſuch as maſters uſe for tuning of harpſichords; it was the common *London* pitch.

A neighbour of mine, who is ſaid to have a nice ear, remarks that the owls about this village hoot in three different keys, in G flat, or F ſharp, in B flat and A flat. He heard two hooting to each other, the one in A flat, and the other in B flat. *Query:* Do theſe different notes proceed from different ſpecies, or only from various individuals? The ſame perſon finds upon trial that the note of the cuckoo (of which we have but one ſpecies) varies in different individuals; for, about *Selborne* wood, he found they were moſtly in D: he heard two ſing together, the one in D, the other in D ſharp, who made a diſagreeable concert: he afterwards heard one in D ſharp, and about *Woolmer-foreſt* ſome in C. As to nightingales, he ſays that their notes are ſo ſhort, and their tranſitions ſo rapid, that he cannot well aſcertain their key. Perhaps in a cage, and in a room, their notes may be more diſtinguiſhable. This perſon has tried to ſettle the notes of a ſwift, and of ſeveral other ſmall birds, but cannot bring them to any criterion.

As

As I have often remarked that redwings are some of the first birds that suffer with us in severe weather, it is no wonder at all that they retreat from *Scandinavian* winters: and much more the *ordo* of *grallæ*, who, all to a bird, forsake the northern parts of *Europe* at the approach of winter. " *Grallæ tanquam conjuratæ unanimiter in* " *fugam se conjiciunt; ne earum unicam quidem inter nos habitantem in-* " *venire possimus; ut enim æstate in australibus degere nequeunt ob defectum* " *lumbricorum, terramque siccam ; ita nec in frigidis ob eandem causam*", says *Ekmarck* the *Swede*, in his ingenious little treatise called *Migrationes Avium*, which by all means you ought to read while your thoughts run on the subject of migration. See *Amænitates Academicæ*, vol. 4, p. 565.

Birds may be so circumstanced as to be obliged to migrate in one country and not in another: but the *grallæ*, (which procure their food from marshes and boggy grounds) must in winter forsake the more northerly parts of *Europe*, or perish for want of food.

I am glad you are making inquiries from *Linnæus* concerning the woodcock : it is expected of him that he should be able to account for the motions and manner of life of the animals of his own *Fauna*.

Faunists, as you observe, are too apt to acquiesce in bare descriptions, and a few synonyms : the reason is plain ; because all that may be done at home in a man's study, but the investigation of the life and conversation of animals, is a concern of much more trouble and difficulty, and is not to be attained but by the active and inquisitive, and by those that reside much in the country.

Foreign systematics are, I observe, much too vague in their specific differences ; which are almost universally constituted by one or two particular marks, the rest of the description running in general terms. But our countryman, the excellent Mr. *Ray*, is

the

4

the only defcriber that conveys fome precife idea in every term or word, maintaining his fuperiority over his followers and imitators in fpite of the advantage of frefh difcoveries and modern inform-ation.

At this diftance of years it is not in my power to recollect at what periods woodcocks ufed to be fluggifh or alert when I was a fportfman : but, upon my mentioning this circumftance to a friend, he thinks he has obferved them to be remarkably liftlefs againft fnowy foul weather : if this fhould be the cafe, then the inaptitude for flying arifes only from an eagernefs for food; as fheep are ob-ferved to be very intent on grazing againft ftormy wet evenings.

I am, &c. &c.

LETTER XI.

TO THE SAME.

DEAR SIR, SELBORNE, Feb. 8, 1772.

WHEN I ride about in the winter, and fee fuch prodigious flocks of various kinds of birds, I cannot help admiring at thefe congre-gations, and wifhing that it was in my power to account for thofe appearances almoft peculiar to the feafon. The two great motives which regulate the proceedings of the brute creation are love and hunger ; the former incites animals to perpetuate their kind, the

U latter

latter induces them to preferve individuals : whether either of thefe fhould feem to be the ruling paffion in the matter of congregating is to be confidered. As to love, that is out of the queftion at a time of the year when that foft paffion is not indulged : befides, during the amorous feafon, fuch a jealoufy prevails between the male birds that they can hardly bear to be together in the fame hedge or field. Moft of the finging and elation of fpirits of that time feem to me to be the effect of rivalry and emulation : and it is to this fpirit of jealoufy that I chiefly attribute the equal difperfion of birds in the fpring over the face of the country.

Now as to the bufinefs of food : as thefe animals are actuated by inftinct to hunt for neceffary food, they fhould not, one would fup- pofe, crowd together in purfuit of fuftenance at a time when it is moft likely to fail; yet fuch affociations do take place in hard weather chiefly, and thicken as the feverity increafes. As fome kind of felf- intereft and felf-defence is no doubt the motive for the proceeding, may it not arife from the helpleffnefs of their ftate in fuch rigorous feafons; as men crowd together, when under great calamities, though they know not why? Perhaps approximation may difpel fome de- gree of cold ; and a crowd may make each individual appear fafer from the ravages of birds of prey and other dangers.

If I admire when I fee how much congenerous birds love to con- gregate, I am the more ftruck when I fee incongruous ones in fuch ftrict amity. If we do not much wonder to fee a flock of rooks ufually attended by a train of daws, yet it is ftrange that the former fhould fo frequently have a flight of ftarlings for their fatellites. Is it becaufe rooks have a more difcerning fcent than their attend- dants, and can lead them to fpots more productive of food? Anato- mifts fay that rooks, by reafon of two large nerves which run down between the eyes into the upper mandible, have a more delicate feeling in their beaks than other round-billed birds, and can grope

for

for their meat when out of fight. Perhaps then their affociates attend them on the motive of intereft, as greyhounds wait on the motions of their finders; and as lions are faid to do on the yelpings of jackalls. Lapwings and ftarlings fometimes affociate.

LETTER XII.

TO THE SAME.

DEAR SIR,   March 9, 1772.

As a gentleman and myfelf were walking on the fourth of laft *November* round the fea-banks at *Newhaven*, near the mouth of the *Lewes* river, in purfuit of natural knowledge, we were furprifed to fee three houfe-fwallows gliding very fwiftly by us. That morning was rather chilly, with the wind at north-weft; but the tenor of the weather for fome time before had been delicate, and the noons remarkably warm. From this incident, and from repeated accounts which I meet with, I am more and more induced to believe that many of the fwallow kind do not depart from this ifland; but lay themfelves up in holes and caverns; and do, infect-like and bat-like, come forth at mild times, and then retire again to their *latebræ*. Nor make I the leaft doubt but that, if I lived at *Newhaven, Seaford, Brighthelmftone*, or any of thofe towns near the chalk-cliffs of the *Suffex* coaft, by proper obfervations, I fhould fee fwallows ftirring at periods of the winter, when the noons were foft and inviting, and the fun warm and invigorating. And I am the more of this

opinion

opinion from what I have remarked during some of our late springs, that though some swallows did make their appearance about the usual time, *viz.* the thirteenth or fourteenth of *April*, yet meeting with an harsh reception, and blustering cold north-east winds, they immediately withdrew, absconding for several days, till the weather gave them better encouragement.

———————

LETTER XIII.

TO THE SAME.

DEAR SIR,
                                                    April 12, 1772.

WHILE I was in *Suffex* last autumn my residence was at the village near *Lewes*, from whence I had formerly the pleasure of writing to you. On the first of *November* I remarked that the old tortoise, formerly mentioned, began first to dig the ground in order to the forming it's hybernaculum, which it had fixed on just beside a great tuft of hepaticas. It scrapes out the ground with it's fore-feet, and throws it up over it's back with it's hind; but the motion of it's legs is ridiculously slow, little exceeding the hour-hand of a clock; and suitable to the composure of an animal said to be a whole month in performing one feat of copulation. Nothing can be more assiduous than this creature night and day in scooping the earth, and forcing it's great body into the cavity; but, as the noons of that season proved unusually warm and sunny, it was continually interrupted, and called forth by the heat in the middle of the day;

                                                         and

and though I continued there till the thirteenth of *November*, yet the work remained unfinished. Harsher weather, and frosty mornings, would have quickened it's operations. No part of it's behaviour ever struck me more than the extreme timidity it always expresses with regard to rain; for though it has a shell that would secure it against the wheel of a loaded cart, yet does it discover as much solicitude about rain as a lady dressed in all her best attire, shuffling away on the first sprinklings, and running it's head up in a corner. If attended to, it becomes an excellent weather-glass; for as sure as it walks elate, and as it were on tiptoe, feeding with great earnestness in a morning, so sure will it rain before night. It is totally a diurnal animal, and never pretends to stir after it becomes dark. The tortoise, like other reptiles, has an arbitrary stomach as well as lungs; and can refrain from eating as well as breathing for a great part of the year. When first awakened it eats nothing; nor again in the autumn before it retires: through the height of the summer it feeds voraciously, devouring all the food that comes in it's way. I was much taken with it's sagacity in discerning those that do it kind offices: for, as soon as the good old lady comes in sight who has waited on it for more than thirty years, it hobbles towards it's benefactress with aukward alacrity; but remains inattentive to strangers. Thus not only " *the ox knoweth his owner, and the ass his master's crib* ᵇ," but the most abject reptile and torpid of beings distinguishes the hand that feeds it, and is touched with the feelings of gratitude !

I am, &c. &c.

P. S. In about three days after I left *Sussex* the tortoise retired into the ground under the hepatica.

Isaiah i. 3.

LETTER

LETTER XIV.

TO THE SAME.

DEAR SIR, SELBORNE, March 26, 1773.

THE more I reflect on the στοργη of animals, the more I am asto-
nished at it's effects. Nor is the violence of this affection more
wonderful than the shortness of it's duration. Thus every hen is
in her turn the virago of the yard, in proportion to the helplessness
of her brood; and will fly in the face of a dog or a sow in defence
of those chickens, which in a few weeks she will drive before her
with relentless cruelty.

This affection sublimes the passions, quickens the invention, and
sharpens the sagacity of the brute creation. Thus an hen, just become a mother, is no longer that placid bird she used to be, but
with feathers standing an end, wings hovering, and clocking note,
she runs about like one possessed. Dams will throw themselves in
the way of the greatest danger in order to avert it from their pro-
geny. Thus a partridge will tumble along before a sportsman in
order to draw away the dogs from her helpless covey. In the time
of nidification the most feeble birds will assault the most rapacious.
All the hirundines of a village are up in arms at the sight of an hawk,
whom they will persecute till he leaves that district. A very exact
observer has often remarked that a pair of ravens nesting in the rock
of *Gibraltar* would suffer no vulture or eagle to rest near their station,
but would drive them from the hill with an amazing fury: even the

blue

blue thrush at the feafon of breeding would dart out from the clefts of the rocks to chafe away the keftril, or the fparrow-hawk. If you ftand near the neft of a bird that has young, fhe will not be induced to betray them by an inadvertent fondnefs, but will wait about at a diftance with meat in her mouth for an hour together.

Should I farther corroborate what I have advanced above by fome anecdotes which I probably may have mentioned before in converfation, yet you will, I truft, pardon the repetition for the fake of the illuftration.

The flycatcher of the Zoology (the *ftoparola* of *Ray*), builds every year in the vines that grow on the walls of my houfe. A pair of thefe little birds had one year inadvertently placed their neft on a naked bough, perhaps in a fhady time, not being aware of the inconvenience that followed. But an hot funny feafon coming on before the brood was half fledged, the reflection of the wall became infupportable, and muft inevitably have deftroyed the tender young, had not affection fuggefted an expedient, and prompted the parent-birds to hover over the neft all the hotter hours, while with wings expanded, and mouths gaping for breath, they fcreened off the heat from their fuffering offspring.

A farther inftance I once faw of notable fagacity in a willow-wren, which had built in a bank in my fields. This bird a friend and myfelf had obferved as fhe fat in her neft; but were particularly careful not to difturb her, though we faw fhe eyed us with fome degree of jealoufy. Some days after as we paffed that way we were defirous of remarking how this brood went on; but no neft could be found, till I happened to take up a large bundle of long green mofs, as it were, careleffly thrown over the neft in order to dodge the eye of any impertinent intruder.

A ftill

A ſtill more remarkable mixture of ſagacity and inſtinct occurred to me one day as my people were pulling off the lining of an hotbed, in order to add ſome freſh dung. From out of the ſide of this bed leaped an animal with great agility that made a moſt groteſque figure; nor was it without great difficulty that it could be taken; when it proved to be a large white-bellied field-mouſe with three or four young clinging to her teats by their mouths and feet. It was amazing that the deſultory and rapid motions of this dam ſhould not oblige her litter to quit their hold, eſpecially when it appeared that they were ſo young as to be both naked and blind!

To theſe inſtances of tender attachment, many more of which might be daily diſcovered by thoſe that are ſtudious of nature, may be oppoſed that rage of affection, that monſtrous perverſion of the ϛοϱγη, which induces ſome females of the brute creation to devour their young becauſe their owners have handled them too freely, or removed them from place to place! Swine, and ſometimes the more gentle race of dogs and cats, are guilty of this horrid and prepoſterous murder. When I hear now and then of an abandoned mother that deſtroys her offspring, I am not ſo much amazed; ſince reaſon per-verted, and the bad paſſions let looſe, are capable of any enormity: but why the parental feelings of brutes, that uſually flow in one moſt uniform tenor, ſhould ſometimes be ſo extravagantly diverted, I leave to abler philoſophers than myſelf to determine.

I am, &c.

LETTER

LETTER XV.

TO THE SAME.

DEAR SIR,   SELBORNE, July 8, 1773.

SOME young men went down lately to a pond on the verge of *Wolmer-forest* to hunt flappers, or young wild-ducks, many of which they caught, and, among the reft, fome very minute yet well-fledged wild-fowls alive, which upon examination I found to be teals. I did not know till then that teals ever bred in the fouth of *England*, and was much pleafed with the difcovery: this I look upon as a great ftroke in natural hiftory.

We have had, ever fince I can remember, a pair of white owls that conftantly breed under the eaves of this church. As I have paid good attention to the manner of life of thefe birds during their feafon of breeding, which lafts the fummer through, the following remarks may not perhaps be unacceptable:—About an hour before funfet (for then the mice begin to run) they fally forth in queft of prey, and hunt all round the hedges of meadows and fmall enclofures for them, which feem to be their only food. In this irregular country we can ftand on an eminence and fee them beat the fields over like a fetting-dog, and often drop down in the grafs or corn. I have minuted thefe birds with my watch for an hour together, and have found that they return to their neft, the one or the other of them, about once in five minutes; reflecting at the fame time on the adroitnefs that every animal is poffeffed of as far as regards the well being of itfelf and off-

X   fpring.

spring. But a piece of addrefs, which they fhew when they return loaded, fhould not, I think, be paffed over in filence.—As they take their prey with their claws, fo they carry it in their claws to their neft: but, as the feet are neceffary in their afcent under the tiles, they conftantly perch firft on the roof of the chancel, and fhift the moufe from their claws to their bill, that the feet may be at liberty to take hold of the plate on the wall as they are rifing under the eaves.

White owls feem not (but in this I am not pofitive) to hoot at all: all that clamorous hooting appears to me to come from the wood kinds. The white owl does indeed fnore and hifs in a tremendous manner; and thefe menaces well anfwer the intention of intimidating: for I have known a whole village up in arms on fuch an occafion, imagining the church-yard to be full of goblins and fpectres. White owls alfo often fcream horribly as they fly along; from this fcreaming probably arofe the common people's imaginary fpecies of *fcreech-owl*, which they fuperftitioufly think attends the windows of dying perfons. The plumage of the remiges of the wings of every fpecies of owl that I have yet examined is remarkably foft and pliant. Perhaps it may be neceffary that the wings of thefe birds fhould not make much refiftance or rufhing, that they may be enabled to fteal through the air unheard upon a nimble and watchful quarry.

While I am talking of owls, it may not be improper to mention what I was told by a gentleman of the county of *Wilts*. As they were grubbing a vaft hollow pollard-afh that had been the manfion of owls for centuries, he difcovered at the bottom a mafs of matter that at firft he could not account for. After fome examination, he found that it was a congeries of the bones of mice (and perhaps of birds and bats) that had been heaping together

for

for ages, being caft up in pellets out of the crops of many genera-
tions of inhabitants. For owls caft up the bones, fur, and feathers,
of what they devour, after the manner of hawks. He believes,
he told me, that there were bufhels of this kind of fubftance.

When brown owls hoot their throats fwell as big as an hen's egg.
I have known an owl of this fpecies live a full year without any
water. Perhaps the cafe may be the fame with all birds of prey.
When owls fly they ftretch out their legs behind them as a balance
to their large heavy heads . for as moft nocturnal birds have large
eyes and ears they muft have large heads to contain them. Large
eyes I prefume are neceffary to collect every ray of light, and
large concave ears to command the fmalleft degree of found
or noife. I am, &c.

It will be proper to premife here that the fixteenth, eighteenth, twentieth,
and twenty-firft letters have been publifhed already in the Philofophical Tranfactions: but as nicer obfervation has furnifhed feveral corrections and additions, it is
hoped that the republication of them will not give offence; efpecially as thefe
fheets would be very imperfect without them, and as they will be new to
many readers who had no opportunity of feeing them when they made their firft
appearance.

The *hirundines* are a moft inoffenfive, harmlefs, entertaining,
focial, and ufeful tribe of birds : they touch no fruit in our gardens; delight, all except one fpecies, in attaching themfelves to
our houfes; amufe us with their migrations, fongs, and marvellous
agility; and clear our outlets from the annoyances of gnats and
other troublefome infects. Some diftricts in the fouth feas, near

Guiaquil,

Guiaquil[c], are defolated, it feems, by the infinite fwarms of venom-ous mofquitoes, which fill the air, and render thofe coafts infup-portable. It would be worth inquiring whether any fpecies of *hirundines* is found in thofe regions. Whoever contemplates the myriads of infects that fport in the fun-beams of a fummer evening in this country, will foon be convinced to what a degree our atmofphere would be choaked with them was it not for the friendly interpofition of the fwallow tribe.

Many fpecies of birds have their peculiar *lice*; but the *hirundines* alone feem to be annoyed with *dipterous* infects, which infeft every fpecies, and are fo large, in proportion to themfelves, that they muft be extremely irkfome and injurious to them. Thefe are the *hippobofcæ hirundinis*, with narrow fubulated wings, abounding in every neft; and are hatched by the warmth of the bird's own body during incubation, and crawl about under it's feathers.

A *fpecies* of them is familiar to horfemen in the fouth of *England* under the name of *foreft-fly*; and to fome of *fide-fly*, from it's run-ing fideways like a crab. It creeps under the tails, and about the groins, of horfes, which, at their firft coming out of the north, are rendered half frantic by the tickling fenfation; while our own breed little regards them.

The curious *Reaumur* difcovered the large eggs, or rather *pupæ*, of thefe flies as big as the flies themfelves, which he hatched in his own bofom. Any perfon that will take the trouble to examine the old nefts of either fpecies of fwallows may find in them the black fhining cafes or fkins of the *pupæ* of thefe infects : but for other particulars, too long for this place, we refer the reader to *l' Hiftoire d' Infectes* of that admirable entomologift. Tom. iv, pl. 11.

c See *Ulloa*'s Travels.

LETTER

LETTER XVI.

TO THE SAME.

DEAR SIR,   SELBORNE, Nov. 20. 1773.

In obedience to your injunctions I sit down to give you some account of the house-martin, or martlet; and, if my monography of this little domestic and familiar bird should happen to meet with your approbation, I may probably soon extend my inquiries to the rest of the *British hirundines*—the swallow, the swift, and the bank-martin.

A few house-martins begin to appear about the sixteenth of *April*; usually some few days later than the swallow. For some time after they appear the hirundines in general pay no attention to the business of nidification, but play and sport about, either to recruit from the fatigue of their journey, if they do migrate at all, or else that their blood may recover it's true tone and texture after it has been so long benumbed by the severities of winter. About the middle of *May*, if the weather be fine, the martin begins to think in earnest of providing a mansion for it's family. The crust or shell of this nest seems to be formed of such dirt or loam as comes most readily to hand, and is tempered and wrought together with little bits of broken straws to render it tough and tenacious. As this bird often builds against a perpendicular wall without any projecting ledge under, it requires it's utmost efforts to get the first foundation firmly fixed, so that it may safely carry the superstructure. On this occasion the bird not only clings with it's claws,

but

but partly fupports itfelf by ftrongly inclining it's tail againft the wall, making that a fulcrum; and thus fteadied it works and plafters the materials into the face of the brick or ftone. But then, that this work may not, while it is foft and green, pull itfelf down by it's own weight, the provident architect has prudence and for-bearance enough not to advance her work too faft; but by building only in the morning, and by dedicating the reft of the day to food and amufement, gives it fufficient time to dry and harden. About half an inch feems to be a fufficient layer for a day. Thus careful workmen when they build mud-walls (informed at firft perhaps by this little bird) raife but a moderate layer at a time, and then defift; left the work fhould become top-heavy, and fo be ruined by it's own weight. By this method in about ten or twelve days is formed an hemifpheric neft with a fmall aperture towards the top, ftrong, compact, and warm; and perfectly fitted for all the purpofes for which it was intended. But then nothing is more common than for the houfe-fparrow, as foon as the fhell is finifhed, to feize on it as it's own, to eject the owner, and to line it after it's own manner.

After fo much labour is beftowed in erecting a manfion, as Nature feldom works in vain, martins will breed on for feveral years together in the fame neft, where it happens to be well fhelter-ed and fecure from the injuries of weather. The fhell or cruft of the neft is a fort of ruftic-work full of knobs and protuberances on the outfide: nor is the infide of thofe that I have examined fmoothed with any exactnefs at all; but is rendered foft and warm, and fit for incubation, by a lining of fmall ftraws, graffes, and fea-thers; and fometimes by a bed of mofs interwoven with wool. In this neft they tread, or engender, frequently during the time of building; and the hen lays from three to five white eggs.

At

At firſt when the young are hatched, and are in a naked and helpleſs condition, the parent birds, with tender aſſiduity, carry out what comes away from their young. Was it not for this affectionate cleanlineſs the neſtlings would ſoon be burnt up, and deſtroyed in ſo deep and hollow a neſt, by their own cauſtic excrement. In the quadruped creation the ſame neat precaution is made uſe of; particularly among dogs and cats, where the dams lick away what proceeds from their young. But in birds there ſeems to be a particular proviſion, that the dung of neſtlings is enveloped in a tough kind of jelly, and therefore is the eaſier conveyed off without ſoiling or daubing. Yet, as nature is cleanly in all her ways, the young perform this office for themſelves in a little time by thruſting their tails out at the aperture of their neſt. As the young of ſmall birds preſently arrive at their ἡλικία, or full growth, they ſoon become impatient of confinement, and ſit all day with their heads out at the orifice, where the dams, by clinging to the neſt, ſupply them with food from morning to night. For a time the young are fed on the wing by their parents; but the feat is done by ſo quick and almoſt imperceptible a flight, that a perſon muſt have attended very exactly to their motions before he would be able to perceive it. As ſoon as the young are able to ſhift for themſelves, the dams immediately turn their thoughts to the buſineſs of a ſecond brood: while the firſt flight, ſhaken off and rejected by their nurſes, congregate in great flocks, and are the birds that are ſeen cluſtering and hovering on ſunny mornings and evenings round towers and ſteeples, and on the roofs of churches and houſes. Theſe congregatings uſually begin to take place about the firſt week in *Auguſt*; and therefore we may conclude that by that time the firſt flight is pretty well over. The young of this ſpecies do not quit their abodes all together; but the more forward birds get abroad ſome days

before

before the reft. Thefe approaching the eaves of buildings, and playing about before them, make people think that feveral old ones
attend one neft.   They are often capricious in fixing on a nefting-
place, beginning many edifices, and leaving them unfinifhed ; but
when once a neft is completed in a fheltered place, it ferves for
feveral feafons. Thofe which breed in a ready finifhed houfe get the
ftart in hatching of thofe that build new by ten days or a fortnight.
Thefe induftrious artificers are at their labours in the long days before four in the morning : when they fix their materials they plafter
them on with their chins, moving their heads with a quick vibratory
motion.   They dip and wafh as they fly fometimes in very hot
weather, but not fo frequently as fwallows.   It has been obferved
that martins ufually build to a north-eaft or north-weft afpect, that
the heat of the fun may not crack and deftroy their nefts : but inftances are alfo remembered where they bred for many years in vaft
abundance in an hot ftifled inn-yard, againft a wall facing to the
fouth.

Birds in general are wife in their choice of fituation : but in this
neighbourhood every fummer is feen a ftrong proof to the contrary
at an houfe without eaves in an expofed diftrict, where fome martins build year by year in the corners of the windows. But, as the
corners of thefe windows (which face to the fouth-eaft and fouth-
weft) are too fhallow, the nefts are wafhed down every hard rain ;
and yet thefe birds drudge on to no purpofe from fummer to
fummer, without changing their afpect or houfe.   It is a piteous
fight to fee them labouring when half their neft is wafhed away and
bringing dirt - - - " *generis lapfi farcire ruinas*".   Thus is inftinct a
moft wonderful unequal faculty; in fome inftances fo much above
reafon, in other refpects fo far below it ! Martins love to frequent
towns, efpecially if there are great lakes and rivers at hand ; nay
                                                                they

they even affect the close air of *London*. And I have not only seen them nesting in the *Borough*, but even in the *Strand* and *Fleet-street*; but then it was obvious from the dinginess of their aspect that their feathers partook of the filth of that sooty atmosphere. Martins are by far the least agile of the four species; their wings and tails are short, and therefore they are not capable of such surprising turns and quick and glancing evolutions as the swallow. Accordingly they make use of a placid easy motion in a middle region of the air, seldom mounting to any great height, and never sweeping long together over the surface of the ground or water. They do not wander far for food, but affect sheltered districts, over some lake, or under some hanging wood, or in some hollow vale, especially in windy weather. They breed the latest of all the swallow kind: in 1772 they had nestlings on to *October* the twenty-first, and are never without unfledged young as late as *Michaelmas*.

As the summer declines the congregating flocks increase in numbers daily by the constant accession of the second broods; till at last they swarm in myriads upon myriads round the villages on the *Thames*, darkening the face of the sky as they frequent the aits of that river, where they roost. They retire, the bulk of them I mean, in vast flocks together about the beginning of *October*: but have appeared of late years in a considerable flight in this neighbourhood, for one day or two, as late as *November* the third and sixth, after they were supposed to have been gone for more than a fortnight. They therefore withdraw with us the latest of any species. Unless these birds are very short-lived indeed, or unless they do not return to the district where they are bred, they must undergo vast devastations some how, and some where; for the birds that return yearly bear no manner of proportion to the birds that retire.

Y

House-

House-martins are diſtinguiſhed from their congeners by having their legs covered with ſoft downy feathers down to their toes. They are no ſongſters; but twitter in a pretty inward ſoft manner in their neſts. During the time of breeding they are often greatly moleſted with fleas.

I am, &c.

LETTER XVII.

TO THE SAME.

DEAR SIR,   RINGMER, near LEWES, Dec. 9, 1773.

I RECEIVED your laſt favour juſt as I was ſetting out for this place; and am pleaſed to find that my monography met with your approbation. My remarks are the reſult of many years obſervation; and are, I truſt, true in the whole: though I do not pretend to ſay that they are perfectly void of miſtake, or that a more nice obſerver might not make many additions, ſince ſubjects of this kind are inexhauſtible.

If you think my letter worthy the notice of your reſpectable ſociety, you are at liberty to lay it before them; and they will conſider it, I hope, as it was intended, as an humble attempt to promote a more minute inquiry into natural hiſtory; into the life and converſation of animals. Perhaps hereafter I may be induced to take the houſe-ſwallow under conſideration; and from that proceed to the reſt of the *Britiſh hirundines*.

Though

Though I have now travelled the *Suffex-downs* upwards of thirty years, yet I ftill inveftigate that chain of majeftic mountains with frefh admiration year by year; and think I fee new beauties every time I traverfe it. This range, which runs from *Chichefter* eaftward as far as *Eaft-Bourn*, is about fixty miles in length, and is called *The South Downs*, properly fpeaking, only round *Lewes*. As you pafs along you command a noble view of the wild, or weald, on one hand, and the broad downs and fea on the other. Mr. *Ray* ufed to vifit a family[d] juft at the foot of thefe hills, and was fo ravifhed with the profpect from *Plumpton-plain*, near *Lewes*, that he mentions thofe fcapes in his " Wifdom of God in the " Works of the Creation" with the utmoft fatisfaction, and thinks them equal to any thing he had feen in the fineft parts of *Europe*.

For my own part, I think there is fomewhat peculiarly fweet and amufing in the fhapely figured afpect of chalk-hills in preference to thofe of ftone, which are rugged, broken, abrupt, and fhapelefs.

Perhaps I may be fingular in my opinion, and not fo happy as to convey to you the fame idea; but I never contemplate thefe mountains without thinking I perceive fomewhat analogous to growth in their gentle fwellings and fmooth fungus-like protuber-ances, their fluted fides, and regular hollows and flopes, that carry at once the air of vegetative dilation and expanfion - - - - - -

- - - Or was there ever a time when thefe immenfe maffes of calcarious matter were thrown into fermentation by fome adventi-tious moifture; were raifed and leavened into fuch fhapes by fome plaftic power; and fo made to fwell and heave their broad backs into the fky fo much above the lefs animated clay of the wild below?

[d] *Mr. Courthope of Danny.*

By

By what I can gueſs from the admeaſurements of the hills that have been taken round my houſe, I ſhould ſuppoſe that theſe hills ſurmount the wild at an average at about the rate of five hundred feet.

One thing is very remarkable as to the ſheep: from the weſtward till you get to the river *Adur* all the flocks have horns, and ſmooth white faces, and white legs; and a hornleſs ſheep is rarely to be ſeen: but as ſoon as you paſs that river eaſtward, and mount *Beeding-hill*, all the flocks at once become hornleſs, or, as they call them, poll-ſheep; and have moreover black faces with a white tuft of wool on their foreheads, and ſpeckled and ſpotted legs: ſo that you would think that the flocks of *Laban* were paſturing on one ſide of the ſtream, and the variegated breed of his ſon-in-law *Jacob* were cantoned along on the other. And this diverſity holds good reſpectively on each ſide from the valley of *Bramber* and *Beeding* to the eaſtward, and weſtward all the whole length of the downs. If you talk with the ſhepherds on this ſubject, they tell you that the caſe has been ſo from time immemorial; and ſmile at your ſimplicity if you aſk them whether the ſituation of theſe two different breeds might not be reverſed? However, an intelligent friend of mine near *Chicheſter* is determined to try the experiment; and has this autumn, at the hazard of being laughed at, introduced a parcel of black-faced hornleſs rams among his horned weſtern ewes. The black-faced poll-ſheep have the ſhorteſt legs and the fineſt wool.

As I had hardly ever before travelled theſe downs at ſo late a ſeaſon of the year, I was determined to keep as ſharp a look-out as poſſible ſo near the ſouthern coaſt, with reſpect to the ſummer ſhort-winged birds of paſſage. We make great inquiries concerning the withdrawing of the ſwallow kind, without examining enough into

the

the caufes why *this tribe* is never to be feen in winter : for, *entre nous*, the difappearing of the latter is more marvellous than that of the former, and much more unaccountable. The *hirundines*, if they pleafe, are certainly capable of migration ; and yet no doubt are often found in a torpid ftate : but redftarts, nightingales, white-throats, black-caps, &c. &c. are very ill provided for long flights ; have never been once found, as I ever heard of, in a torpid ftate, and yet can never be fuppofed, in fuch troops, from year to year to dodge and elude the eyes of the curious and inquifitive, which from day to day difcern the other fmall birds that are known to abide our winters. But, notwithftanding all my care, I faw nothing like a fummer bird of paffage : and, what is more ftrange, not one wheat-ear, though they abound fo in the autumn as to be a confiderable perquifite to the fhepherds that take them ; and though many are to be feen to my knowledge all the winter through in many parts of the fouth of *England*. The moft intelligent fhepherds tell me that fome few of thefe birds appear on the downs in *March*, and then withdraw to breed probably in warrens and ftone-quarries : now and then a neft is plowed up in a fallow on the downs under a furrow, but it is thought a rarity. At the time of wheat-harveft they begin to be taken in great numbers ; are fent for fale in vaft quantities to *Brighthelmftone* and *Tunbridge* ; and appear at the tables of all the gentry that entertain with any degree of elegance. About *Michaelmas* they retire and are feen no more till *March*. Though thefe birds are, when in feafon, in great plenty on the fouth downs round *Lewes*, yet at *Eaft-Bourn*, which is the eaftern extremity of thofe downs, they abound much more. One thing is very remarkable—that though in the height of the feafon fo many hundreds of dozens are taken, yet they never are feen to flock ; and it is a rare thing to fee more than three or four at a time : fo that there

muft

muſt be a perpetual flitting and conſtant progreſſive ſucceſſion. It does not appear that any wheat-ears are taken to the weſtward of *Houghton-bridge*, which ſtands on the river *Arun*.

I did not fail to look particularly after my new migration of *ring-ouſels*; and to take notice whether they continued on the downs to this ſeaſon of the year; as I had formerly remarked them in the month of *October* all the way from *Chicheſter* to *Lewes* where-ever there were any ſhrubs and covert: but not one bird of this ſort came within my obſervation. I only ſaw a few *larks* and *whin-chats*, ſome *rooks*, and ſeveral *kites* and *buzzards*.

About *Midſummer* a flight of *croſs-bills* comes to the pine-groves about this houſe, but never makes any long ſtay.

The old tortoiſe, that I have mentioned in a former letter, ſtill continues in this garden; and retired under ground about the twentieth of *November*, and came out again for one day on the thirtieth: it lies now buried in a wet ſwampy border under a wall facing to the ſouth, and is enveloped at preſent in mud and mire!

Here is a large rookery round this houſe, the inhabitants of which ſeem to get their livelihood very eaſily; for they ſpend the greateſt part of the day on their neſt-trees when the weather is mild. Theſe rooks retire every evening all the winter from this rookery, where they only call by the way, as they are going to rooſt in deep woods: at the dawn of day they always reviſit their neſt-trees, and are preceded a few minutes by a flight of daws, that act, as it were, as their harbingers.

I am, &c.

LETTER

LETTER XVIII.

TO THE SAME.

DEAR SIR, SELBORNE, Jan. 29, 1774.

THE houfe-fwallow, or chimney-fwallow, is undoubtedly the firft comer of all the *Britifh hirundines*; and appears in general on or about the thirteenth of *April*, as I have remarked from many years obfervation. Not but now and then a ftraggler is feen much earlier: and, in particular, when I was a boy I obferved a fwallow for a whole day together on a funny warm *Shrove Tuefday*; which day could not fall out later than the middle of *March*, and often happened early in *February*.

It is worth remarking that thefe birds are feen firft about lakes and mill-ponds; and it is alfo very particular, that if thefe early vifiters happen to find froft and fnow, as was the cafe of the two dreadful fprings of 1770 and 1771, they immediately withdraw for a time. A circumftance this much more in favour of hiding than migration; fince it is much more probable that a bird fhould retire to it's hybernaculum juft at hand, than return for a week or two only to warmer latitudes.

The fwallow, though called the chimney-fwallow, by no means builds altogether in chimnies, but often within barns and out-houfes againft the rafters; and fo fhe did in *Virgil*'s time:

— — — — " Antè
" *Garrula quàm tignis nidos fufpendat hirundo.*"

In

In *Sweden* she builds in barns, and is called *ladu swala*, the barn-swallow. Besides, in the warmer parts of *Europe* there are no chimnies to houses, except they are *English-built*: in these countries she constructs her nest in porches, and gate-ways, and galleries, and open halls.

Here and there a bird may affect some odd, peculiar place; as we have known a swallow build down the shaft of an old well, through which chalk had been formerly drawn up for the purpose of manure: but in general with us this *hirundo* breeds in chimnies; and loves to haunt those stacks where there is a constant fire, no doubt for the sake of warmth. Not that it can subsist in the immediate shaft where there is a fire; but prefers one adjoining to that of the kitchen, and disregards the perpetual smoke of that funnel, as I have often observed with some degree of wonder.

Five or six or more feet down the chimney does this little bird begin to form her nest about the middle of *May*, which consists, like that of the house-martin, of a crust or shell composed of dirt or mud, mixed with short pieces of straw to render it tough and permanent; with this difference, that whereas the shell of the martin is nearly hemispheric, that of the swallow is open at the top, and like half a deep dish: this nest is lined with fine grasses, and feathers which are often collected as they float in the air.

Wonderful is the address which this adroit bird shews all day long in ascending and descending with security through so narrow a pass. When hovering over the mouth of the funnel, the vibrations of her wings acting on the confined air occasion a rumbling like thunder. It is not improbable that the dam submits to this inconvenient situation so low in the shaft, in order to secure her broods from rapacious birds, and particularly from owls, which

frequently

frequently fall down chimnies, perhaps in attempting to get at these neftlings.

The fwallow lays from four to fix white eggs, dotted with red fpecks; and brings out her firft brood about the laft week in *June*, or the firft week in *July*. The progreffive method by which the young are introduced into life is very amufing: firft, they emerge from the fhaft with difficulty enough, and often fall down into the rooms below: for a day or fo they are fed on the chimney-top, and then are conducted to the dead leaflefs bough of fome tree, where, fitting in a row, they are attended with great affiduity, and may then be called *perchers*. In a day or two more they become *flyers*, but are ftill unable to take their own food; therefore they play about near the place where the dams are hawking for flies; and, when a mouthful is collected, at a certain fignal given, the dam and the neftling advance, rifing towards each other, and meeting at an angle; the young one all the while uttering fuch a little quick note of gratitude and complacency, that a perfon muft have paid very little regard to the wonders of Nature that has not often remarked this feat.

The dam betakes herfelf immediately to the bufinefs of a fecond brood as foon as fhe is difengaged from her firft; which at once affociates with the firft broods of *houfe-martins*; and with them congregates, cluftering on funny roofs, towers, and trees. This hirundo brings out her fecond brood towards the middle and end of *Auguft*.

All the fummer long is the fwallow a moft inftructive pattern of unwearied induftry and affection; for, from morning to night, while there is a family to be fupported, fhe fpends the whole day in fkimming clofe to the ground, and exerting the moft fudden turns and quick evolutions. Avenues, and long walks under

Z   hedges,

hedges, and pasture-fields, and mown meadows where cattle graze, are her delight, especially if there are trees intersperfed; because in such spots insects most abound. When a fly is taken a smart snap from her bill is heard, resembling the noise at the shutting of a watch-case; but the motion of the mandibles are too quick for the eye.

The swallow, probably the male bird, is the *excubitor* to house-martins, and other little birds, announcing the approach of birds of prey. For as soon as an hawk appears, with a shrill alarming note he calls all the swallows and martins about him; who pursue in a body, and buffet and strike their enemy till they have driven him from the village, darting down from above on his back, and rising in a perpendicular line in perfect security. This bird also will sound the alarm, and strike at cats when they climb on the roofs of houses, or otherwise approach the nests. Each species of hirundo drinks as it flies along, sipping the surface of the water; but the swallow alone, in general, *washes* on the wing, by dropping into a pool for many times together: in very hot weather house-martins and bank-martins dip and wash a little.

The swallow is a delicate songster, and in soft sunny weather sings both perching and flying; on trees in a kind of concert, and on chimney tops: is also a bold flyer, ranging to distant downs and commons even in windy weather, which the other species seem much to dislike; nay, even frequenting exposed sea-port towns, and making little excursions over the salt water. Horsemen on wide downs are often closely attended by a little party of swallows for miles together, which plays before and behind them, sweeping around, and collecting all the sculking insects that are roused by the trampling of the horses feet: when the wind blows hard, without this expedient, they are often forced to settle to pick up their lurking prey.   This

This species feeds much on little *coleoptera*, as well as on gnats and flies; and often settles on dug ground, or paths, for gravels to grind and digest it's food. Before they depart, for some weeks, to a bird, they forsake houses and chimnies, and roost in trees; and usually withdraw about the beginning of *October*; though some few stragglers may appear on at times till the first week in *November*.

Some few pairs haunt the new and open streets of *London* next the fields, but do not enter, like the house-martin, the close and crowded parts of the city.

Both male and female are distinguished from their congeners by the length and forkedness of their tails. They are undoubtedly the most nimble of all the species: and when the male pursues the female in amorous chase, they then go beyond their usual speed, and exert a rapidity almost too quick for the eye to follow.

After this circumstantial detail of the life and discerning στοργη of the swallow, I shall add, for your farther amusement, an anecdote or two not much in favour of her sagacity:—

A certain swallow built for two years together on the handles of a pair of garden-shears, that were stuck up against the boards in an out-house, and therefore must have her nest spoiled whenever that implement was wanted: and, what is stranger still, another bird of the same species built it's nest on the wings and body of an owl that happened by accident to hang dead and dry from the rafter of a barn. This owl, with the nest on it's wings, and with eggs in the nest, was brought as a curiosity worthy the most elegant private museum in *Great-Britain*. The owner, struck with the oddity of the sight, furnished the bringer with a large shell, or conch, desiring him to fix it just where the owl hung: the person did as he was ordered, and the following year a pair, pro-

bably

bably the fame pair, built their neft in the conch, and laid their eggs.

The owl and the conch make a ftrange grotefque appearance, and are not the leaft curious fpecimens in that wonderful collection of art and nature [f].

Thus is inftinct in animals, taken the leaft out of it's way, an undiftinguishing, limited faculty; and blind to every circumftance that does not immediately refpect felf-prefervation, or lead at once to the propagation or fupport of their fpecies.

I am,

With all refpect, &c. &c.

LETTER XIX.

TO THE SAME.

DEAR SIR,   SELBORNE, Feb. 14, 1774.

I RECEIVED your favour of the eighth, and am pleafed to find that you read my little hiftory of the fwallow wtih your ufual candour: nor was I the lefs pleafed to find that you made objections where you faw reafon.

[f] Sir *Afhton Lever's* Mufæum.

As

As to the quotations, it is difficult to fay precifely which fpecies of hirundo *Virgil* might intend in the lines in queftion, fince the ancients did not attend to fpecific differences like modern natu-ralifts : yet fomewhat may be gathered, enough to incline me to fuppofe that in the two paffages quoted the poet had his eye on the fwallow.

In the firft place the epithet *garrula* fuits the *fwallow* well, who is a great fongfter ; and not the *martin*, which is rather a mute bird ; and when it fings is fo inward as fcarce to be heard. Befides, if *tignum* in that place fignifies a *rafter* rather than a *beam*, as it feems to me to do, then I think it muft be the *fwallow* that is alluded to, and not the *martin* ; fince the former does frequently build *within the roof* againft the *rafters* ; while the latter always, as far as I have been able to obferve, builds *without the roof* againft eaves and cornices.

As to the *fimile*, too much ftrefs muft not be laid on it : yet the epithet *nigra* fpeaks plainly in favour of the fwallow, whofe back and wings are very black ; while the rump of the martin is milk-white, it's back and wings blue, and all it's under part white as fnow. Nor can the clumfy motions (comparatively clumfy) of the martin well reprefent the fudden and artful evolutions and quick turns which *Juturna* gave to her brother's chariot, fo as to elude the eager purfuit of the enraged *Æneas*. The verb *fonat* alfo feems to imply a bird that is fomewhat loquacious [g].

" [g] *Nigra* velut magnas domini cum divitis ædes
" Pervolat, et pennis alta atria luftrat hirundo,
" Pabula parva legens, nidifque loquacibus efcas :
" Et nunc porticibus vacuis, nunc humida circum
" Stagna *fonat* " — — — —

We

We have had a very wet autumn and winter, so as to raise the springs to a pitch beyond any thing since 1764; which was a remarkable year for floods and high waters. The land-springs, which we call *lavants*, break out much on the downs of *Suffex*, *Hampfhire*, and *Wiltfhire*. The country people say when the *lavants* rise corn will always be dear; meaning that when the earth is so glutted with water as to send forth springs on the downs and uplands, that the corn-vales must be drowned: and so it has proved for thefe ten or eleven years past. For land-springs have never obtained more since the memory of man than during that period; nor has there been known a greater fcarcity of all forts of grain, confidering the great improvements of modern hufbandry. Such a run of wet feafons a century or two ago would, I am perfuaded, have occafioned a famine. Therefore pamphlets and newfpaper letters, that talk of combinations, tend to inflame and miflead; since we must not expect plenty till Providence fends us more favourable feafons.

The wheat of last year, all round this diftrict, and in the county of *Rutland*, and elfewhere, yields remarkably bad: and our wheat on the ground, by the continual late fudden viciffitudes from fierce froft to pouring rains, looks poorly; and the turnips rot very faft.

I am, &c.

LETTER

LETTER XX.

TO THE SAME.

DEAR SIR,   SELBORNE, Feb. 26, 1774.

THE sand-martin, or bank-martin, is by much the least of any of the *British hirundines*; and, as far as we have ever seen, the smallest known hirundo : though *Brisson* asserts that there is one much smaller, and that is the *hirundo esculenta*.

But it is much to be regretted that it is scarce possible for any observer to be so full and exact as he could wish in reciting the circumstances attending the life and conversation of this little bird, since it is *fera naturâ*, at least in this part of the kingdom, disclaiming all domestic attachments, and haunting wild heaths and commons where there are large lakes : while the other species, especially the swallow and house-martin, are remarkably gentle and domesticated, and never seem to think themselves safe but under the protection of man.

Here are in this parish, in the sand-pits and banks of the lakes of *Woolmer-forest*, several colonies of these birds; and yet they are never seen in the village; nor do they at all frequent the cottages that are scattered about in that wild district. The only instance I ever remember where this species haunts any building is at the town of *Bishop's Waltham*, in this county, where many sand-martins nestle and breed in the scaffold-holes of the back-wall of *William* of *Wykeham's* stables : but then this wall stands in a very sequestered and

retired

retired enclofure, and faces upon a large and beautiful lake. And indeed this fpecies feems fo to delight in large waters, that no inftance occurs of their abounding, but near vaft pools or rivers: and in particular it has been remarked that they fwarm in the banks of the *Thames* in fome places below *London-bridge*.

It is curious to obferve with what different degrees of architectonic fkill Providence has endowed birds of the fame genus, and fo nearly correfpondent in their general mode of life! for while the fwallow and the houfe-martin difcover the greateft addrefs in raifing and fecurely fixing crufts or fhells of loam as cunabula for their young, the bank-martin terebrates a round and regular hole in the fand or earth, which is ferpentine, horizontal, and about two feet deep. At the inner end of this burrow does this bird depofit, in a good degree of fafety, her rude neft, confifting of fine graffes and feathers, ufually goofe-feathers, very inartificially laid together.

Perfeverance will accomplifh any thing: though at firft one would be difinclined to believe that this weak bird, with her foft and tender bill and claws, fhould ever be able to bore the ftubborn fand-bank without entirely difabling herfelf: yet with thefe feeble inftruments have I feen a pair of them make great difpatch: and could remark how much they had fcooped that day by the frefh fand which ran down the bank, and was of a different colour from that which lay loofe and bleached in the fun.

In what fpace of time thefe little artifts are able to mine and finifh thefe cavities I have never been able to difcover, for reafons given above; but it would be a matter worthy of obfervation, where it falls in the way of any naturalift to make his remarks. This I have often taken notice of, that feveral holes of different depths are left unfinifhed at the end of fummer. To imagine that

thefe

thefe beginnings were intentionally made in order to be in the greater forwardnefs for next fpring, is allowing perhaps too much forefight and *rerum prudentia* to a fimple bird. May not the caufe of thefe *latebræ* being left unfinifhed arife from their meeting in thofe places with ftrata too harfh, hard, and folid, for their purpofe, which they relinquifh, and go to a frefh fpot that works more freely? Or may they not in other places fall in with a foil as much too loofe and mouldering, liable to flounder, and threatening to overwhelm them and their labours?

One thing is remarkable — that, after fome years, the old holes are forfaken and new ones bored; perhaps becaufe the old habitations grow foul and fetid from long ufe, or becaufe they may fo abound with fleas as to become untenantable. This fpecies of fwallow moreover is ftrangely annoyed with fleas: and we have feen fleas, bed-fleas (*pulex irritans*), fwarming at the mouths of thefe holes, like bees on the ftools of their hives.

The following circumftance fhould by no means be omitted — that thefe birds do *not* make ufe of their caverns by way of hybernacula, as might be expected; fince banks fo perforated have been dug out with care in the winter, when nothing was found but empty nefts.

The fand-martin arrives much about the fame time with the fwallow, and lays, as fhe does, from four to fix white eggs. But as this fpecies is *cryptogame*, carrying on the bufinefs of nidification, incubation, and the fupport of it's young in the dark, it would not be fo eafy to afcertain the time of breeding, were it not for the coming forth of the broods, which appear much about the time, or rather fomewhat earlier than thofe of the fwallow. The neftlings are fupported in common like thofe of their congeners, with gnats and other fmall infects; and fometimes they are fed with *libellulæ*

A a (dragon-

(dragon-flies) almoſt as long as themſelves. In the laſt week in
June we have ſeen a row of theſe ſitting on a rail near a great pool
as *perchers*; and ſo young and helpleſs, as eaſily to be taken by
hand: but whether the dams ever feed them on the wing, as
ſwallows and houſe-martins do, we have never yet been able to
determine; nor do we know whether they purſue and attack birds
of prey.

When they happen to breed near hedges and encloſures, they
are diſpoſſeſſed of their breeding holes by the houſe-ſparrow, which
is on the ſame account a fell adverſary to houſe-martins.

Theſe *hirundines* are no ſongſters, but rather mute, making only
a little harſh noiſe when a perſon approaches their neſts. They
ſeem not to be of a ſociable turn, never with us congregating with
their congeners in the autumn. Undoubtedly they breed a ſecond
time, like the houſe-martin and ſwallow; and withdraw about
Michaelmas.

Though in ſome particular diſtricts they may happen to abound,
yet in the whole, in the ſouth of *England* at leaſt, is this much the
rareſt ſpecies. For there are few towns or large villages but what
abound with houſe-martins; few churches, towers, or ſteeples, but
what are haunted by ſome ſwifts; ſcarce a hamlet or ſingle cottage-
chimney that has not it's ſwallow; while the bank-martins, ſcattered
here and there, live a ſequeſtered life among ſome abrupt ſand-hills,
and in the banks of ſome few rivers.

Theſe birds have a peculiar manner of flying; flitting about
with odd jerks, and vacillations, not unlike the motions of a
butterfly. Doubtleſs the flight of all *hirundines* is influenced by,
and adapted to, the peculiar ſort of inſects which furniſh their food.
Hence it would be worth inquiry to examine what particular genus
of inſects affords the principal food of each reſpective ſpecies of
ſwallow.   Notwithſtanding

Notwithstanding what has been advanced above, some few sand-martins, I see, haunt the skirts of *London*, frequenting the dirty pools in *Saint George's-Fields*, and about *White-Chapel*. The question is where these build, since there are no banks or bold shores in that neighbourhood: perhaps they nestle in the scaffold holes of some old or new deserted building. They dip and wash as they fly sometimes, like the house-martin and swallow.

Sand-martins differ from their congeners in the diminutiveness of their size, and in their colour, which is what is usually called a mouse-colour. Near *Valencia*, in *Spain*, they are taken, says *Willughby*, and sold in the markets for the table; and are called by the country people, probably from their desultory jerking manner of flight, *Papilion de Montagna*.

LETTER XXI.

TO THE SAME.

DEAR SIR,   SELBORNE, Sept. 28, 1774.

As the *swift* or *black-martin* is the largest of the *British hirundines*, so is it undoubtedly the latest comer. For I remember but one instance of it's appearing before the last week in *April*: and in some of our late frosty, harsh springs, it has not been seen till the beginning of *May*. This species usually arrives in pairs.

A a 2   The

The fwift, like the fand-martin, is very defective in architecture, making no cruft, or fhell, for it's neft; but forming it of dry graffes and feathers, very rudely and inartificially put together. With all my attention to thefe birds, I have never been able once to difcover one in the act of collecting or carrying in materials: fo that I have fufpected (fince their nefts are exactly the fame) that they fome-times ufurp upon the houfe-fparrows, and expel them, as fparrows do the houfe and fand-martin; well remembering that I have feen them fquabbling together at the entrance of their holes; and the fparrows up in arms, and much-difconcerted at thefe intruders. And yet I am affured, by a nice obferver in fuch matters, that they do collect feathers for their nefts in *Andalufia*; and that he has fhot them with fuch materials in their mouths.

Swifts, like fand-martins, carry on the bufinefs of nidification quite in the dark, in crannies of caftles, and towers, and fteeples, and upon the tops of the walls of churches under the roof; and therefore cannot be fo narrowly watched as thofe fpecies that build more openly: but, from what I could ever obferve, they begin nefting about the middle of *May*; and I have remarked, from eggs taken, that they have fat hard by the ninth of *June*. In general they haunt tall buildings, churches, and fteeples, and breed only in fuch: yet in this village fome pairs frequent the loweft and meaneft cottages, and educate their young under thofe thatched roofs. We remember but one inftance where they breed out of buildings; and that is in the fides of a deep chalkpit near the town of *Odiham*, in this county, where we have feen many pairs entering the crevices, and fkimming and fqueaking round the precipices.

As I have regarded thefe amufive birds with no fmall attention, if I fhould advance fomething new and peculiar with refpect to them, and different from all other birds, I might perhaps be cre-
dited;

dited; especially as my affertion is the refult of many years exact obfervation. The fact that I would advance is, that fwifts *tread*, or copulate, on the wing: and I would wifh any nice obferver, that is ftartled at this fuppofition, to ufe his own eyes, and I think he will foon be convinced. In another *clafs* of animals, *viz.* the *infect*, nothing is fo common as to fee the different fpecies of many genera in conjunction as they fly. The fwift is almoft continually on the wing; and as it never fettles on the ground, on trees, or roofs, would feldom find opportunity for amorous rites, was it not enabled to indulge them in the air. If any perfon would watch thefe birds of a fine morning in *May*, as they are failing round at a great height from the ground, he would fee, every now and then, one drop on the back of another, and both of them fink down together for many fathoms with a loud piercing fhriek. This I take to be the juncture when the bufinefs of generation is carrying on.

As the fwift eats, drinks, collects materials for it's neft, and, as it feems, propagates on the wing; it appears to live more in the air than any other bird, and to perform all functions there fave thofe of fleeping and incubation.

This *hirundo* differs widely from it's congeners in laying invariably but *two* eggs at a time, which are milk-white, long, and peaked at the fmall end; whereas the other fpecies lay at each brood from *four* to *fix*. It is a moft alert bird, rifing very early, and retiring to rooft very late; and is on the wing in the height of fummer at leaft fixteen hours. In the longeft days it does not withdraw to reft till a quarter before nine in the evening, being the lateft of all day birds. Juft before they retire whole groups of them affemble high in the air, and fqueak, and fhoot about with wonderful rapidity. But this bird is never fo much alive as in fultry thundry weather, when it expreffes great alacrity, and calls

forth

forth all it's powers. In hot mornings feveral, getting together in little parties, dafh round the fteeples and churches, fqueaking as they go in a very clamorous manner : thefe, by nice obfervers, are fuppofed to be males ferenading their fitting hens; and not without reafon, fince they feldom fqueak till they come clofe to the walls or eaves, and fince thofe within utter at the fame time a little inward note of complacency.

When the hen has fat hard all day, fhe rufhes forth juft as it is almoft dark, and ftretches and relieves her weary limbs, and fnatches a fcanty meal for a few minutes, and then returns to her duty of incubation. Swifts, when wantonly and cruelly fhot while they have young, difcover a little lump of infects in their mouths, which they pouch and hold under their tongue. In general they feed in a much higher diftrict than the other fpecies; a proof that gnats and other infects do alfo abound to a confiderable height in the air : they alfo range to vaft diftances; fince loco-motion is no labour to them, who are endowed with fuch wonderful powers of wing. Their powers feem to be in proportion to their leavers; and their wings are longer in proportion than thofe of almoft any other bird. When they mute, or cafe themfelves in flight, they raife their wings, and make them meet over their backs.

At fome certain times in the fummer I had remarked that fwifts were hawking very low for hours together over pools and ftreams; and could not help inquiring into the object of their purfuit that induced them to defcend fo much below their ufual range. After fome trouble, I found that they were taking *phryganeæ*, *ephemeræ*, and *libellulæ* (cadew-flies, may-flies, and dragon-flies) that were juft emerged out of their aurelia ftate. I then no longer wondered that they fhould be fo willing to ftoop for a prey that afforded them fuch plentiful and fucculent nourifhment.

They

They bring out their young about the middle or latter end of *July:* but as thefe never become perchers, nor, that ever I could difcern, are fed on the wing by their dams, the coming forth of the young is not fo notorious as in the other fpecies.

On the thirtieth of laft *June* I untiled the eaves of an houfe where many pairs build, and found in each neft only *two* fquab, naked *pulli :* on the eighth of *July* I repeated the fame inquiry, and found they had made very little progrefs towards a fledged ftate, but were ftill naked and helplefs. From whence we may conclude that birds whofe way of life keeps them perpetually on the wing would not be able to quit their neft till the end of the month. Swallows and martins, that have numerous families, are continually feeding them every two or three minutes ; while fwifts, that have but two young to maintain, are much at their leifure, and do not attend on their nefts for hours together.

Sometimes they purfue and ftrike at hawks that come in their way; but not with that vehemence and fury that fwallows exprefs on the fame occafion. They are out all day long in wet days, feeding about, and difregarding ftill rain : from whence two things may be gathered ; firft, that many infects abide high in the air, even in rain ; and next, that the feathers of thefe birds muft be well preened to refift fo much wet. Windy, and particularly windy weather with heavy fhowers, they diflike ; and on fuch days withdraw, and are fcarce ever feen.

There is a circumftance refpecting the *colour* of fwifts, which feems not to be unworthy our attention. When they arrive in the fpring they are all over of a gloffy, dark foot-colour, except their chins, which are white ; but, by being all day long in the fun and air, they become quite weather-beaten and bleached before they depart, and yet they return gloffy again in the fpring. Now, if

they

they purfue the fun into lower latitudes, as fome fuppofe, in order to enjoy a perpetual fummer, why do they not return bleached? Do they not rather perhaps retire to reft for a feafon, and at that juncture moult and change their feathers, fince all other birds are known to moult foon after the feafon of breeding?

Swifts are very anomalous in many particulars, diffenting from all their congeners not only in the number of their young, but in breeding but *once* in a fummer; whereas all the other *Britiſh hirundines* breed invariably *twice*. It is paft all doubt that fwifts can breed but once, fince they withdraw in a fhort time after the flight of their young, and fome time before their congeners bring out their fecond broods. We may here remark, that, as fwifts breed but *once* in a fummer, and only *two* at a time, and the other *hirundines twice*, the latter, who lay from four to fix eggs, increafe at an average five times as faft as the former.

But in nothing are fwifts more fingular than in their early retreat. They retire, as to the main body of them, by the tenth of *Auguft*, and fometimes a few days fooner: and every ftraggler invariably withdraws by the twentieth, while their congeners, all of them, ftay till the beginning of *October*; many of them all through that month, and fome occafionally to the beginning of *November*. This early retreat is myfterious and wonderful, fince that time is often the fweeteft feafon in the year. But, what is more extraordinary, they begin to retire ftill earlier in the moft foutherly parts of *Andalufia*, where they can be no ways influenced by any defect of heat; or, as one might fuppofe, defect of food. Are they regulated in their motions with us by a failure of food, or by a propenfity to moulting, or by a difpofition to reft after fo rapid a life, or by what? This is one of thofe incidents in natural hiftory that not only baffles our fearches, but almoft eludes our gueffes!

Thefe

Thefe hirundines never perch on trees or roofs, and fo never congregate with their congeners. They are fearlefs while haunting their nefting places, and are not to be fcared with a gun; and are often beaten down with poles and cudgels as they ftoop to go under the eaves. Swifts are much infefted with thofe pefts to the genus called *hippobofcæ hirundinis*; and often wriggle and fcratch themfelves, in their flight, to get rid of that clinging annoyance.

Swifts are no fongfters, and have only one harfh fcreaming note; yet there are ears to which it is not difpleafing, from an agreeable affociation of ideas, fince that note never occurs but in the moft lovely fummer weather.

They never fettle on the ground but through accident; and when down can hardly rife, on account of the fhortnefs of their legs and the length of their wings: neither can they walk, but only crawl; but they have a ftrong grafp with their feet, by which they cling to walls. Their bodies being flat they can enter a very narrow crevice; and where they cannot pafs on their bellies they will turn up edgewife.

The particular formation of the foot difcriminates the fwift from all the *Britifh hirundines*; and indeed from all other known birds, the *hirundo melba*, or great white-bellied fwift of *Gibraltar*, excepted; for it is fo difpofed as to carry " *omnes quatuor digitos anticos*" all it's four toes forward; befides the leaft toe, which fhould be the back-toe, confifts of one bone alone, and the other three only of two apiece. A conftruction moft rare and peculiar, but nicely adapted to the purpofes in which their feet are employed. This, and fome peculiarities attending the noftrils and under mandible, have induced a difcerning [h] naturalift to fuppofe that this *fpecies* might conftitute a *genus per fe*.

[h] John Antony Scopoli, of Carniola, M. D.

B b

In

In *London* a party of swifts frequents the *Tower*, playing and feeding over the river just below the bridge: others haunt some of the churches of the *Borough* next the fields; but do not venture, like the *house-martin*, into the close crowded part of the town.

The *Swedes* have bestowed a very pertinent name on this swallow, calling it *ring swala*, from the perpetual *rings* or circles that it takes round the scene of it's nidification.

Swifts feed on *coleoptera*, or small beetles with hard cases over their wings, as well as on the softer insects; but it does not appear how they can procure gravel to grind their food, as swallows do, since they never settle on the ground. Young ones, over-run with *hippoboscæ*, are sometimes found, under their nests, fallen to the ground; the number of vermin rendering their abode insupportable any longer. They frequent in this village several abject cottages; yet a succession still haunts the same unlikely roofs: a good proof this that the same birds return to the same spots. As they must stoop very low to get up under these humble eaves, cats lie in wait, and sometimes catch them on the wing.

On the fifth of *July*, 1775, I again untiled part of a roof over the nest of a swift. The dam sat in the nest; but so strongly was she affected by natural στοργη for her brood, which she supposed to be in danger, that, regardless of her own safety, she would not stir, but lay sullenly by them, permitting herself to be taken in hand. The squab young we brought down and placed on the grass-plot, where they tumbled about, and were as helpless as a new-born child. While we contemplated their naked bodies, their unwieldy disproportioned abdomina, and their heads, too heavy for their necks to support, we could not but wonder when we reflected that these shiftless beings in a little more than a fortnight would be able to dash through the air almost with the

inconceivable

inconceivable fwiftnefs of a meteor; and perhaps, in their emigration, muft traverfe vaft continents and oceans as diftant as the equator. So foon does Nature advance fmall birds to their ἡλικια, or ftate of perfection; while the progreffive growth of men and large quadrupeds is flow and tedious!

I am, &c.

LETTER XXII.

TO THE SAME.

DEAR SIR,   SELBORNE, Sept. 13, 1774.

By means of a ftraight cottage-chimney I had an opportunity this fummer of remarking, at my leifure, how fwallows afcend and defcend through the fhaft: but my pleafure, in contemplating the addrefs with which this feat was performed to a confiderable depth in the chimney, was fomewhat interrupted by apprehenfions left my eyes might undergo the fame fate with thofe of *Tobit.*[i]

Perhaps it may be fome amufement to you to hear at what times the different fpecies of hirundines arrived this fpring in three very diftant counties of this kingdom. With us the

i Tobit 2. 10.

fwallow

swallow was seen first on *April* the 4th, the swift on *April* the 24th, the bank-martin on *April* the 12th, and the house-martin not till *April* the 30th. At *South Zele, Devonshire,* swallows did not arrive till *April* the 25th; swifts, in plenty, on *May* the 1st; and house-martins not till the middle of *May.* At *Blackburn,* in *Lancashire,* swifts were seen *April* the 28th, swallows *April* the 29th, house-martins *May* the 1st. Do these different dates, in such distant districts, prove any thing for or against migration?

A farmer, near *Weyhill,* fallows his land with two teams of asses; one of which works till noon, and the other in the afternoon. When these animals have done their work, they are penned all night, like sheep, on the fallow. In the winter they are confined and foddered in a yard, and make plenty of dung.

Linnæus says that hawks " *pacifcuntur inducias cum avibus, quamdiv cuculus cuculat :*" but it appears to me that, during that period, many little birds are taken and destroyed by birds of prey, as may be seen by their feathers left in lanes and under hedges.

The *missel-thrush* is, while breeding, fierce and pugnacious, driving such birds as approach its nest, with great fury, to a distance. The *Welch* call it *pen y llwyn,* the head or master of the coppice. He suffers no magpie, jay, or blackbird, to enter the garden where he haunts; and is, for the time, a good guard to the new-sown legumens. In general he is very successful in the defence of his family: but once I observed in my garden, that several magpies came determined to storm the nest of a missel-thrush : the dams defended their mansion with great vigour, and fought resolutely *pro aris & focis* ; but numbers at last prevailed, they tore the nest to pieces, and swallowed the young alive.

In

In the feafon of nidification the wildeft birds are comparatively tame. Thus the *ring-dove* breeds in my fields, though they are continually frequented; and the miffel-thrufh, though moft fhy and wild in the autumn and winter, builds in my garden clofe to a walk where people are paffing all day long.

Wall-fruit abounds with me this year; but my grapes, that ufed to be forward and good, are at prefent backward beyond all precedent: and this is not the worft of the ftory; for the fame ungenial weather, the fame black cold folftice, has injured the more neceffary fruits of the earth, and difcoloured and blighted our wheat. The crop of hops promifes to be very large.

Frequent returns of deafnefs incommode me fadly, and half difqualify me for a naturalift; for, when thofe fits are upon me, I lofe all the pleafing notices and little intimations arifing from rural founds; and *May* is to me as filent and mute with refpect to the notes of birds, &c. as *Auguft*. My eyefight is, thank God, quick and good; but with refpect to the other fenfe, I am, at times, difabled:

" And Wifdom at one entrance quite fhut out."

LETTER

LETTER XXIII.

TO THE SAME.

DEAR SIR, SELBORNE, June 8, 1775.

ON *September* the 21st, 1741, being then on a visit, and intent on field-diversions, I rose before daybreak: when I came into the enclosures, I found the stubbles and clover-grounds matted all over with a thick coat of cobweb, in the meshes of which a copious and heavy dew hung so plentifully that the whole face of the country seemed, as it were, covered with two or three setting-nets drawn one over another. When the dogs attempted to hunt, their eyes were so blinded and hoodwinked that they could not proceed, but were obliged to lie down and scrape the incumbrances from their faces with their fore-feet, so that, finding my sport interrupted, I returned home musing in my mind on the oddness of the occurrence.

As the morning advanced the sun became bright and warm, and the day turned out one of those most lovely ones which no season but the autumn produces; cloudless, calm, serene, and worthy of the South of *France* itself.

About nine an appearance very unusual began to demand our attention, a shower of cobwebs falling from very elevated regions, and continuing, without any interruption, till the close of the day. These webs were not single filmy threads, floating in the air in all directions, but perfect flakes or rags; some near an

inch

inch broad, and five or six long, which fell with a degree of velocity that shewed they were considerably heavier than the atmosphere.

On every side as the observer turned his eyes might he behold a continual succession of fresh flakes falling into his sight, and twinkling like stars as they turned their sides towards the sun.

How far this wonderful shower extended would be difficult to say; but we know that it reached *Bradley*, *Selborne*, and *Alresford*, three places which lie in a sort of a triangle, the shortest of whose sides is about eight miles in extent.

At the second of those places there was a gentleman (for whose veracity and intelligent turn we have the greatest veneration) who observed it the moment he got abroad; but concluded that, as soon as he came upon the hill above his house, where he took his morning rides, he should be higher than this meteor, which he imagined might have been blown, like *Thistle-down*, from the common above: but, to his great astonishment, when he rode to the most elevated part of the down, 300 feet above his fields, he found the webs in appearance still as much above him as before; still descending into sight in a constant succession, and twinkling in the sun, so as to draw the attention of the most incurious.

Neither before nor after was any such fall observed; but on this day the flakes hung in the trees and hedges so thick, that a diligent person sent out might have gathered baskets full.

The remark that I shall make on these cobweb-like appearances, called *gossamer*, is, that, strange and superstitious as the notions about them were formerly, nobody in these days doubts but that they are the real production of small spiders, which swarm in the fields in fine weather in autumn, and have a power of shooting out webs from their tails so as to render themselves buoyant, and

lighter

lighter than air. But why thefe apterous infects fhould *that day* take fuch a wonderful aërial excurfino, and why their webs fhould at once become fo grofs and material as to be confiderably more weighty than air, and to defcend with precipitation, is a matter beyond my fkill. If I might be allowed to hazard a fuppofition, I fhould imagine that thofe filmy threads, when firft fhot, might be entangled in the rifmg dew, and fo drawn up, fpiders and all, by a brifk evaporation into the regions where clouds are formed : and if the fpiders have a power of coiling and thickening their webs in the air, as Dr. *Lifter* fays they have, [fee his Letters to Mr. *Ray*] then, when they were become heavier than the air, they muft fall.

Every day in fine weather, in autumn chiefly, do I fee thofe fpiders fhooting out their webs and mounting aloft : they will go off from your finger if you will take them into your hand. Laft fummer one alighted on my book as I was reading in the parlour; and, running to the top of the page, and fhooting out a web, took it's departure from thence. But what I moft wondered at was, that it went off with confiderable velocity in a place where no air was ftirring; and I am fure that I did not affift it with my breath. So that thefe little crawlers feem to have, while mounting, fome loco-motive power without the ufe of wings, and to move in the air fafter than the air itfelf.

LETTER

LETTER XXIV.

TO THE SAME.

DEAR SIR,   SELBORNE, Aug. 15, 1775.

THERE is a wonderful spirit of sociality in the brute creation, independent of sexual attachment: the congregating of gregarious birds in the winter is a remarkable instance.

Many horses, though quiet with company, will not stay one minute in a field by themselves: the strongest fences cannot restrain them. My neighbour's horse will not only not stay by himself abroad, but he will not bear to be left alone in a strange stable without discovering the utmost impatience, and endeavouring to break the rack and manger with his fore feet. He has been known to leap out at a stable-window, through which dung was thrown, after company; and yet in other respects is remarkably quiet. Oxen and cows will not fatten by themselves; but will neglect the finest pasture that is not recommended by society. It would be needless to instance in sheep, which constantly flock together.

But this propensity seems not to be confined to animals of the same species; for we know a doe, still alive, that was brought up from a little fawn with a dairy of cows; with them it goes a-field, and with them it returns to the yard. The dogs of the house take no notice of this deer, being used to her; but, if strange dogs come by, a chase ensues; while the master smiles to see his favourite securely leading her pursuers over hedge, or

C c   gate,

gate, or ftile, till fhe returns to the cows, who, with fierce lowings and menacing horns, drive the affailants quite out of the pafture.

Even great difparity of kind and fize does not always prevent focial advances and mutual fellowfhip. For a very intelligent and obfervant perfon has affured me that, in the former part of his life, keeping but one horfe, he happened alfo on a time to have but one folitary hen. Thefe two incongruous animals fpent much of their time together in a lonely orchard, where they faw no creature but each other. By degrees an apparent regard began to take place between thefe two fequeftered individuals. The fowl would approach the quadruped with notes of complacency, rubbing herfelf gently againft his legs : while the horfe would look down with fatisfaction, and move with the greateft caution and circumfpection, left he fhould trample on his dimunitive companion. Thus, by mutual good offices, each feemed to confole the vacant hours of the other : fo that *Milton*, when he puts the following fentiment in the mouth of *Adam*, feems to be fomewhat miftaken :

> " Much lefs can *bird* with *beaft*, or fifh with fowl,
> " So well converfe, nor with the ox the ape."

I am, &c.

LETTER

LETTER XXV.

TO THE SAME.

DEAR SIR, SELBORNE, Oct. 2, 1775.

W E have two gangs or hordes of gypfies which infeft the fouth
and weft of *England*, and come round in their circuit two or three
times in the year. One of thefe tribes calls itfelf by the noble name
of *Stanley*, of which I have nothing particular to fay; but the other
is diftinguifhed by an appellative fomewhat remarkable—As far as
their harfh gibberifh can be underftood, they feem to fay that the
name of their clan is *Curleople:* now the termination of this word is
apparently *Grecian:* and as *Mezeray* and the graveft hiftorians all
agree that thefe vagrants did certainly migrate from *Egypt* and the
Eaft, two or three centuries ago, and fo fpread by degrees over
Europe, may not this family-name, a little corrupted, be the very
name they brought with them from the *Levant?* It would be
matter of fome curiofity, could one meet with an intelligent perfon
among them, to inquire whether, in their jargon, they ftill retain
any *Greek* words : the *Greek* radicals will appear in hand, foot,
head, water, earth, &c. It is poffible that amidft their cant and
corrupted dialect many mutilated remains of their native language
might ftill be difcovered.

With regard to thofe peculiar people, the gypfies, one thing
is very remarkable, and efpecially as they came from warmer
climates; and that is, that while other beggars lodge in barns,

C c 2 ftables,

ſtables, and cow-houſes, theſe ſturdy ſavages ſeem to pride themſelves in braving the ſeverities of winter, and in living *ſub dio* the
whole year round. Laſt *September* was as wet a month as ever was
known ; and yet during thoſe deluges did a young gypſy-girl lie-in
in the midſt of one of our hop-gardens, on the cold ground, with
nothing over her but a piece of a blanket extended on a few
hazel-rods bent hoop faſhion, and ſtuck into the earth at each end,
in circumſtances too trying for a cow in the ſame condition : yet
within this garden there was a large hop-kiln, into the chambers of
which ſhe might have retired, had ſhe thought ſhelter an object
worthy her attention.

Europe itſelf, it ſeems, cannot ſet bounds to the rovings of theſe
vagabonds ; for Mr. *Bell*, in his return from *Peking*, met a gang
of theſe people on the confines of *Tartary*, who were endeavouring
to penetrate thoſe deſerts and try their fortune in *China* [k].

Gypſies are called in *French*, *Bohemiens*; in *Italian* and modern
Greek, *Zingani*.

I am, &c.

[k] See Bell's Travels in China.

LETTER

LETTER XXVI.

TO THE SAME.

DEAR SIR,   SELBORNE, Nov. 1, 1775.

> " Hîc - - - - tædæ pingues, hîc plurimus ignis
> " Semper, et assiduâ postes fuligine nigri."

I SHALL make no apology for troubling you with the detail of a
very simple piece of domestic œconomy, being satisfied that you
think nothing beneath your attention that tends to utility: the
matter alluded to is the use of *rushes* instead of candles, which
I am well aware prevails in many districts besides this; but as
I know there are countries also where it does not obtain, and as
I have considered the subject with some degree of exactness, I shall
proceed in my humble story, and leave you to judge of the
expediency.

The proper species of *rush* for this purpose seems to be the
juncus effusus, or common soft rush, which is to be found in most
moist pastures, by the sides of streams, and under hedges. These
rushes are in best condition in the height of summer; but may be
gathered, so as to serve the purpose well, quite on to autumn.
It would be needless to add that the largest and longest are best.
Decayed labourers, women, and children, make it their business
to procure and prepare them. As soon as they are cut they must
be flung into water, and kept there; for otherwise they will dry
and shrink, and the peel will not run. At first a person would
find

find it no eafy matter to diveft a rufh of it's peel or rind, fo as to leave one regular, narrow, even rib from top to bottom that may fupport the pith : but this, like other feats, foon becomes familiar even to children ; and we have feen an old woman, ftone-blind, performing this bufinefs with great difpatch, and feldom failing to ftrip them with the niceft regularity. When thefe *junci* are thus far prepared, they muft lie out on the grafs to be bleached, and take the dew for fome nights, and afterwards be dried in the fun.

Some addrefs is required in dipping thefe rufhes in the fcalding fat or greafe ; but this knack alfo is to be attained by practice. The careful wife of an induftrious *Hampfhire* labourer obtains all her fat for nothing ; for fhe faves the fcummings of her bacon-pot for this ufe ; and, if the greafe abounds with falt, fhe caufes the falt to precipitate to the bottom, by fetting the fcummings in a warm oven. Where hogs are not much in ufe, and efpecially by the fea-fide, the coarfer animal-oils will come very cheap. A pound of common greafe may be procured for four pence ; and about fix pounds of greafe will dip a pound of rufhes ; and one pound of rufhes may be bought for one fhilling : fo that a pound of rufhes, medicated and ready for ufe, will coft three fhillings. If men that keep bees will mix a little wax with the greafe, it will give it a confiftency, and render it more cleanly, and make the rufhes burn longer : mutton-fuet would have the fame effect.

A good rufh, which meafured in length two feet four inches and an half, being minuted, burnt only three minutes fhort of an hour : and a rufh ftill of greater length has been known to burn one hour and a quarter.

Thefe rufhes give a good clear light. Watch-lights (coated with tallow), it is true, fhed a difmal one, " darknefs vifible ;" but

then

then the wick of thofe have *two* ribs of the rind, or peel, to fupport the pith, while the wick of the dipped rufh has but *one*. The *two* ribs are intended to impede the progrefs of the flame and make the candle laft.

In a pound of dry rufhes, avoirdupois, which I caufed to be weighed and numbered, we found upwards of one thoufand fix hundred individuals. Now fuppofe each of thefe burns, one with another, only half an hour, then a poor man will purchafe eight hundred hours of light, a time exceeding thirty-three entire days, for three fhillings. According to this account each rufh, before dipping, cofts $\frac{1}{33}$ of a farthing, and $\frac{1}{11}$ afterwards. Thus a poor family will enjoy $5\frac{1}{2}$ hours of comfortable light for a farthing. An experienced old houfekeeper affures me that one pound and an half of rufhes completely fupplies his family the year round, fince working people burn no candle in the long days, becaufe they rife and go to bed by daylight.

Little farmers ufe rufhes much in the fhort days, both morning and evening, in the dairy and kitchen; but the very poor, who are always the worft œconomifts, and therefore muft continue very poor, buy an halfpenny candle every evening, which, in their blowing open rooms, does not burn much more than two hours. Thus have they only two hours light for their money inftead of eleven.

While on the fubject of rural œconomy, it may not be improper to mention a pretty implement of houfewifery that we have feen no where elfe; that is, little neat befoms which our forefters make from the ftalks of the *polytricum commune*, or *great golden maiden-hair*, which they call *filk-wood*, and find plenty in the bogs. When this mofs is well combed and dreffed, and divefted of it's outer fkin, it becomes of a beautiful bright-chefnut colour; and, being

soft

foot and pliant, is very proper for the dusting of beds, curtains, carpets, hangings, &c. If these besoms were known to the brush-makers in town, it is probable they might come much in use for the purpose above-mentioned [1].

I am, &c.

LETTER XXVII.

TO THE SAME.

DEAR SIR,   SELBORNE, Dec. 12, 1775.

WE had in this village more than twenty years ago an idiot boy, whom I well remember, who, from a child, shewed a strong propensity to bees; they were his food, his amusement, his sole object. And as people of this cast have seldom more than one point in view, so this lad exerted all his few faculties on this one pursuit. In the winter he dosed away his time, within his father's house, by the fire side, in a kind of torpid state, seldom departing from the chimney-corner; but in the summer he was all alert, and in quest of his game in the fields, and on sunny banks. Honey-bees, humble-bees, and wasps, were his prey wherever he found them: he had no apprehensions from their stings, but would seize

[1] A besom of this sort is to be seen in Sir *Ashton Lever*'s Museum.

them

seize them *nudis manibus*, and at once difarm them of their weapons, and fuck their bodies for the fake of their honey-bags. Sometimes he would fill his bofom between his fhirt and his fkin with a number of thefe captives; and fometimes would confine them in bottles. He was a very *merops apiafter*, or *bee-bird*; and very injurious to men that kept bees; for he would flide into their bee-gardens, and, fitting down before the ftools, would rap with his finger on the hives, and fo take the bees as they came out. He has been known to overturn hives for the fake of honey, of which he was paffionately fond. Where metheglin was making he would linger round the tubs and veffels, begging a draught of what he called *bee-wine*. As he ran about he ufed to make a humming noife with his lips, refembling the buzzing of bees. This lad was lean and fallow, and of a cadaverous complexion; and, except in his favourite purfuit, in which he was wonderfully adroit, difcovered no manner of underftanding. Had his capacity been better, and directed to the fame object, he had perhaps abated much of our wonder at the feats of a more modern exhibiter of bees: and we may juftly fay of him now,

> " — — — — — — — — Thou,
> " Had thy prefiding ftar propitious fhone,
> " Should'ft *Wildman* be — — — —."

When a tall youth he was removed from hence to a diftant village, where he died, as I underftand, before he arrived at manhood.

I am, &c.

Dd

LETTER

LETTER XXVIII.

TO THE SAME.

DEAR SIR,　　　　　　　　　SELBORNE, Jan. 8, 1776.

IT is the hardest thing in the world to shake off superstitious pre-judices: they are sucked in as it were with our mother's milk; and, growing up with us at a time when they take the fastest hold and make the most lasting impressions, become so interwoven into our very constitutions, that the strongest good sense is required to disengage ourselves from them. No wonder therefore that the lower people retain them their whole lives through, since their minds are not invigorated by a liberal education, and therefore not enabled to make any efforts adequate to the occasion.

Such a preamble seems to be necessary before we enter on the superstitions of this district, lest we should be suspected of exag-geration in a recital of practices too gross for this enlightened age.

But the people of *Tring*, in *Hertfordshire*, would do well to re-member, that no longer ago than the year 1751, and within twenty miles of the capital, they seized on two superannuated wretches, crazed with age, and overwhelmed with infirmities, on a suspicion of witchcraft; and, by trying experiments, drowned them in a horse-pond.

In a farm-yard near the middle of this village stands, at this day, a row of pollard-ashes, which, by the seams and long cicatrices down their sides, manifestly shew that, in former times, they have been cleft asunder. These trees, when young and flexible, were severed and held open by wedges, while ruptured children, strip-

ped

ped naked, were pushed through the apertures, under a persuasion that, by such a process, the poor babes would be cured of their infirmity. As soon as the operation was over, the tree, in the suffering part, was plastered with loam, and carefully swathed up. If the parts coalesced and soldered together, as usually fell out, where the feat was performed with any adroitness at all, the party was cured; but, where the cleft continued to gape, the operation, it was supposed, would prove ineffectual. Having occasion to enlarge my garden not long since, I cut down two or three such trees, one of which did not grow together.

We have several persons now living in the village, who, in their childhood, were supposed to be healed by this superstitious ceremony, derived down perhaps from our *Saxon* ancestors, who practised it before their conversion to Christianity.

At the south corner of the *Plestor*, or area, near the church, there stood, about twenty-years ago, a very old grotesque hollow pollard-ash, which for ages had been looked on with no small veneration as a *shrew-ash*. Now a shrew-ash is an ash whose twigs or branches, when gently applied to the limbs of cattle, will immediately relieve the pains which a beast suffers from the running of a *shrew-mouse* over the part affected: for it is supposed that a shrew-mouse is of so baneful and deleterious a nature, that wherever it creeps over a beast, be it horse, cow, or sheep, the suffering animal is afflicted with cruel anguish, and threatened with the loss of the use of the limb. Against this accident, to which they were continually liable, our provident fore-fathers always kept a shrew-ash at hand, which, when once medicated, would maintain it's virtue for ever. A shrew-ash was made thus [m] :—Into the body of the

m For a similar practice, see *Plot's Staffordshire.*

tree

tree a deep hole was bored with an auger, and a poor devoted shrew-mouse was thrust in alive, and plugged in, no doubt, with several quaint incantations long since forgotten. As the ceremonies necessary for such a consecration are no longer understood, all succession is at an end, and no such tree is known to subsist in the manor, or hundred.

As to that on the *Plestor*

"The late vicar stubb'd and burnt it."

when he was way-warden, regardless of the remonstrances of the by-standers, who interceded in vain for it's preservation, urging it's power and efficacy, and alledging that it had been

"Religione patrum multos servata per annos."

I am, &c.

LETTER XXIX.

TO THE SAME.

DEAR SIR,   SELBORNE, Feb. 7, 1776.

In heavy fogs, on elevated situations especially, trees are perfect alembics; and no one that has not attended to such matters can imagine how much water one tree will distil in a night's time, by condensing the vapour, which trickles down the twigs and boughs, so as to make the ground below quite in a float. In *Newton-lane*,

in

in *October* 1775, on a mifty day, a particular oak in leaf dropped fo faft that the cart-way ftood in puddles and the ruts ran with water, though the ground in general was dufty.

In fome of our fmaller iflands in the *Weft-Indies*, if I miftake not, there are no fprings or rivers; but the people are fupplied with that neceffary element, water, merely by the dripping of fome large tall trees, which, ftanding in the bofom of a mountain, keep their heads conftantly enveloped with fogs and clouds, from which they difpenfe their kindly never-ceafing moifture; and fo render thofe diftricts habitable by condenfation alone.

Trees in leaf have fuch a vaft proportion more of furface than thofe that are naked, that, in theory, their condenfations fhould greatly exceeed thofe that are ftripped of their leaves; but, as the former *imbibe* alfo a great quantity of moifture, it is difficult to fay which drip moft: but this I know, that deciduous trees that are entwined with much ivy feem to diftil the greateft quantity. Ivy-leaves are fmooth, and thick, and cold, and therefore condenfe very faft; and befides ever-greens imbibe very little. Thefe facts may furnifh the intelligent with hints concerning what forts of trees they fhould plant round fmall ponds that they would wifh to be perennial; and fhew them how advantageous fome trees are in preference to others.

Trees perfpire profufely, condenfe largely, and check evaporation fo much, that woods are always moift; no wonder therefore that they contribute much to pools and ftreams.

That trees are great promoters of lakes and rivers appears from a well known fact in *North-America*; for, fince the woods and forefts have been grubbed and cleared, aii bodies of water are much diminifhed; fo that fome ftreams, that were very confiderable a

century

century ago, will not now drive a common mill[n]. Besides, most woodlands, forests, and chases, with us abound with pools and morasses; no doubt for the reason given above.

To a thinking mind few phenomena are more strange than the state of little ponds on the summits of chalk-hills, many of which are never dry in the most trying droughts of summer. On *chalk-hills* I say, because in many rocky and gravelly soils springs usually break out pretty high on the sides of elevated grounds and mountains; but no person acquainted with chalky districts will allow that they ever saw springs in such a soil but in vallies and bottoms, since the waters of so pervious a stratum as chalk all lie on one dead level, as well-diggers have assured me again and again.

Now we have many such little round ponds in this district; and one in particular on our sheep-down, three hundred feet above my house; which, though never above three feet deep in the middle, and not more than thirty feet in diameter, and containing perhaps not more than two or three hundred hogsheads of water, yet never is known to fail, though it affords drink for three hundred or four hundred sheep, and for at least twenty head of large cattle beside. This pond, it is true, is over-hung with two moderate beeches, that, doubtless, at times afford it much supply: but then we have others as small, that, without the aid of trees, and in spite of evaporation from sun and wind, and perpetual consumption by cattle, yet constantly maintain a moderate share of water, without overflowing in the wettest seasons, as they would do if supplied by springs. By my journal of *May*, 1775, it appears that " the small " and even considerable ponds in the vales are now dried up, while " the small ponds on the very tops of hills are but little affected." Can this difference be accounted for from evaporation alone, which certainly is more prevalent in bottoms? or rather have not those

[n] Vide *Kalm's* Travels to *North-America*.

elevated

elevated pools fome unnoticed recruits, which in the night time counterbalance the wafte of the day; without which the cattle alone muft foon exhauft them? And here it will be neceffary to enter more minutely into the caufe. Dr. *Hales*, in his Vegetable Statics, advances, from experiment, that " the moifter the earth is the more " dew falls on it in a night : and more than a *double* quantity of " dew falls on a furface of *water* than there does on an equal " furface of moift earth." Hence we fee that water, by it's coolnefs, is enabled to affimilate to itfelf a large quantity of moifture nightly by condenfation; and that the air, when loaded with fogs and vapours, and even with copious dews, can alone advance a confiderable and never-failing refource. Perfons that are much abroad, and travel early and late; fuch as fhepherds, fifhermen, &c. can tell what prodigious fogs prevail in the night on elevated downs, even in the hotteft parts of fummer; and how much the furfaces of things are drenched by thofe fwimming vapours, though, to the fenfes, all the while, little moifture feems to fall.

I am, &c.

LETTER

LETTER XXX.

TO THE SAME.

DEAR SIR,   SELBORNE, April 3, 1776.

MONSIEUR HERISSANT, a *French* anatomist, seems persuaded that he has discovered the reason why cuckoos do not hatch their own eggs; the impediment, he supposes, arises from the internal structure of their parts, which incapacitates them for incubation. According to this gentleman, the crop, or craw, of a cuckoo does not lie before the sternum at the bottom of the neck, as in the *gallinæ, columbæ,* &c. but immediately behind it, on and over the bowels, so as to make a large protuberance in the belly[o].

Induced by this assertion, we procured a cuckoo; and, cutting open the breast-bone, and exposing the intestines to sight, found the crop lying as mentioned above. This stomach was large and round, and stuffed hard like a pincushion with food, which, upon nice examination, we found to consist of various insects; such as small scarabs, spiders, and dragon-flies; the last of which we have seen cuckoos catching on the wing as they were just emerging out of the aurelia state. Among this farrago also were to be seen maggots, and many seeds, which belonged either to gooseberries, currants, cranberries, or some such fruit; so that these birds ap-

[o] *Histoire de l'Academie Royale,* 1752.

parently

parently subsist on insects and fruits: nor was there the least appearance of bones, feathers, or fur, to support the idle notion of their being birds of prey.

The sternum in this bird seemed to us to be remarkably short, between which and the anus lay the crop, or craw, and immediately behind that the bowels against the back-bone.

It must be allowed, as this anatomist observes, that the crop placed just upon the bowels must, especially when full, be in a very uneasy situation during the business of incubation; yet the test will be to examine whether birds that are actually known to sit for certain are not formed in a similar manner. This inquiry I proposed to myself to make with a *fern-owl*, or goat-sucker, as soon as opportunity offered: because, if their formation proves the same, the reason for incapacity in the cuckoo will be allowed to have been taken up somewhat hastily.

Not long after a fern-owl was procured, which, from it's habit and shape, we suspected might resemble the cuckoo in it's internal construction. Nor were our suspicions ill-grounded; for, upon the dissection, the crop, or craw, also lay behind the sternum, immediately on the viscera, between them and the skin of the belly. It was bulky, and stuffed hard with large *phalænæ*, moths of several sorts, and their eggs, which no doubt had been forced out of those insects by the action of swallowing.

Now as it appears that this bird, which is so well known to practise incubation, is formed in a similar manner with cuckoos, Monsieur *Herissant*'s conjecture, that cuckoos are incapable of incubation from the disposition of their intestines, seems to fall to the ground; and we are still at a loss for the cause of that strange and singular peculiarity in the instance of the *cuculus canorus*.

E e

We

We found the case to be the same with the ring-tail hawk, in respect to formation; and, as far as I can recollect, with the swift; and probably it is so with many more sorts of birds that are not granivorous.

I am, &c.

————————

LETTER XXXI.

TO THE SAME.

DEAR SIR,   SELBORNE, April 29, 1776.

On *August* the 4th, 1775, we surprised a large viper, which seemed very heavy and bloated, as it lay in the grass basking in the sun. When we came to cut it up, we found that the abdomen was crowded with young, fifteen in number; the shortest of which measured full seven inches, and were about the size of full-grown earth-worms. This little fry issued into the world with the true viper-spirit about them, shewing great alertness as soon as disengaged from the belly of the dam: they twisted and wriggled about, and set themselves up, and gaped very wide when touched with a stick, shewing manifest tokens of menace and defiance, though as yet they had no manner of fangs that we could find, even with the help of our glasses.

To

To a thinking mind nothing is more wonderful than that early inſtinct which impreſſes young animals with the notion of the ſituation of their natural weapons, and of uſing them properly in their own defence, even before thoſe weapons ſubſiſt or are formed. Thus a young cock will ſpar at his adverſary before his ſpurs are grown ; and a calf or a lamb will puſh with their heads before their horns are ſprouted. In the ſame manner did theſe young adders attempt to bite before their fangs were in being. The dam however was furniſhed with very formidable ones, which we lifted up (for they fold down when not uſed) and cut them off with the point of our ſciſſars.

There was little room to ſuppoſe that this brood had ever been in the open air before; and that they were taken in for refuge, at the mouth of the dam, when ſhe perceived that danger was approaching; becauſe then probably we ſhould have found them ſomewhere in the neck, and not in the abdomen.

LETTER

LETTER XXXII.

TO THE SAME.

Castration has a ſtrange effect: it emaſculates both man, beaſt, and bird, and brings them to a near reſemblance of the other ſex. Thus eunuchs have ſmooth unmuſcular arms, thighs, and legs; and broad hips, and beardleſs chins, and ſqueaking voices. Gelt-ſtags and bucks have hornleſs heads, like hinds and does. Thus wethers have ſmall horns, like ewes; and oxen large bent horns, and hoarſe voices when they low, like cows: for bulls have ſhort ſtraight horns; and though they mutter and grumble in a deep tremendous tone, yet they low in a ſhrill high key. Capons have ſmall combs and gills, and look pallid about the head, like pullets; they alſo walk without any parade, and hover chickens like hens. Barrow-hogs have alſo ſmall tuſks like ſows.

Thus far it is plain that the deprivation of *maſculine vigour* puts a ſtop to the growth of thoſe parts or appendages that are looked upon as it's inſignia. But the ingenious Mr. *Liſle*, in his book on huſbandry, carries it much farther; for he ſays that the loſs of thoſe inſignia alone has ſometimes a ſtrange effect on the ability itſelf: he had a boar ſo fierce and venereous, that, to prevent miſchief, orders were given for his tuſks to be broken off. No ſooner had the beaſt ſuffered this injury than his powers forſook him, and he neglected thoſe females to whom before he was paſſionately attached, and from whom no fences could reſtrain him.

LETTER XXXIII.

TO THE SAME.

THE natural term of an hog's life is little known, and the reaſon is plain—becauſe it is neither profitable nor convenient to keep that turbulent animal to the full extent of it's time : however, my neighbour, a man of ſubſtance, who had no occaſion to ſtudy every little advantage to a nicety, kept an half bred Bantam-ſow, who was as thick as ſhe was long, and whoſe belly ſwept on the ground till ſhe was advanced to her ſeventeenth year ; at which period ſhe ſhewed ſome tokens of age by the decay of her teeth and the decline of her fertility.

For about ten years this prolific mother produced two litters in the year of about ten at a time, and once above twenty at a litter ; but, as there were near double the number of pigs to that of teats, many died. From long experience in the world this female was grown very ſagacious and artful :—when ſhe found occaſion to converſe with a boar ſhe uſed to open all the intervening gates, and march, by herſelf, up to a diſtant farm where one was kept ; and when her purpoſe was ſerved would return by the ſame means. At the age of about fifteen her litters began to be reduced to four or five ; and ſuch a litter ſhe exhibited when in her fatting-pen. She proved, when fat, good bacon, juicy, and tender ; the rind, or ſward, was remarkably thin. At a moderate computation ſhe

was

was allowed to have been the fruitful parent of three hundred pigs: a prodigious inftance of fecundity in fo large a quadruped! She was killed in fpring 1775.

I am, &c.

————————

LETTER XXXIV.

TO THE SAME.

DEAR SIR,   SELBORNE, May 9, 1776.

"— — — — — admorunt ubera tigres."

WE have remarked in a former letter how much incongruous animals, in a lonely ftate, may be attached to each other from a fpirit of fociality; in this it may not be amifs to recount a different motive which has been known to create as ftrange a fondnefs.

My friend had a little helplefs *leveret* brought to him, which the fervants fed with milk in a fpoon, and about the fame time his cat kittened and the young were difpatched and buried. The hare was foon loft, and fuppofed to be gone the way of moft fond-lings, to be killed by fome dog or cat. However, in about a fortnight, as the mafter was fitting in his garden in the dufk of the evening, he obferved his cat, with tail erect, trotting towards him, and calling with little fhort inward notes of complacency,

such

such as they use towards their kittens, and something gamboling after, which proved to be the leveret that the cat had supported with her milk, and continued to support with great affection.

Thus was a graminivorous animal nurtured by a carnivorous and predaceous one!

Why so cruel and sanguinary a beast as a cat, of the ferocious genus of *Feles*, the *murium leo*, as *Linnæus* calls it, should be affected with any tenderness towards an animal which is it's natural prey, is not so easy to determine.

This strange affection probably was occasioned by that desiderium, those tender maternal feelings, which the loss of her kittens had awakened in her breast; and by the complacency and ease she derived to herself from the procuring her teats to be drawn, which were too much distended with milk, till, from habit, she became as much delighted with this foundling as if it had been her real offspring.

This incident is no bad solution of that strange circumstance which grave historians as well as the poets assert, of exposed children being sometimes nurtured by female wild beasts that probably had lost their young. For it is not one whit more marvellous that *Romulus* and *Remus*, in their infant state, should be nursed by a she-wolf, than that a poor little sucking leveret should be fostered and cherished by a bloody grimalkin.

— — — — " viridi fœtam Mavortis in antro
" Procubuisse lupam: geminos huic ubera circum
" Ludere pendentes pueros, et lambere matrem
" Impavidos: illam tereti cervice reflexam
" Mulcere alternos, et corpora fingere linguâ."

LETTER

LETTER XXXV.

TO THE SAME.

DEAR SIR,
SELBORNE, May 20, 1777.

Lands that are fubject to frequent inundations are always poor; and probably the reafon may be becaufe the worms are drowned. The moft infignificant infects and reptiles are of much more confequence, and have much more influence in the œconomy of Nature, than the incurious are aware of; and are mighty in their effect, from their minutenefs, which renders them lefs an object of attention; and from their numbers and fecundity. Earth-worms, though in appearance a fmall and defpicable link in the chain of Nature, yet, if loft, would make a lamentable chafm. For, to fay nothing of half the birds, and fome quadrupeds which are almoft entirely fupported by them, worms feem to be the great promoters of vegetation, which would proceed but lamely without them, by boring, perforating, and loofening the foil, and rendering it pervious to rains and the fibres of plants, by drawing ftraws and ftalks of leaves and twigs into it; and, moft of all, by throwing up fuch infinite numbers of lumps of earth called worm-cafts, which, being their excrement, is a fine manure for grain and grafs. Worms probably provide new foil for hills and flopes where the rain wafhes the earth away; and they affect flopes, probably to avoid being flooded. Gardeners and farmers exprefs their deteftation of worms; the former becaufe they render their walks unfightly, and make them much work: and the latter becaufe, as they think, worms eat their

green

green corn. But thefe men would find that the earth without worms would foon become cold, hard-bound, and void of fermentation; and confequently fteril: and befides, in favour of worms, it fhould be hinted that green corn, plants, and flowers, are not fo much injured by them as by many fpecies of *coleoptera* (fcarabs), and *tipulæ* (long-legs) in their larva, or grub-ftate; and by unnoticed myriads of fmall fhell-lefs fnails, called flugs, which filently and imperceptibly make amazing havoc in the field and garden q.

Thefe hints we think proper to throw out in order to fet the inquifitive and difcerning to work.

A good monography of worms would afford much entertainment and information at the fame time, and would open a large and new field in natural hiftory. Worms work moft in the fpring; but by no means lie torpid in the dead months; are out every mild, night in the winter, as any perfon may be convinced that will take the pains to examine his grafs-plots with a candle; are hermaphrodites, and much addicted to venery, and confequently very prolific.

I am, &c.

q Farmer *Young*, of *Norton-farm*, fays that this fpring (1777) about four acres of his wheat in one field was entirely deftroyed by *flugs*, which fwarmed on the blades of corn, and devoured it as faft as it fprang.

Ff   LETTER

LETTER XXXVI.

TO THE SAME.

DEAR SIR,   SELBORNE, Nov. 22, 1777.

YOU cannot but remember that the twenty-fixth and twenty-
feventh of laft *March* were very hot days; fo fultry that every
body complained and were reftlefs under thofe fenfations to which
they had not been reconciled by gradual approaches.

This fudden fummer-like heat was attended by many fummer
coincidences; for on thofe two days the thermometer rofe to
fixty-fix in the fhade; many fpecies of infects revived and came
forth; fome bees fwarmed in this neighbourhood; the old tortoife,
near *Lewes*, in *Suffex*, awakened and came forth out of it's dormitory; and, what is moft to my prefent purpofe, many *houfe-fwallows*
appeared and were very alert in many places, and particularly at
Cobham, in *Surrey*.

But as that fhort warm period was fucceeded as well as preceded
by harfh fevere weather, with frequent frofts and ice, and cutting
winds, the infects withdrew, the tortoife retired again into the
ground, and the fwallows were feen no more until the tenth of
April, when, the rigour of the fpring abating, a fofter feafon began
to prevail.

Again; it appears by my journals for many years paft that *houfe-
martins* retire, to a bird, about the beginning of *October*; fo that a
perfon not very obfervant of fuch matters would conclude that they

had

had taken their laſt farewell: but then it may be ſeen in my diaries alſo that conſiderable flocks have diſcovered themſelves again in the firſt week of *November*, and often on the fourth day of that month only *for one day*; and that not as if they were in actual migration, but playing about at their leiſure and feeding calmly, as if no enterprize of moment at all agitated their ſpirits. And this was the caſe in the beginning of this very month; for, on the fourth of *November*, more than twenty houſe-martins, which, in appearance, had all departed about the ſeventh of *October*, were ſeen again, for that *one morning only*, ſporting between my fields and *the Hanger*, and feaſting on inſects which ſwarmed in that ſheltered diſtrict. The preceding day was wet and bluſtering, but the fourth was dark and mild, and ſoft, the wind at ſouth-weſt, and the ther-mometer at 58½; a pitch not common at that ſeaſon of the year. Moreover, it may no tbe amiſs to add in this place, that whenever the thermometer is above 50 the bat comes flitting out in every autumnal and winter-month.

From all theſe circumſtances laid together, it is obvious that torpid inſects, reptiles, and quadrupeds, are awakened from their profoundeſt ſlumbers by a little untimely warmth; and therefore that nothing ſo much promotes this death-like ſtupor as a defect of heat. And farther, it is reaſonable to ſuppoſe that two whole ſpecies, or at leaſt many individuals of thoſe two ſpecies, of *Britiſh hirundines*, do never leave this iſland at all, but partake of the ſame benumbed ſtate: for we cannot ſuppoſe that, after a month's abſence, houſe-martins can return from ſouthern regions to appear for *one* morning in *November*, or that houſe-ſwallows ſhould leave the diſ-tricts of *Africa* to enjoy, in *March*, the tranſient ſummer of a *couple* of days.

I am, &c.

LETTER

LETTER XXXVII.

TO THE SAME.

DEAR SIR, SELBORNE, Jan. 8, 1778.

THERE was in this village several years ago a miferable pauper, who, from his birth, was afflicted with a leprofy, as far as we are aware of a fingular kind, fince it affected only the palms of his hands and the foles of his feet. This fcaly eruption ufually broke out twice in the year, at the fpring and fall; and, by peeling away, left the fkin fo thin and tender that neither his hands or feet were able to perform their functions; fo that the poor object was half his time on crutches, incapable of employ, and languifhing in a tirefome ftate of indolence and inactivity. His habit was lean, lank, and cadaverous. In this fad plight he dragged on a miferable exiftence, a burden to himfelf and his parifh, which was obliged to fupport him till he was relieved by death at more than thirty years of age.

The good women, who love to account for every defect in children by the doctrine of longing, faid that his mother felt a violent propenfity for oyfters, which fhe was unable to gratify; and that the black rough fcurf on his hands and feet were the fhells of that fifh. We knew his parents, neither of which were lepers; his father in particular lived to be far advanced in years.

In all ages the leprofy has made dreadful havock among mankind. The *Ifraelites* feem to have been greatly afflicted with it from

from the moſt remote times; as appears from the peculiar and repeated injunctions given them in the *Levitical* law [r]. Nor was the rancour of this foul diſorder much abated in the laſt period of their commonwealth, as may be ſeen in many paſſages of the New Teſtament.

Some centuries ago this horrible diſtemper prevailed all *Europe* over; and our forefathers were by no means exempt, as appears by the large proviſion made for objects labouring under this calamity. There was an hoſpital for female lepers in the dioceſe of *Lincoln*, a noble one near *Durham*, three in *London* and *Southwark*, and perhaps many more in or near our great towns and cities. Moreover, ſome crowned heads, and other wealthy and charitable perſonages, bequeathed large legacies to ſuch poor people as languiſhed under this hopeleſs infirmity.

It muſt therefore, in theſe days, be, to an humane and thinking perſon, a matter of equal wonder and ſatisfaction, when he contemplates how nearly this peſt is eradicated, and obſerves that a leper now is a rare ſight. He will, moreover, when engaged in ſuch a train of thought, naturally inquire for the reaſon. This happy change perhaps may have originated and been continued from the much ſmaller quantity of ſalted meat and fiſh now eaten in theſe kingdoms; from the uſe of linen next the ſkin; from the plenty of better bread; and from the profuſion of fruits, roots, legumes, and greens, ſo common in every family. Three or four centuries ago, before there were any encloſures, ſown-graſſes, field-turnips, or field-carrots, or hay, all the cattle which had grown fat in ſummer, and were not killed for winter-uſe, were turned out ſoon after *Michaelmas* to ſhift as they could through the

[r] See Leviticus, chap. xiii. and xiv.

dead

dead months; so that no fresh meat could be had in winter or spring. Hence the marvellous account of the vast stores of salted flesh found in the larder of the eldest *Spencer*[s] in the days of *Edward* the Second, even so late in the spring as the third of *May*. It was from magazines like these that the turbulent barons supported in idleness their riotous swarms of retainers ready for any disorder or mischief. But agriculture is now arrived at such a pitch of perfection, that our best and fattest meats are killed in the winter; and no man need eat salted flesh, unless he prefers it, that has money to buy fresh.

One cause of this distemper might be, no doubt, the quantity of wretched fresh and salt fish consumed by the commonalty at all seasons as well as in lent; which our poor now would hardly be persuaded to touch.

The use of linen changes, shirts or shifts, in the room of sordid and filthy woollen, long worn next the skin, is a matter of neatness comparatively modern; but must prove a great means of preventing cutaneous ails. At this very time woollen instead of linen prevails among the poorer *Welch*, who are subject to foul eruptions.

The plenty of good wheaten bread that now is found among all ranks of people in the south, instead of that miserable sort which used in old days to be made of barley or beans, may contribute not a little to the sweetening their blood and correcting their juices; for the inhabitants of mountainous districts, to this day, are still liable to the itch and other cutaneous disorders, from a wretchedness and poverty of diet.

As to the produce of a garden, every middle-aged person of observation may perceive, within his own memory, both in town

---

[s] *Viz.* Six hundred bacons, eighty carcasses of beef, and six hundred muttons.

and

and country, how vaftly the confumption of vegetables is increafed. Green-ftalls in cities now fupport multitudes in a comfortable ftate, while gardeners get fortunes. Every decent labourer alfo has his garden, which is half his fupport, as well as his delight; and common farmers provide plenty of beans, peas, and greens, for their hinds to eat with their bacon; and thofe few that do not are defpifed for their fordid parfimony, and looked upon as regardlefs of the welfare of their dependants. Potatoes have prevailed in this little diftrict, by means of premiums, within thefe twenty years only; and are much efteemed here now by the poor, who would fcarce have ventured to tafte them in the laft reign.

Our *Saxon* anceftors certainly had fome fort of cabbage, becaufe they call the month of *February fprout-cale*; but, long after their days, the cultivation of gardens was little attended to. The religious, being men of leifure, and keeping up a conftant correfpondence with *Italy*, were the firft people among us that had gardens and fruit-trees in any perfection, within the walls of their abbies [t] and priories. The *barons* neglected every purfuit that did not lead to war or tend to the pleafure of the chafe.

It was not till gentlemen took up the ftudy of horticulture themfelves that the knowledge of gardening made fuch hafty advances. Lord *Cobham*, Lord *Ila*, and Mr. *Waller* of *Beaconsfield*, were fome of the firft people of rank that promoted the elegant fcience of ornamenting without defpifing the fuperintendence of the kitchen quarters and fruit walls.

[t] " In *monafteries* the lamp of knowledge continued to burn, however dimly. In " them men of bufinefs were formed for the ftate: the art of writing was cultivated by " the *monks*; they were the only proficients in mechanics, *gardening*, and architecture." See *Dalrymple's* Annals of *Scotland*.

A remark

A remark made by the excellent Mr. *Ray* in his Tour of *Europe* at once furprifes us, and corroborates what has been advanced above; for we find him obferving, fo late as his days, that " the " *Italians* ufe feveral herbs for fallets, which *are not yet* or have " not been but *lately* ufed in *England,* viz. *felleri* (celery) which is " nothing elfe but the fweet fmallage; the young fhoots whereof, " with a little of the head of the root cut off, they eat raw with " oil and pepper." and farther he adds " *curled endive* blanched " is much ufed beyond feas; and, for a raw fallet, feemed to excel " lettuce itfelf." Now this journey was undertaken no longer ago than in the year 1663.

I am, &c.

LETTER XXXVIII.

TO THE SAME.

" Fortè puer, comitum feductus ab agmine fido,
" Dixerat, ecquis adeft? et, adeft, refponderat echo.
" Hic ftupet; utque aciem partes divifit in omnes;
" Voce, veni, clamat magnâ. Vocat illa vocantem."

DEAR SIR, SELBORNE, Feb. 12, 1778.

In a diftrict fo diverfified as this, fo full of hollow vales and hanging woods, it is no wonder that echoes fhould abound. Many we have difcovered that return the cry of a pack of dogs, the notes of a hunting-horn, a tunable ring of bells, or the melody of birds, very agreeably: but we were ftill at a lofs for a polyfyllabical,

articulate

articulate echo, till a young gentleman, who had parted from his company in a fummer evening walk, and was calling after them, ftumbled upon a very curious one in a fpot where it might leaft be expected.   At firft he was much furprifed, and could not be per-fuaded but that he was mocked by fome boy; but, repeating his trials in feveral languages, and finding his refpondent to be a very adroit polyglot, he then difcerned the deception.

This echo in an evening, before rural noifes ceafe, would re-peat ten fyllables moft articulately and diftinctly, efpecially if quick dactyls were chofen.   The laft fyllables of

" Tityre, tu patulæ recubans - - -"

were as audibly and intelligibly returned as the firft : and there is no doubt, could trial have been made, but that at midnight, when the air is very elaftic, and a dead ftillnefs prevails, one or two fyllables more might have been obtained; but the diftance rendered fo late an experiment very inconvenient.

Quick dactyls, we obferved, fucceeded beft; for when we came to try it's powers in flow, heavy, embaraffed fpondees of the fame number of fyllables,

" Monftrum horrendum, informe, ingens - - -"

we could perceive a return but of four or five.

All echoes have fome one place to which they are returned ftronger and more diftinct than to any other ; and that is always the place that lies at right angles with the object of repercuffion, and is not too near, nor too far off. Buildings, or naked rocks, re-echo much more articulately than hanging wood or vales ; becaufe in the latter the voice is as it were entangled, and embaraffed in the covert, and weakened in the rebound.

G g   The

The true object of this echo, as we found by various experiments, is the ftone-built, tiled hop-kiln in *Gally-lane*, which meafures in front 40 feet, and from the ground to the eaves 12 feet. The true *centrum phonicum*, or juft diftance, is one particular fpot in the *King's-field*, in the path to *Nore-hill*, on the very brink of the fteep balk above the hollow cart way.   In this cafe there is no choice of diftance ; but the path, by meer contingency, happens to be the lucky, the identical fpot, becaufe the ground rifes or falls fo immediately, if the fpeaker either retires or advances, that his mouth would at once be above or below the object.

We meafured this polyfyllabical echo with great exactnefs, and found the diftance to fall very fhort of Dr. *Plot's* rule for diftinct articulation : for the Doctor, in his hiftory of *Oxfordfhire*, allows 120 feet for the return of each fyllable diftinctly : hence this echo, which gives ten diftinct fyllables, ought to meafure 400 yards, or 120 feet to each fyllable ; whereas our diftance is only 258 yards, or near 75 feet, to each fyllable.   Thus our meafure falls fhort of the Doctor's, as five to eight : but then it muft be acknowledged that this candid philofopher was convinced afterwards, that fome latitude muft be admitted of in the diftance of echoes according to time and place.

When experiments of this fort are making, it fhould always be remembered that weather and the time of day have a vaft influence on an echo ; for a dull, heavy, moift air deadens and clogs the found ; and hot funfhine renders the air thin and weak, and deprives it of all it's fpringinefs ; and a ruffling wind quite defeats the whole.   In a ftill, clear, dewy evening the air is moft elaftic ; and perhaps the later the hour the more fo.

Echo has always been so amusing to the imagination, that the poets have personified her; and in their hands she has been the occasion of many a beautiful fiction. Nor need the gravest man be ashamed to appear taken with such a phænomenon, since it may become the subject of philosophical or mathematical inquiries.

One should have imagined that echoes, if not entertaining, must at least have been harmless and inoffensive; yet *Virgil* advances a strange notion, that they are injurious to bees. After enumerating some probable and reasonable annoyances, such as prudent owners would wish far removed from their bee-gardens, he adds

" — — — — — — aut ubi concava pulsu
" Saxa sonant, vocisque offensa resultat imago."

This wild and fanciful assertion will hardly be admitted by the philosophers of these days; especially as they all now seem agreed that insects are not furnished with any organs of hearing at all. But if it should be urged, that though they cannot *hear* yet perhaps they may *feel* the repercussions of sounds, I grant it is possible they may. Yet that these impressions are distasteful or hurtful, I deny, because bees, in good summers, thrive well in my outlet, where the echoes are very strong: for this village is another *Anathoth*, a place of *responses* or *echoes*. Besides, it does not appear from experiment that bees are in any way capable of being affected by sounds: for I have often tried my own with a large speaking-trumpet held close to their hives, and with such an exertion of voice as would have haled a ship at the distance of a mile, and still these insects pursued their various employments undisturbed, and without shewing the least sensibility or resentment.

G g 2

Some

Some time fince it's difcovery this echo is become totally filent, though the object, or hop-kiln, remains : nor is there any myftery in this defect ; for the field between is planted as an hop-garden, and the voice of the fpeaker is totally abforbed and loft among the poles and entangled foliage of the hops. And when the poles are removed in autumn the difappointment is the fame ; becaufe a tall quick-fet hedge, nurtured up for the purpofe of fhelter to the hop ground, entirely interrupts the impulfe and repercuffion of the voice : fo that till thofe obftructions are removed no more of it's garrulity can be expected.

Should any gentleman of fortune think an echo in his park or outlet a pleafing incident, he might *build* one at little or no ex-penfe. For whenever he had occafion for a new barn, ftable, dog-kennel, or the like ftructure, it would be only needful to erect this building on the gentle declivity of an hill, with a like rifing oppo-fite to it, at a few hundred yards diftance ; and perhaps fuccefs might be the eafier enfured could fome canal, lake, or ftream, intervene. From a feat at the *centrum phonicum* he and his friends might amufe themfelves fometimes of an evening with the prattle of this loquacious nymph ; of whofe complacency and decent re-ferve more may be faid than can with truth of every individual of her fex ; fince fhe is — — — — — — — — — —

　　"— — — — — — — quæ nec *reticere* loquenti,
　"Nec *prior* ipfa *loqui* didicit refonabilis echo."

I am, &c.

P. S.

P. S. The claffic reader will, I truft, pardon the following lovely quotation, fo finely defcribing echoes, and fo poetically accounting for their caufes from popular fuperftition:

" Quæ benè quom videas, rationem reddere poffis
" Tute tibi atque aliis, quo pacto per loca fola
" Saxa pareis formas verborum ex ordine reddant,
" Palanteis comites quom monteis inter opacos
" Quærimus, et magnâ difperfos voce ciemus.
" Sex etiam, aut feptem loca vidi reddere voces
" Unam quom jaceres : ita colles collibus ipfis
" Verba repulfantes iterabant dicta referre.
" Hæc loca capripedes Satyros, Nymphafque tenere
" Finitimi fingunt, et Faunos effe loquuntur ;
" Quorum noctivago ftrepitu, ludoque jocanti
" Adfirmant volgo taciturna filentia rumpi,
" Chordarumque fonos fieri, dulceifque querelas,
" Tibia quas fundit digitis pulfata canentum :
" Et genus agricolûm latè fentifcere, quom Pan
" Pinea femiferi capitis velamina quaffans,
" Unco fæpe labro calamos percurrit hianteis,
" Fiftula filveftrem ne ceffet fundere mufam."

Lucretius, Lib. iv. 1. 576.

LETTER

LETTER XXXIX.

TO THE SAME.

DEAR SIR,
SELBORNE, May 13, 1778.

AMONG the many fingularities attending thofe amufing birds the *fwifts*, I am now confirmed in the opinion that we have every year the fame number of pairs invariably; at leaft the refult of my inquiry has been exactly the fame for a long time paft. The fwallows and martins are fo numerous, and fo widely diftributed over the village, that it is hardly poffible to recount them; while the fwifts, though they do not all build in the church, yet fo frequently haunt it, and play and rendezvous round it, that they are eafily enumerated. The number that I conftantly find are *eight pairs*; about half of which refide in the church, and the reft build in fome of the loweft and meaneft thatched cottages. Now as thefe eight pairs, allowance being made for accidents, breed yearly eight pairs more, what becomes annually of this increafe; and what determines every fpring which pairs fhall vifit us, and reoccupy their ancient haunts?

Ever fince I have attended to the fubject of ornithology, I have always fuppofed that that fudden reverfe of affection, that ftrange αντιστοργη, which immediately fucceeds in the feathered kind to the moft paffionate fondnefs, is the occafion of an equal difperfion of birds over the face of the earth. Without this provifion one favourite diftrict would be crowded with inhabitants, while others

would

would be deftitute and forfaken. But the parent birds feem to maintain a jealous fuperiority, and to oblige the young to feek for new abodes: and the rivalry of the males, in many kinds, prevents their crowding the one on the other. Whether the fwallows and houfe-martins return in the fame exact number annually is not eafy to fay, for reafons given above: but it is apparent, as I have re-marked before in my Monographies, that the numbers returning bear no manner of proportion to the numbers retiring.

LETTER XL.

TO THE SAME.

DEAR SIR,   SELBORNE, June 2, 1778.

THE ftanding objection to botany has always been, that it is a purfuit that amufes the fancy and exercifes the memory, with-out improving the mind or advancing any real knowledge: and, where the fcience is carried no farther than a mere fyftematic claffification, the charge is but too true. But the botanift that is defirous of wiping off this afperfion fhould be by no means con-tent with a lift of names; he fhould ftudy plants philofophically, fhould inveftigate the laws of vegetation, fhould examine the powers and virtues of efficacious herbs, fhould promote their cultivation;

cultivation; and graft the gardener, the planter, and the hufband-man, on the phytologift. Not that fyftem is by any means to be thrown afide; without fyftem the field of Nature would be a pathlefs wildernefs; but fyftem fhould be fubfervient to, not the main object of, purfuit.

Vegetation is highly worthy of our attention; and in itfelf is of the utmoft confequence to mankind, and productive of many of the greateft comforts and elegancies of life. To plants we owe timber, bread, beer, honey, wine, oil, linen, cotton, &c. what not only ftrengthens our hearts, and exhilerates our fpirits, but what fecures us from inclemencies of weather and adorns our per-fons. Man, in his true ftate of nature, feems to be fubfifted by fpontaneous vegetation: in middle climes, where graffes prevail, he mixes fome animal food with the produce of the field and gar-den: and it is towards the polar extremes only that, like his kin-dred bears and wolves, he gorges himfelf with flefh alone, and is driven, to what hunger has never been known to compel the very beafts, to prey on his own fpecies. [u]

The productions of vegetation have had a vaft influence on the commerce of nations, and have been the great promoters of navi-gation, as may be feen in the articles of fugar, tea, tobacco, opium, ginfeng, betel, paper, &c. As every climate has it's peculiar pro-duce, our natural wants bring on a mutual intercourfe; fo that by means of trade each diftant part is fupplied with the growth of every latitude. But, without the knowledge of plants and their culture, we muft have been content with our hips and haws, with-out enjoying the delicate fruits of *India* and the falutiferous drugs of *Peru*.

[u] See the late Voyages to the fouth-feas.

Inftead

Inftead of examining the minute diftinctions of every various fpecies of each obfcure genus, the botanift fhould endeavour to make himfelf acquainted with thofe that are ufeful. You fhall fee a man readily afcertain every herb of the field, yet hardly know wheat from barley, or at leaft one fort of wheat or barley from another.

But of all forts of vegetation the *graffes* feem to be moft neglected; neither the farmer nor the grazier feem to diftinguifh the annual from the perennial, the hardy from the tender, nor the fucculent and nutritive from the dry and juicelefs.

The ftudy of graffes would be of great confequence to a northerly, and grazing kingdom. The botanift that could improve the fwerd of the diftrict where he lived would be an ufeful member of fociety: to raife a thick turf on a naked foil would be worth volumes of fyftematic knowledge; and he would be the beft commonwealth's man that could occafion the growth of " *two blades* " *of grafs* where *one* alone was feen before."

I am, &c.

H h LETTER

LETTER XLI.

TO THE SAME.

DEAR SIR,   SELBORNE, July 3, 1778.

In a diftrict fo diverfified with fuch a variety of hill and dale, afpects, and foils, it is no wonder that great choice of plants fhould be found. Chalks, clays, fands, fheep-walks and downs, bogs, heaths, woodlands, and champaign fields, cannot but furnifh an ample *Flora*. The deep rocky lanes abound with *filices*, and the paftures and moift woods with *fungi*. If in any branch of botany we may feem to be wanting, it muft be in the large aquatic plants, which are not to be expected on a fpot far removed from rivers, and lying up amidft the hill country at the fpring heads. To enumerate all the plants that have been difcovered within our limits would be a needlefs work; but a fhort lift of the more rare, and the fpots where they are to be found, may be neither unacceptable nor unentertaining :—

Helleborus fœtidus, ftinking hellebore, bear's foot, or fetterworth, all over the *High-wood* and *Coney-croft-hanger :* this continues a great branching plant the winter through, bloffoming about *January*, and is very ornamental in fhady walks and fhrubberies. The good women give the leaves powdered to children troubled with worms; but it is a violent remedy, and ought to be adminiftered with caution.

Helleborus

Helleborus viridis, green hellebore,—in the deep ſtony lane on the left hand juſt before the turning to *Norton-farm,* and at the top of *Middle Dorton* under the hedge : this plant dies down to the ground early in autumn, and ſprings again about *February,* flowering almoſt as ſoon as it appears above ground.

Vaccinium oxycoccos, creeping bilberries, or cranberries,—in the bogs of *Bin's-pond ;*

Vaccinium myrtillus, whortle, or bleaberries,—on the dry hillocks of *Woolmer-foreſt ;*

Droſera rotundifolia,	round-leaved ſundew.	} In the bogs
————*longifolia,*	long-leaved ditto.	} of *Bin's-pond.*

Comarum paluſtre, purple comarum, or marſh cinque foil,—in the bogs of *Bin's-pond ;*

Hypericum androſæmum, Tutſan, St. John's Wort,—in the ſtony, hollow lanes ;

Vinca minor, leſs periwinkle,—in *Selborne-hanger* and *Shrub-wood ;*

Monotropa hypopithys, yellow monotropa, or birds' neſt,—in *Selborne-hanger* under the ſhady beeches, to whoſe roots it ſeems to be paraſitical—at the north-weſt end of the *Hanger ;*

Chlora perfoliata, Blackſtonia perfoliata, Hudſoni, perfoliated yellow-wort,—on the banks in the *King's-field ;*

Paris quadrifolia, herb Paris, true-love, or one-berry,—in the *Church-litten-coppice ;*

Chryſoſplenium oppoſitifolium, oppoſite golden ſaxifrage,—in the dark and rocky hollow lanes ;

Gentiana amarella, autumnal gentian, or fellwort,—on the *Zig-zag* and *Hanger ;*

Lathræa

Lathræa squammaria, tooth-wort,—in the *Church-litten-coppice* under some hazels near the foot-bridge, in *Trimming's* garden hedge, and on the dry wall opposite *Grange-yard*;

Dipsacus pilosus, small teasel,—in the *Short* and *Long Lith.*

Lathyrus sylvestris, narrow-leaved, or wild lathyrus,—in the bushes at the foot of the *Short Lith,* near the path;

Ophrys spiralis, ladies traces,—in the *Long Lith,* and towards the south-corner of the common;

Ophrys nidus avis, birds' nest ophrys,—in the *Long Lith* under the shady beeches among the dead leaves; in *Great Dorton* among the bushes, and on the *Hanger* plentifully;

Serapias latifolia, helleborine,—in the *High-wood* under the shady beeches;

Daphne laureola, spurge laurel,—in *Selborne-Hanger* and the *High-wood*;

Daphne mezereum, the mezereon,—in *Selborne-Hanger* among the shrubs at the south-east end above the cottages.

Lycoperdon tuber, truffles,—in the *Hanger* and *High-wood.*

Sambucus ebulus, dwarf elder, walwort, or danewort,—among the rubbish and ruined foundations of the *Priory*.

LETTER

LETTER XLII.

TO THE SAME.

" Omnibus animalibus reliquis certus et uniufmodi, et in fuo cuique genere
" inceffus: eft:aves folæ vario meatu feruntur, et in terrâ, et in äere."
PLIN. Hift. Nat. lib. x. cap. 38.

DEAR SIR,   SELBORNE, Aug. 7, 1778.

A GOOD ornithologift fhould be able to diftinguifh birds by their
air as well as by their colours and fhape; on the ground as well
as on the wing, and in the bufh as well as in the hand. For,
though it muft not be faid that every *fpecies* of birds has a manner
peculiar to itfelf, yet there is fomewhat in moft *genera* at leaft, that
at firft fight difcriminates them, and enables a judicious obferver
to pronounce upon them with fome certainty. Put a bird in
motion

" — — Et verâ inceffu patuit — — — —"

Thus *kites* and *buzzards* fail round in circles with wings expanded and motionlefs; and it is from their gliding manner that the
former are ftill called in the north of *England gleads*, from the *Saxon*
verb *glidan*, to glide. The *keftrel*, or *wind-hover*, has a peculiar mode
of hanging in the air in one place, his wings all the while being
brifkly agitated. *Hen-harriers* fly low over heaths or fields of corn,
and beat the ground regularly like a pointer or fetting-dog. *Owls*
move in a buoyant manner, as if lighter than the air; they feem
to want ballaft. There is a peculiarity belonging to *ravens* that
muft draw the attention even of the moft incurious—they fpend all
their

their leisure time in striking and cuffing each other on the wing in
a kind of playful skirmish; and, when they move from one place
to another, frequently turn on their backs with a loud croak, and
seem to be falling to the ground. When this odd gesture betides
them, they are scratching themselves with one foot, and thus
lose the center of gravity. *Rooks* sometimes dive and tumble in
a frolicksome manner; *crows* and *daws* swagger in their walk;
wood-peckers fly *volatu undoso*, opening and closing their wings at
every stroke, and so are always rising or falling in curves. All
of this genus use their tails, which incline downward, as a support
while they run up trees. *Parrots*, like all other hooked-clawed
birds, walk aukwardly, and make use of their bill as a third foot,
climbing and descending with ridiculous caution. All the *gallinæ*
parade and walk gracefully, and run nimbly; but fly with difficulty, with an impetuous whirring, and in a straight line. *Magpies* and *jays* flutter with powerless wings, and make no dispatch;
herons seem incumbered with too much sail for their light bodies;
but these vast hollow wings are necessary in carrying burdens, such
as large fishes, and the like; *pigeons*, and particularly the sort
called *smiters*, have a way of clashing their wings the one against
the other over their backs with a loud snap; another variety called
tumblers turn themselves over in the air. Some birds have move-
ments peculiar to the season of love: thus *ring-doves*, though strong
and rapid at other times, yet in the spring hang about on the wing
in a toying and playful manner; thus the *cock-snipe*, while breeding, forgetting his former flight, fans the air like the wind-hover;
and the *green-finch* in particular exhibits such languishing and
faultering gestures as to appear like a wounded and dying bird;
the *king-fisher* darts along like an arrow; *fern-owls*, or *goat-suckers*,
glance in the dusk over the tops of trees like a meteor; *starlings*

as

as it were swim along, while *missel-thrushes* use a wild and desultory flight; *swallows* sweep over the surface of the ground and water, and distinguish themselves by rapid turns and quick evolutions; *swifts* dash round in circles; and the *bank-martin* moves with frequent vacillations like a butterfly. Most of the small birds fly by jerks, rising and falling as they advance. Most small birds hop; but *wagtails* and *larks* walk, moving their legs alternately. *Skylarks* rise and fall perpendicularly as they sing; *woodlarks* hang poised in the air; and *titlarks* rise and fall in large curves, singing in their descent. The *white-throat* uses odd jerks and gesticulations over the tops of hedges and bushes. All the *duck-kind* waddle; *divers* and *auks* walk as if fettered, and stand erect on their tails: these are the *compedes* of *Linnæus*. *Geese* and *cranes*, and most wild-fowls, move in figured flights, often changing their position. The secondary *remiges* of *Tringæ*, *wild-ducks*, and some others, are very long, and give their wings, when in motion, an hooked appearance. *Dabchicks*, *moor-hens*, and *coots*, fly erect, with their legs hanging down, and hardly make any dispatch; the reason is plain, their wings are placed too forward out of the true center of gravity; as the legs of *auks* and *divers* are situated too backward.

LETTER

LETTER XLIII.

TO THE SAME.

DEAR SIR,   SELBORNE, Sept. 9, 1778.

From the motion of birds, the tranfition is natural enough to
their notes and language, of which I fhall fay fomething. Not
that I would pretend to underftand their language like the *vizier*;
who, by the recital of a converfation which paffed between two
owls, reclaimed a fultan,[x] before delighting in conqueft and de-
vaftation; but I would be thought only to mean that many of
the winded tribes have various founds and voices adapted to exprefs their various paffions, wants, and feelings; fuch as anger,
fear, love, hatred, hunger, and the like. All fpecies are not
equally eloquent; fome are copious and fluent as it were in
their utterance, while others are confined to a few important
founds: no bird, like the fifh kind, is quite mute, though fome
are rather filent. The language of birds is very ancient, and, like
other ancient modes of fpeech, very elliptical; little is faid, but
much is meant and underftood.

The notes of the eagle-kind are fhrill and piercing; and about
the feafon of nidification much diverfified, as I have been often
affured by a curious obferver of Nature, who long refided at
Gibraltar, where eagles abound. The notes of our *hawks* much
refemble thofe of the king of birds. *Owls* have very expreffive
notes; they hoot in a fine vocal found, much refembling the *vox*

---

[x] See Spectator, Vol. VII, N°. 512.

humana

humana, and reducible by a pitch-pipe to a mufical key. This note feems to exprefs complacency and rivalry among the males: they ufe alfo a quick call and an horrible fcream; and can fnore and hifs when they mean to menace. *Ravens,* befides their loud croak, can exert a deep and folemn note that makes the woods to echo; the amorous found of a *crow* is ftrange and ridiculous; *rooks,* in the breeding feafon, attempt fometimes in the gaiety of their hearts to fing, but with no great fuccefs; the *parrot*-kind have many modulations of voice, as appears by their aptitude to learn human founds; *doves* coo in an amorous and mournful manner, and are emblems of defpairing lovers; the *woodpecker* fets up a fort of loud and hearty laugh; the *fern-owl,* or *goat-fucker,* from the dufk till day-break, ferenades his mate with the clattering of caftanets. All the tuneful *paſſeres* exprefs their complacency by fweet modulations, and a variety of melody. The *fwallow,* as has been obferved in a former letter, by a fhrill alarm befpeaks the attention of the other *hirundines,* and bids them be aware that the hawk is at hand. Aquatic and gregarious birds, efpecially the nocturnal, that fhift their quarters in the dark, are very noify and loquacious; as cranes, wild-geefe, wild-ducks, and the like: their perpetual clamour prevents them from difperfing and lofing their companions.

In fo extenfive a fubject, fketches and outlines are as much as can be expected; for it would be endlefs to inftance in all the infinite variety of the feathered nation. We fhall therefore confine the remainder of this letter to the few domeftic fowls of our yards, which are moft known, and therefore beft underftood. And firft the *peacock,* with his gorgeous train, demands our attention; but, like moft of the gaudy birds, his notes are grating and fhocking to the ear: the yelling of cats, and the braying of an afs, are not

I i   more

more difguftful. The voice of the *goofe* is trumpet-like, and clanking; and once faved the Capitol at *Rome*, as grave hiftorians affert: the hifs alfo of the *gander* is formidable and full of menace, and " protective of his young." Among *ducks* the fexual dif-tinction of voice is remarkable; for, while the *quack* of the female is loud and fonorous, the voice of the *drake* is inward and harfh, and feeble, and fcarce difcernible. The cock *turkey* ftruts and gobbles to his miftrefs in a moft uncouth manner; he hath alfo a pert and petulant note when he attacks his adverfary. When a hen *turkey* leads forth her young brood fhe keeps a watchful eye; and if a bird of prey appear, though ever fo high in the air, the careful mother announces the enemy with a little inward moan, and watches him with a fteady and attentive look; but, if he approach, her note becomes earneft and alarming, and her outcries are redoubled.

No inhabitants of a yard feem poffeffed of fuch a variety of expreffion and fo copious a language as common poultry. Take a chicken of four or five days old, and hold it up to a window where there are flies, and it will immediately feize it's prey, with little twitterings of complacency; but if you tender it a wafp or a bee, at once it's note becomes harfh, and expreffive of difapprobation and a fenfe of danger. When a pullet is ready to lay fhe intimates the event by a joyous and eafy foft note. Of all the occurrences of their life that of *laying* feems to be the moft important; for no fooner has a hen difburdened herfelf, than fhe rufhes forth with a clamorous kind of joy, which the cock and the reft of his miftreffes immediately adopt. The tumult is not confined to the family concerned, but catches from yard to yard, and fpreads to every homeftead within hearing, till at laft the whole village is in an uproar. As foon as a hen becomes a mother

her

her new relation demands a new language; she then runs clocking and screaming about, and seems agitated as if possessed. The father of the flock has also a considerable vocabulary; if he finds food, he calls a favourite concubine to partake; and if a bird of prey passes over, with a warning voice he bids his family beware. The gallant *chanticleer* has, at command, his amorous phrases and his terms of defiance. But the sound by which he is best known is his *crowing*: by this he has been distinguished in all ages as the countryman's clock or larum, as the watchman that proclaims the divisions of the night. Thus the poet elegantly styles him:

" —— —— —— the crested cock, whose clarion sounds
" The silent hours."

A neighbouring gentleman one summer had lost most of his chickens by a sparrow-hawk, that came gliding down between a faggot pile and the end of his house to the place where the coops stood. The owner, inwardly vexed to see his flock thus diminishing, hung a setting net adroitly between the pile and the house, into which the caitif dashed, and was entangled. Resentment suggested the law of retaliation; he therefore clipped the hawk's wings, cut off his talons, and, fixing a cork on his bill, threw him down among the brood-hens. Imagination cannot paint the scene that ensued; the expressions that fear, rage, and revenge, inspired, were new, or at least such as had been unnoticed before: the exasperated matrons upbraided, they execrated, they insulted, they triumphed. In a word, they never desisted from buffeting their adversary till they had torn him in an hundred pieces.

LETTER XLIV.

TO THE SAME.

SELBORNE.

" —— —— ᴗ —— monftrent"
" —— —— —— —— —— ——"
" Quid tantum Oceano properent fe tingere foles"
" Hyberni; vel quæ tardis mora noctibus obftet."

Gentlemen who have outlets might contrive to make ornament fubfervient to utility: a pleafing eye-trap might alfo contribute to promote fcience: an obelifk in a garden or park might
be both an embellifhment and an *heliotrope*.

Any perfon that is curious, and enjoys the advantage of a good
horizon, might, with little trouble, make two *heliotropes*; the one
for the *winter*, the other for the *fummer* folftice: and thefe two
erections might be conftructed with very little expenfe; for two
pieces of timber frame-work, about ten or twelve feet high, and
four feet broad at the bafe, and clofe lined with plank, would
anfwer the purpofe.

The erection for the former fhould, if poffible, be placed
within fight of fome window in the common fitting parlour; becaufe men, at that dead feafon of the year, are ufually within
doors at the clofe of the day; while that for the latter might be
fixed for any given fpot in the garden or outlet: whence the owner
might contemplate, in a fine fummer's evening, the utmoft extent
that the fun makes to the *northward* at the feafon of the longeft
days.

days. Now nothing would be neceffary but to place thefe two *objects* with fo much exactnefs, that the *wefterly* limb of the fun, at fetting, might but juft clear the *winter heliotrope* to the *weft* of it on the *fhorteft* day; and that the *whole* difc of the fun, at the *longeft* day, might exactly at fetting alfo clear the *fummer heliotrope* to the *north* of it.

By this fimple expedient it would foon appear that there is no fuch thing, ftrictly fpeaking, as a folftice; for, from the *fhorteft* day, the owner would, every clear evening, fee the difc *advancing*, at it's fetting, to the *weftward* of the object; and, from the *longeft* day, obferve the fun *retiring* backwards every evening at it's fet-ing, towards the object *weftward*, till, in a few nights, it would fet quite behind it, and fo by degrees to the *weft* of it: for when the fun comes near the fummer folftice, the whole difc of it would at firft fet behind the object; after a time the *northern* limb would firft appear, and fo every night gradually more, till at length the whole diameter would fet *northward* of it for about three nights; but on the middle night of the three, fenfibly more remote than the former or following. When beginning it's recefs from the fummer tropic, it would continue more and more to be hidden every night, till at length it would defcend quite behind the object again; and fo nightly more and more to the *weftward*.

LETTER

LETTER XLV.

TO THE SAME.

SELBORNE.

" —— —— —— Mugire videbis
" Sub pedibus terram, et defcendere montibus ornos."

WHEN I was a boy I ufed to read, with aftonifhment and implicit affent, accounts in *Baker's Chronicle* of walking hills and travelling mountains. *John Philips*, in his *Cyder*, alludes to the credit that was given to fuch ftories with a delicate but quaint vein of humour peculiar to the author of the *Splendid Shilling*.

" I nor advife, nor reprehend the choice
" Of *Marcley Hill*; the apple no where finds
" A kinder mould; yet 'tis unfafe to truft
" Deceitful ground: who knows but that once more
" This mount may journey, and his prefent fite
" Forfaken, to thy neighbour's bounds transfer
" Thy goodly plants, affording matter ftrange
" For law debates!"

But, when I came to confider better, I began to fufpect that though our hills may never have journeyed far, yet that the ends of many of them have flipped and fallen away at diftant periods, leaving the cliffs bare and abrupt. This feems to have been the cafe with *Nore* and *Whetham Hills*; and efpecially with the ridge between *Harteley Park* and *Ward-le-ham*, where the ground has flid into vaft fwellings and furrows; and lies ftill in fuch romantic confufion confusion as cannot be accounted for from any other cause. A strange event, that happened not long since, justifies our suspicions; which, though it befell not within the limits of this parish, yet as it was within the hundred of *Selborne*, and as the circumstances were singular, may fairly claim a place in a work of this nature.

The months of *January* and *February*, in the year 1774, were remarkable for great melting snows and vast gluts of rain; so that by the end of the latter month the land-springs, or *lavants*, began to prevail, and to be near as high as in the memorable winter of 1764. The beginning of *March* also went on in the same tenor; when, in the night between the 8th and 9th of that month, a considerable part of the great woody hanger at *Hawkley* was torn from it's place, and fell down, leaving a high free-stone cliff naked and bare, and resembling the steep side of a chalk-pit. It appears that this huge fragment, being perhaps sapped and undermined by waters, foundered, and was ingulfed, going down in a perpendicular direction; for a gate which stood in the field, on the top of the hill, after sinking with it's posts for thirty or forty feet, remained in so true and upright a position as to open and shut with great exactness, just as in it's first situation. Several oaks also are still standing, and in a state of vegetation, after taking the same desperate leap. That great part of this prodigious mass was absorbed in some gulf below, is plain also from the inclining ground at the bottom of the hill, which is free and unincumbered; but would have been buried in heaps of rubbish, had the fragment parted and fallen forward. About an hundred yards from the foot of this hanging coppice stood a cottage by the side of a lane; and two hundred yards lower, on the other side of the lane, was a farm-house, in which lived a labourer and

his

his family; and, juft by, a ftout new barn. The cottage was inhabited by an old woman and her fon, and his wife. Thefe people in the evening, which was very dark and tempeftuous, obferved that the brick floors of their kitchens began to heave and part; and that the walls feemed to open, and the roofs to crack: but they all agree that no tremor of the ground, indicating an earthquake, was ever felt; only that the wind continued to make a moft tremendous roaring in the woods and hangers. The miferable inhabitants, not daring to go to bed, remained in the utmoft folicitude and confufion, expecting every moment to be buried under the ruins of their fhattered edifices. When day-light came they were at leifure to contemplate the devaftations of the night: they then found that a deep rift, or chafm, had opened under their houfes, and torn them, as it were, in two; and that one end of the barn had fuffered in a fimilar manner; that a pond near the cottage had undergone a ftrange reverfe, becoming deep at the fhallow end, and fo *vice verfa*; that many large oaks were removed out of their perpendicular, fome thrown down, and fome fallen into the heads of neighbouring trees; and that a gate was thruft forward, with it's hedge, full fix feet, fo as to require a new track to be made to it. From the foot of the cliff the general courfe of the ground, which is pafture, inclines in a moderate defcent for half a mile, and is interfperfed with fome hillocks, which were rifted, in every direction, as well towards the great woody hanger, as from it. In the firft pafture the deep clefts began; and running acrofs the lane, and under the buildings, made fuch vaft fhelves that the road was impaffable for fome time; and fo over to an arable field on the other fide, which was ftrangely torn and difordered. The fecond pafture field, being more foft and fpringy, was protruded forward without many fiffures in the turf, which

was

was raifed in long ridges refembling graves, lying at right angles to the motion. At the bottom of this enclofure the foil and turf rofe many feet againft the bodies of fome oaks that obftructed their farther courfe and terminated this awful commotion.

The perpendicular height of the precipice, in general, is twenty-three yards; the length of the lapfe, or flip, as feen from the fields below, one hundred and eighty-one; and a partial fall, concealed in the coppice, extends feventy yards more: fo that the total length of this fragment that fell was two hundred and fifty-one yards. About fifty acres of land fuffered from this violent convulfion; two houfes were entirely deftroyed; one end of a new barn was left in ruins, the walls being cracked through the very ftones that compofed them; a hanging coppice was changed to a naked rock; and fome grafs grounds and an arable field fo broken and rifted by the chafms as to be rendered, for a time, neither fit for the plough or fafe for pafturage, till confiderable labour and expenfe had been beftowed in levelling the furface and filling in the gaping fiffures.

K k   LETTER

LETTER XLVI.

TO THE SAME.

SELBORNE.

" — — — refonant arbufta — — — — "

THERE is a fteep abrupt pafture field interfperfed with furze clofe to the back of this village, well known by the name of the *Short Lithe,* confifting of a rocky dry foil, and inclining to the afternoon fun. This fpot abounds with the *gryllus campeftris,* or *field-cricket*; which, though frequent in thefe parts, is by no means a common infect in many other counties.

As their cheerful fummer cry cannot but draw the attention of a naturalift, I have often gone down to examine the œconomy of thefe *grylli,* and ftudy their mode of life : but they are fo fhy and cautious that it is no eafy matter to get a fight of them; for, feeling a perfon's footfteps as he advances, they ftop fhort in the midft of their fong, and retire backward nimbly into their burrows, where they lurk till all fufpicion of danger is over.

At firft we attempted to dig them out with a fpade, but without any great fuccefs; for either we could not get to the bottom of the hole, which often terminated under a great ftone; or elfe, in breaking up the ground, we inadvertently fqueezed the poor in-fect to death. Out of one fo bruifed we took a multitude of eggs,

which

which were long and narrow, of a yellow colour, and covered with a very tough skin. By this accident we learned to distinguish the male from the female; the former of which is shining black, with a golden stripe across his shoulders; the latter is more dusky, more capacious about the abdomen, and carries a long sword-shaped weapon at her tail, which probably is the instrument with which she deposits her eggs in crannies and safe receptacles.

Where violent methods will not avail, more gentle means will often succeed; and so it proved in the present case; for, though a spade be too boisterous and rough an implement, a pliant stalk of grass, gently insinuated into the caverns, will probe their windings to the bottom, and quickly bring out the inhabitant; and thus the humane inquirer may gratify his curiosity without injuring the object of it. It is remarkable that, though these insects are furnished with long legs behind, and brawny thighs for leaping, like grasshoppers; yet when driven from their holes they shew no activity, but crawl along in a shiftless manner, so as easily to be taken: and again, though provided with a curious apparatus of wings, yet they never exert them when there seems to be the greatest occasion. The males only make that shrilling noise perhaps out of rivalry and emulation, as is the case with many animals which exert some sprightly note during their breeding time: it is raised by a brisk friction of one wing against the other. They are solitary beings, living singly male or female, each as it may happen; but there must be a time when the sexes have some intercourse, and then the wings may be useful perhaps during the hours of night. When the males meet they will fight fiercely, as I found by some which I put into the crevices of a dry stone wall, where I should have been glad to have made them settle. For though they seemed distressed by being taken out of their know-

ledge,

ledge, yet the first that got possession of the chinks would seize on any that were obtruded upon them with a vast row of serrated fangs. With their strong jaws, toothed like the shears of a lobster's claws, they perforate and round their curious regular cells, having no fore-claws to dig, like the mole-cricket. When taken in hand I could not but wonder that they never offered to defend themselves, though armed with such formidable weapons. Of such herbs as grow before the mouths of their burrows they eat indiscriminately; and on a little platform, which they make just by, they drop their dung; and never, in the day time, seem to stir more than two or three inches from home. Sitting in the entrance of their caverns they chirp all night as well as day from the middle of the month of *May* to the middle of *July*; and in hot weather, when they are most vigorous, they make the hills echo; and, in the stiller hours of darkness, may be heard to a considerable distance. In the beginning of the season their notes are more faint and inward; but become louder as the summer advances, and so die away again by degrees.

Sounds do not always give us pleasure according to their sweetness and melody; nor do harsh sounds always displease. We are more apt to be captivated or disgusted with the associations which they promote, than with the notes themselves. Thus the shrilling of the *field-cricket*, though sharp and stridulous, yet marvellously delights some hearers, filling their minds with a train of summer ideas of every thing that is rural, verdurous, and joyous.

About the tenth of *March* the crickets appear at the mouths of their cells, which they then open and bore, and shape very elegantly. All that ever I have seen at that season were in their pupa state, and had only the rudiments of wings, lying under a skin or

coat,

coat, which muft be caft before the infect can arrive at it's perfect ftate[y]; from whence I fhould fuppofe that the old ones of laft year do not always furvive the winter. In *Auguft* their holes begin to be obliterated, and the infects are feen no more till fpring.

Not many fummers ago I endeavoured to tranfplant a colony to the terrace in my garden, by boring deep holes in the floping turf. The new inhabitants ftayed fome time, and fed and fung; but wandered away by degrees, and were heard at a farther diftance every morning; fo that it appears that on this emergency they made ufe of their wings in attempting to return to the fpot from which they were taken.

One of thefe crickets, when confined in a paper cage and fet in the fun, and fupplied with plants moiftened with water, will feed and thrive, and become fo merry and loud as to be irkfome in the fame room where a perfon is fitting : if the plants are not wetted it will die.

[y] We have obferved that they caft thefe fkins in *April*, which are then feen lying at the mouths of their holes.

LETTER

LETTER XLVII.

TO THE SAME.

DEAR SIR, SELBORNE.

" Far from all resort of mirth
" Save the cricket on the hearth." MILTON's *Il Penseroso.*

WHILE many other insects must be sought after in fields and woods, and waters, the *gryllus domesticus*, or *house-cricket*, resides altogether within our dwellings, intruding itself upon our notice whether we will or no. This species delights in new-built houses, being, like the spider, pleased with the moisture of the walls; and besides, the softness of the mortar enables them to burrow and mine between the joints of the bricks or stones, and to open communications from one room to another. They are particularly fond of kitchens and bakers' ovens, on account of their perpetual warmth.

Tender insects that live abroad either enjoy only the short period of one summer, or else doze away the cold uncomfortable months in profound slumbers; but these, residing as it were in a torrid zone, are always alert and merry: a good *Christmas* fire is to them like the heats of the dog-days. Though they are frequently heard by day, yet is their natural time of motion only in the night. As soon as it grows dusk, the chirping increases, and they come running forth, and are from the size of a flea to that of their full stature. As one should suppose, from the

burning

burning atmosphere which they inhabit, they are a thirsty race, and shew a great propensity for liquids, being found frequently drowned in pans of water, milk, broth, or the like. Whatever is moist they affect; and therefore often gnaw holes in wet woollen stockings and aprons that are hung to the fire: they are the housewife's barometer, foretelling her when it will rain; and are prognostic sometimes, she thinks, of ill or good luck; of the death of a near relation, or the approach of an absent lover. By being the constant companions of her solitary hours they naturally become the objects of her superstition. These crickets are not only very thirsty, but very voracious; for they will eat the scummings of pots, and yeast, salt, and crumbs of bread; and any kitchen offal or sweepings. In the summer we have observed them to fly, when it became dusk, out of the windows, and over the neighbouring roofs. This feat of activity accounts for the sudden manner in which they often leave their haunts, as it does for the method by which they come to houses where they were not known before. It is remarkable, that many sorts of insects seem never to use their wings but when they have a mind to shift their quarters and settle new colonies. When in the air they move " *volatu undoso*," in waves or curves, like *wood-peckers*, opening and shutting their wings at every stroke, and so are always rising or sinking.

When they increase to a great degree, as they did once in the house where I am now writing, they become noisome pests, flying into the candles, and dashing into people's faces; but may be blasted and destroyed by gunpowder discharged into their crevices and crannies. In families, at such times, they are, like Pharaoh's plague of frogs, — " in their bedchambers, and upon " their beds, and in their ovens, and in their kneading-
" troughs."

" troughs [z]." Their fhrilling noife is occafioned by a brifk attrition of their wings. Cats catch hearth-crickets, and, playing with them as they do with mice, devour them. Crickets may be deftroyed, like wafps, by phials half filled with beer, or any liquid, and fet in their haunts; for, being always eager to drink, they will crowd in till the bottles are full.

LETTER XLVIII.

TO THE SAME.

SELBORNE.

How diverfified are the modes of life not only of incongruous but even of congenerous animals; and yet their fpecific diftinctions are not more various than their propenfities. Thus, while the *field-cricket* delights in funny dry banks, and the *houfe-cricket* rejoices amidft the glowing heat of the kitchen hearth or oven, the *gryllus gryllo talpa* (the *mole-cricket*), haunts moift meadows, and frequents the fides of ponds and banks of ftreams, performing all it's funtions in a fwampy wet foil. With a pair of fore-feet, curioufly adapted to the purpofe, it burrows and works under ground like the mole, raifing a ridge as it proceeds, but feldom throwing up hillocks.

[z] Exod. viii. 3.

As

As *mole-crickets* often infeft gardens by the fides of canals, they are unwelcome guefts to the gardener, raifing up ridges in their fubterraneous progrefs, and rendering the walks unfightly. If they take to the kitchen quarters, they occafion great damage among the plants and roots, by deftroying whole beds of cabbages, young legumes, and flowers. When dug out they feem very flow and helplefs, and make no ufe of their wings by day ; but at night they come abroad, and make long excurfions, as I have been convinced by finding ftragglers, in a morning, in improbable places. In fine weather, about the middle of *April*, and juft at the clofe of day, they begin to folace themfelves with a low, dull, jarring note, continued for a long time without interruption, and not unlike the chattering of the fern-owl, or goat-fucker, but more inward.

About the beginning of *May* they lay their eggs, as I was once an eye-witnefs : for a gardener at an houfe, where I was on a vifit, happening to be mowing, on the 6th of that month, by the fide of a canal, his fcythe ftruck too deep, pared off a large piece of turf, and laid open to view a curious fcene of domeftic œconomy :

" —— —— —— —— ingentem lato dedit ore feneftram :
" Apparet domus intus, et atria longa patefcunt :
" Apparent —— —— —— penetralia."

There were many caverns and winding paffages leading to a kind of chamber, neatly fmoothed and rounded, and about the fize of a moderate fnuff-box. Within this fecret nurfery were depofited near an hundred eggs of a dirty yellow colour, and enveloped in a tough fkin, but too lately excluded to contain any rudiments of young, being full of a vifcous fubftance. The eggs

L l

lay

lay but fhallow, and within the influence of the fun, juft under a little heap of fresh-mowed mould, like that which is raifed by ants.

When *mole-crickets* fly they move " *curfu undofo*," rifing and falling in curves, like the other fpecies mentioned before. In different parts of this kingdom people call them *fen-crickets*, *churr-worms*, and *eve-churrs*, all very appofite names.

Anatomifts, who have examined the inteftines of thefe infects, aftonifh me with their accounts; for they fay that, from the ftructure, pofition, and number of their ftomachs, or maws, there feems to be good reafon to fuppofe that this and the two former fpecies *ruminate* or *chew* the *cud* like many quadrupeds!

LETTER XLIX.

TO THE SAME.

SELBORNE, May 7, 1779.

IT is now more than forty years that I have paid fome attention to the ornithology of this diftrict, without being able to exhauft the fubject : new occurrences ftill arife as long as any inquiries are kept alive.

In the laft week of laft month five of thofe moft rare birds, too uncommon to have obtained an *Englifh* name, but known to natu-
ralifts

Pl. IV.

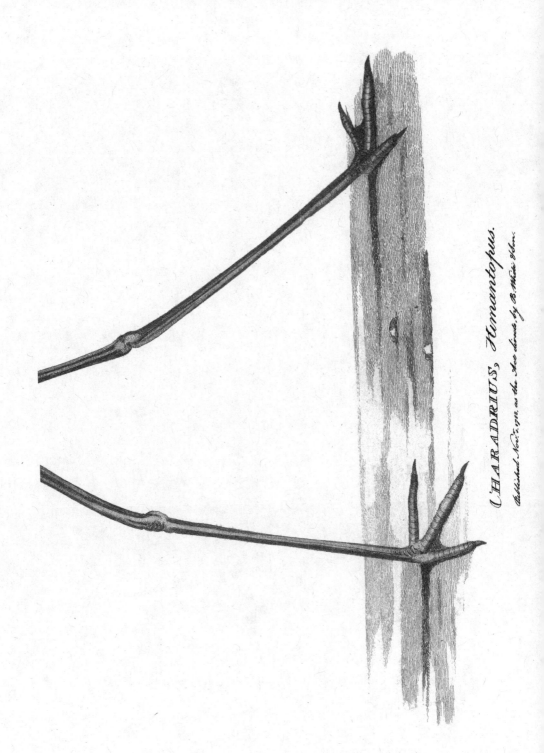

CHARADRIUS, Himantopus.

Published Nov. 2. 1791. as the Act directs, by B. White & Son.

ralifts by the terms of *himantopus*, or *loripes*, and *charadrius himan-topus*, were fhot upon the verge of *Frinfham-pond*, a large lake be-longing to the Bifhop of *Winchefter*, and lying between *Woolmer-foreft*, and the town of *Farnham*, in the county of *Surrey*. The pond keeper fays there were three brace in the flock; but that, after he had fatisfied his curiofity, he fuffered the fixth to remain unmo-lefted. One of thefe fpecimens I procured, and found the length of the legs to be fo extraordinary, that, at firft fight, one might have fuppofed the fhanks had been faftened on to impofe on the credulity of the beholder: they were legs in *caricatura*; and had we feen fuch proportions on a *Chinefe* or *Japan* fcreen we fhould have made large allowances for the fancy of the draughtfman. Thefe birds are of the *plover* family, and might with propriety be called the *ftilt plovers*. *Briffon*, under that idea, gives them the appofite name of *l'echaffe*. My fpecimen, when drawn and ftuffed with pepper, weighed only four ounces and a quarter, though the *naked* part of the thigh meafured three inches and an half, and the legs four inches and an half. Hence we may fafely affert that thefe birds exhibit, weight for inches, incomparably the greateft length of legs of any known bird. The *flamingo*, for inftance, is one of the moft long legged birds, and yet it bears no manner of proportion to the *himantopus*; for a cock *flamingo* weighs, at an average, about four pounds avoirdupois; and his legs and thighs meafure ufually about twenty inches. But four pounds are fifteen times and a fraction more than four ounces, and one quarter; and if four ounces and a quarter have eight inches of legs, four pounds muft have one hundred and twenty inches and a fraction of legs; *viz.* fomewhat more than ten feet; fuch a monftrous proportion as the world never faw! If you fhould try the experiment in ftill larger birds the difparity would ftill increafe. It muft be matter of great

curiofity

curiosity to see the *stilt plover* move; to observe how it can wield such a length of lever with such feeble muscles as the thighs seem to be furnished with. At best one should expect it to be but a bad walker: but what adds to the wonder is, that it has no back toe. Now without that steady prop to support it's steps it must be liable, in speculation, to perpetual vacillations, and seldom able to preserve the true center of gravity.

The old name of *himantopus* is taken from *Pliny*; and, by an aukward metaphor, implies that the legs are as slender and pliant as if cut out of a *thong* of leather. Neither *Willughby* nor *Ray*, in all their curious researches, either at home or abroad, ever saw this bird. Mr. *Pennant* never met with it in all *Great-Britain*, but observed it often in the cabinets of the curious at *Paris*. *Hasselquist* says that it migrates to *Egypt* in the autumn: and a most accurate observer of Nature has assured me that he has found it on the banks of the streams in *Andalusia*.

Our writers record it to have been found only twice in *Great-Britain*. From all these relations it plainly appears that these long legged *plovers* are birds of *South Europe*, and rarely visit our island; and when they do are wanderers and stragglers, and impelled to make so distant and northern an excursion from motives or accidents for which we are not able to account. One thing may fairly be deduced, that these birds come over to us from the continent, since nobody can suppose that a species not noticed once in an age, and of such a remarkable make, can constantly breed unobserved in this kingdom.

LETTER

LETTER L.

TO THE SAME.

DEAR SIR,   SELBORNE, April 21, 1780.

THE old *Suffex* tortoife, that I have mentioned to you fo often, is become my property. I dug it out of it's winter dormitory in *March* laft, when it was enough awakened to exprefs it's refentments by hiffing; and, packing it in a box with earth, carried it eighty miles in poft-chaifes. The rattle and hurry of the journey fo perfectly roufed it that, when I turned it out on a border, it walked twice down to the bottom of my garden; however, in the evening, the weather being cold, it buried itfelf in the loofe mould, and continues ftill concealed.

As it will be under my eye, I fhall now have an opportunity of enlarging my obfervations on it's mode of life, and propenfities; and perceive already that, towards the time of coming forth, it opens a breathing place in the ground near it's head, requiring, I conclude, a freer refpiration as it becomes more alive. This creature not only goes under the earth from the middle of *November* to the middle of *April*, but fleeps great part of the fummer; for it goes to bed in the longeft days at four in the afternoon, and often does not ftir in the morning till late. Befides, it retires to reft for every fhower; and does not move at all in wet days.

When

When one reflects on the state of this strange being, it is a matter of wonder to find that Providence should bestow such a profusion of days, such a seeming waste of longevity, on a reptile that appears to relish it so little as to squander more than two thirds of it's existence in a joyless stupor, and be lost to all sensation for months together in the profoundest of slumbers.

While I was writing this letter, a moist and warm afternoon, with the thermometer at 50, brought forth troops of *shell-snails*; and, at the same juncture, the *tortoise* heaved up the mould and put out it's head; and the next morning came forth, as it were raised from the dead; and walked about till four in the afternoon. This was a curious coincidence! a very amusing occurrence! to see such a similarity of feelings between the two Φερεοικοι! for so the *Greeks* called both the *shell-snail* and the *tortoise*.

Summer birds are, this cold and backward spring, unusually late: I have seen but one swallow yet. This conformity with the weather convinces me more and more that they sleep in the winter.

LETTER

LETTER LI.

TO THE SAME.

SELBORNE, Sept. 3, 1781.

I HAVE now read your miscellanies through with much care and satisfaction; and am to return you my best thanks for the honourable mention made in them of me as a naturalist, which I wish I may deserve.

In some former letters I expressed my suspicions that many of the house-martins do not depart in the winter far from this village. I therefore determined to make some search about the south-east end of the hill, where I imagined they might slumber out the uncomfortable months of winter. But supposing that the examination would be made to the best advantage in the spring, and observing that no martins had appeared by the 11th of *April* last; on that day I employed some men to explore the shrubs and cavities of the suspected spot. The persons took pains, but without any success; however, a remarkable incident occurred in the midst of our pursuit—while the labourers were at work a house-martin, the first that had been seen this year, came down the village in the sight of several people, and went at once into a nest, where it stayed a short time, and then flew over the houses; for some days after no martins were observed, not till the 16th of *April*, and then only a pair. Martins in general were remarkably late this year.

LETTER

LETTER LII.

TO THE SAME.

SELBORNE, Sept. 9, 1781.

I HAVE juft met with a circumftance refpecting fwifts, which furnifhes an exception to the whole tenor of my obfervations ever fince I have beftowed any attention on that fpecies of hirundines. Our fwifts, in general, withdrew this year about the firft day of *Auguft*, all fave one pair, which in two or three days was reduced to a fingle bird. The perfeverance of this individual made me fufpect that the ftrongeft of motives, that of an attach- to her young, could alone occafion fo late a ftay. I watched therefore till the twenty-fourth of *Auguft*, and then difcovered that, under the eaves of the church, fhe attended upon two young, which were fledged, and now put out their white chins from a crevice. Thefe remained till the twenty-feventh, looking more alert every day, and feeming to long to be on the wing. After this day they were miffing at once; nor could I ever obferve them with their dam courfing round the church in the act of learning to fly, as the firft broods evidently do. On the thirty-firft I caufed the eaves to be fearched, but we found in the neft only two callow, dead, ftinking fwifts, on which a fecond neft had been formed. This double neft was full of the black fhining cafes of the *hippobofcæ hirundinis*.

The following remarks on this unufual incident are obvious. The firft is, that though it may be difagreeable to fwifts to remain

beyond

beyond the beginning of *August*, yet that they can subsist longer is undeniable. The second is, that this uncommon event, as it was owing to the loss of the first brood, so it corroborates my former remark, that swifts breed regularly but once; since, was the contrary the case, the occurrence above could neither be new nor rare.

P. S. One swift was seen at *Lyndon*, in the county of *Rutland*, in 1782, so late as the third of *September*.

LETTER LIII.

TO THE SAME.

As I have sometimes known you make inquiries about several kinds of insects, I shall here send you an account of one sort which I little expected to have found in this kingdom. I had often observed that one particular part of a vine growing on the walls of my house was covered in the autumn with a black dust-like appearance, on which the flies fed eagerly; and that the shoots and leaves thus affected did not thrive; nor did the fruit ripen. To this substance I applied my glasses; but could not discover that it had any thing to do with animal life, as I at first expected: but, upon a closer examination behind the larger

M m   boughs,

boughs, we were surprised to find that they were coated over with husky shells, from whose sides proceeded a cotton-like substance, surrounding a multitude of eggs. This curious and uncommon production put me upon recollecting what I have heard and read concerning the *coccus vitis viniferæ* of *Linnæus*, which, in the south of *Europe*, infests many vines, and is an horrid and loathsome pest. As soon as I had turned to the accounts given of this insect, I saw at once that it swarmed on my vine; and did not appear to have been at all checked by the preceding winter, which had been uncommonly severe.

Not being then at all aware that it had any thing to do with *England*, I was much inclined to think that it came from *Gibraltar* among the many boxes and packages of plants and birds which I had formerly received from thence; and especially as the vine infested grew immediately under my study-window, where I usually kept my specimens. True it is that I had received nothing from thence for some years: but as insects, we know, are conveyed from one country to another in a very unexpected manner, and have a wonderful power of maintaining their existence till they fall into a *nidus* proper for their support and increase, I cannot but suspect still that these *cocci* came to me originally from *Andalusia*. Yet, all the while, candour obliges me to confess that Mr. *Lightfoot* has written me word that he once, and but once, saw these insects on a vine at *Weymouth* in *Dorsetshire*; which, it is here to be observed, is a sea-port town to which the *coccus* might be conveyed by shipping.

As many of my readers may possibly never have heard of this strange and unusual insect, I shall here transcribe a passage from a natural history of *Gibraltar*, written by the Reverend *John White*, late vicar of *Blackburn* in *Lancashire*, but not yet published:—

" In

" In the year 1770 a vine, which grew on the eaſt-ſide of my
" houſe, and which had produced the fineſt crops of grapes for
" years paſt, was ſuddenly overſpread on all the woody branches
" with large lumps of a white fibrous ſubſtance reſembling ſpiders
" webs, or rather raw cotton. It was of a very clammy quality,
" ſticking faſt to every thing that touched it, and capable of being
" ſpun into long threads. At firſt I ſuſpected it to be the product
" of ſpiders, but could find none. Nothing was to be ſeen con-
" nected with it but many *brown oval huſky ſhells*, which by no
" means looked like inſects, but rather reſembled bits of the dry
" bark of the vine. The tree had a plentiful crop of grapes ſet,
" when this peſt appeared upon it; but the fruit was manifeſtly
" injured by this foul incumbrance. It remained all the ſummer,
" ſtill increaſing, and loaded the woody and bearing branches to
" a vaſt degree. I often pulled off great quantities by handfuls;
" but it was ſo ſlimy and tenacious that it could by no means be
" cleared. The grapes never filled to their natural perfection,
" but turned watery and vapid. Upon peruſing the works after-
" wards of M. de *Reaumur*, I found this matter perfectly deſcribed
" and accounted for. Thoſe huſky ſhells, which I had obſerved,
" were no other than the *female coccus*, from whoſe ſides this
" cotton-like ſubſtance exſudes, and ſerves as a covering and
" ſecurity for their eggs."

To this account I think proper to add, that, though the female
cocci are ſtationary, and ſeldom remove from the place to which
they ſtick, yet the male is a winged inſect; and that the black
duſt which I ſaw was undoubtedly the excrement of the females,
which is eaten by ants as well as flies. Though the utmoſt ſeverity
of our winter did not deſtroy theſe inſects, yet the attention of the
gardener in a ſummer or two has entirely relieved my vine from
this filthy annoyance.

As

As we have remarked above that infects are often conveyed from one country to another in a very unaccountable manner, I shall here mention an emigration of small *aphides*, which was obferved in the village of *Selborne* no longer ago than *Auguft* the 1ft, 1785.

At about three o'clock in the afternoon of that day, which was very hot, the people of this village were furprifed by a fhower of *aphides*, or *fmother-flies*, which fell in thefe parts. Thofe that were walking in the ftreet at that juncture found themfelves covered with thefe infects, which fettled alfo on the hedges and gardens, blackening all the vegetables where they alighted. My annuals were difcoloured with them, and the ftalks of a bed of onions were quite coated over for fix days after. Thefe armies were then, no doubt, in a ftate of emigration, and fhifting their quarters; and might have come, as far as we know, from the great hop-plantations of *Kent* or *Suffex*, the wind being all that day in the eafterly quarter. They were obferved at the fame time in great clouds about *Farnham*, and all along the vale from *Farnham* to *Alton*[a].

---

[a] For various methods by which feveral infects fhift their quarters, fee *Derham's* Phyfico-Theology.

LETTER

LETTER LIV.

TO THE SAME.

DEAR SIR,

WHEN I happen to vifit a family where *gold* and *filver fifhes* are kept in a glafs bowl, I am always pleafed with the occurrence, becaufe it offers me an opportunity of obferving the actions and propenfities of thofe beings with whom we can be little acquainted in their natural ftate. Not long fince I fpent a fortnight at the houfe of a friend where there was fuch a *vivary*, to which I paid no fmall attention, taking every occafion to remark what paffed within it's narrow limits. It was here that I firft obferved the manner in which fifhes die. As foon as the creature fickens, the head finks lower and lower, and it ftands as it were on it's head; till, getting weaker, and lofing all poife, the tail turns over, and at laft it floats on the furface of the water with it's belly uppermoft. The reafon why fifhes, when dead, fwim in that manner is very obvious; becaufe, when the body is no longer balanced by the fins of the belly, the broad mufcular back preponderates by it's own gravity, and turns the belly uppermoft, as lighter from it's being a cavity, and becaufe it contains the fwimming-bladders, which contribute to render it buoyant. Some that delight in *gold* and *filver fifhes* have adopted a notion that they need no aliment. True it is that they will fubfift for a long time without any apparent food but what they can collect from pure water frequently changed; yet they muft draw fome fupport from animalcula, and other

nourifhment

nourishment supplied by the water; because, though they seem to eat nothing, yet the consequences of eating often drop from them. That they are best pleased with such *jejune* diet may easily be confuted, since if you toss them crumbs they will seize them with great readiness, not to say greediness: however, bread should be given sparingly, lest, turning sour, it corrupt the water. They will also feed on the water-plant called *lemna (duck's meat)*, and also on small fry.

When they want to move a little they gently protrude themselves with their *pinnæ pectorales*; but it is with their strong muscular tails only that they and all fishes shoot along with such inconceivable rapidity. It has been said that the eyes of fishes are immoveable: but these apparently turn them forward or backward in their sockets as their occasions require. They take little notice of a lighted candle, though applied close to their heads, but flounce and seem much frightened by a sudden stroke of the hand against the support whereon the bowl is hung; especially when they have been motionless, and are perhaps asleep. As fishes have no eyelids, it is not easy to discern when they are sleeping or not, because their eyes are always open.

Nothing can be more amusing than a glass bowl containing such fishes: the double refractions of the glass and water represent them, when moving, in a shifting and changeable variety of dimensions, shades, and colours; while the two mediums, assisted by the concavo-convex shape of the vessel, magnify and distort them vastly; not to mention that the introduction of another element and it's inhabitants into our parlours engages the fancy in a very agreeable manner.

Gold

Gold and *silver fishes*, though originally natives of *China* and *Japan*, yet are become so well reconciled to our climate as to thrive and multiply very fast in our ponds and stews. *Linnæus* ranks this species of fish under the genus of *cyprinus*, or *carp*, and calls it *cyprinus auratus*.

Some people exhibit this sort of fish in a very fanciful way; for they cause a glass bowl to be blown with a large hollow space within, that does not communicate with it. In this cavity they put a bird occasionally; so that you may see a goldfinch or a linnet hopping as it were in the midst of the water, and the fishes swimming in a circle round it. The simple exhibition of the fishes is agreeable and pleasant; but in so complicated a way becomes whimsical and unnatural, and liable to the objection due to him,

"Qui variare cupit rem prodigialitèr unam."

I am, &c.

LETTER

LETTER LV.

TO THE SAME.

DEAR SIR,   October 10, 1781.

I THINK I have obferved before that much the moſt confiderable
part of the *houſe-martins* withdraw from hence about the firſt week
in *October*; but that ſome, the latter broods I am now convinced,
linger on till towards the middle of that month : and that at times,
once perhaps in two or three years, a flight, for one day only, has
ſhown itſelf in the firſt week in *November*.

Having taken notice, in *October* 1780, that the laſt flight was
numerous, amounting perhaps to one hundred and fifty ; and that
the ſeaſon was ſoft and ſtill ; I was reſolved to pay uncommon
attention to theſe late birds ; to find, if poſſible, where they rooſted, and to determine the preciſe time of their retreat.   The
mode of life of theſe latter *hirundines* is very favourable to ſuch
a deſign ; for they ſpend the whole day in the ſheltered diſtrict,
between me and the *Hanger*, ſailing about in a placid, eaſy
manner, and feaſting on thoſe inſects which love to haunt a ſpot
ſo ſecure from ruffling winds.   As my principal object was to diſ-
cover the place of their rooſting, I took care to wait on them
before they retired to reſt, and was much pleaſed to find that,
for ſeveral evenings together, juſt at a quarter paſt five in the
afternoon, they all ſcudded away in great haſte towards the ſouth-
eaſt, and darted down among the low ſhrubs above the cottages at
the end of the hill.   This ſpot in many reſpects ſeems to be well
calculated for their winter reſidence : for in many parts it is as

steep

fteep as the roof of any houfe, and therefore fecure from the annoyances of water; and it is moreover clothed with beechen fhrubs, which, being ftunted and bitten by fheep, make the thickeft covert imaginable; and are fo entangled as to be impervious to the fmalleft fpaniel: befides, it is the nature of underwood beech never to caft it's leaf all the winter; fo that, with the leaves on the ground and thofe on the twigs, no fhelter can be more complete. I watched them on to the thirteenth and fourteenth of *October*, and found their evening retreat was exact and uniform; but after this they made no regular appearance. Now and then a ftraggler was feen; and, on the twenty-fecond of *October*, I obferved two in the morning over the village, and with them my remarks for the feafon ended.

From all thefe circumftances put together, it is more than probable that this lingering flight, at fo late a feafon of the year, never departed from the ifland. Had they indulged me that autumn with a *November* vifit, as I much defired, I prefume that, with proper affiftants, I fhould have fettled the matter paft all doubt; but though the third of *November* was a fweet day, and in appearance exactly fuited to my wifhes, yet not a martin was to be feen; and fo I was forced, reluctantly, to give up the purfuit.

I have only to add that were the bufhes, which cover fome acres, and are not my own property, to be grubbed and carefully examined, probably thofe late broods, and perhaps the whole aggregate body of the houfe-martins of this diftrict, might be found there, in different fecret dormitories; and that, fo far from withdrawing into warmer climes, it would appear that they never depart three hundred yards from the village.

———

N n LETTER

LETTER LVI.

TO THE SAME.

THEY who write on natural hiftory cannot too frequentlyadvert to *inftinct*, that wonderful limited faculty, which, in fome inftances, raifes the brute creation as it were above *reafon*, and in others leaves them fo far below it.   Philofophers have defined *inftinct* to be that fecret influence by which every fpecies is impelled naturally to purfue, at all times, the fame way or track, without any teach-ing or example; whereas *reafon*, without inftruction, would often vary and do that by many methods which *inftinct* effects by one alone.   Now this maxim muft be taken in a qualified fenfe; for there are inftances in which *inftinct* does vary and conform to the circumftances of place and convenience.

It has been remarked that every fpecies of bird has a mode of nidification peculiar to itfelf; fo that a fchool-boy would at once pronounce on the fort of neft before him.   This is the cafe among fields and woods, and wilds; but, in the villages round *London*, where moffes and goffamer, and cotton from vegetables, are hardly to be found, the neft of the *chaffinch* has not that elegant finifhed appearance, nor is it fo beautifully ftudded with lichens, as in a more rural diftrict : and the *wren* is obliged to conftruct it's houfe with ftraws and dry graffes, which do not give it that rotundity and compactnefs fo remarkable in the edifices of that little architect.   Again, the regular neft of the *houfe-martin* is

hemifpheric;

hemifpheric; but where a rafter, or a joift, or a cornice, may happen to ftand in the way, the neft is fo contrived as to conform to the obftruction, and becomes flat or oval, or compreffed.

In the following inftances *inftinct* is perfectly uniform and confiftent. There are three creatures, the *fquirrel*, the *field-moufe*, and the bird called the *nut-hatch*, *(fitta Europæa)*, which live much on hazlenuts; and yet they open them each in a different way. The firft, after rafping off the fmall end, fplits the fhell in two with his long fore-teeth, as a man does with his knife; the fecond nibbles a hole with his teeth, fo regular as if drilled with a wimble, and yet fo fmall that one would wonder how the kernel can be extracted through it; while the laft picks an irregular ragged hole with it's bill: but as this artift has no paws to hold the nut firm while he pierces it, like an adroit workman, he fixes it, as it were in a vice, in fome cleft of a tree, or in fome crevice; when, ftanding over it, he perforates the ftubborn fhell. We have often placed nuts in the chink of a gate-poft where *nut-hatches* have been known to haunt, and have always found that thofe birds have readily penetrated them. While at work they make a rapping noife that may be heard at a confiderable diftance.

You that underftand both the theory and practical part of mufic may beft inform us why *harmony* or *melody* fhould fo ftrangely affect fome men, as it were by recollection, for days after a concert is over. What I mean the following paffage will moft readily explain:

" Præhabebat porrò vocibus humanis, inftrumentifque har-
" monicis muficam illam avium : non quod aliâ quoque non
" delectaretur ; fed quod ex muficâ humanâ relinqueretur in
" animo continens quædam, attentionemque et fomnum con-
" turbans agitatio ; dum afcenfus, exfcenfus, tenores, ac muta-

" tiones

" tiones illæ fonorum, et confonantiarum euntque, redeuntque
" per phantafiam:—cum nihil tale relinqui poffit ex modu-
" lationibus avium, quæ, quod non funt perinde a nobis
" imitabiles, non poffunt perinde internam facultatem com-
" movere."   *Gaffendus in Vitâ Peirefkii.*

This curious quotation ftrikes me much by fo well reprefenting my own cafe, and by defcribing what I have fo often felt, but never could fo well exprefs. When I hear fine mufic I am haunted with paffages therefrom night and day; and efpecially at firft waking, which, by their importunity, give me more un-eafinefs than pleafure: elegant leffons ftill teafe my imagination, and recur irrefiftibly to my recollection at feafons, and even when I am defirous of thinking of more ferious matters.

I am, &c.

LETTER

LETTER LVII.

TO THE SAME.

A RARE, and I think a new, little bird frequents my garden, which I have great reason to think is the *pettichaps*: it is common in some parts of the kingdom; and I have received formerly several dead specimens from *Gibraltar*. This bird much resembles the *white-throat*, but has a more white or rather silvery breast and belly; is restless and active, like the *willow-wrens*, and hops from bough to bough, examining every part for food; it also runs up the stems of the *crown-imperials*, and, putting it's head into the bells of those flowers, sips the liquor which stands in the *nectarium* of each petal. Sometimes it feeds on the ground like the *hedge-sparrow*, by hopping about on the grass-plots and mown walks.

One of my neighbours, an intelligent and observing man, informs me that, in the beginning of *May*, and about ten minutes before eight o'clock in the evening, he discovered a great cluster of *house-swallows*, thirty at least he supposes, perching on a willow that hung over the verge of *James Knight's* upper-pond. His attention was first drawn by the twittering of these birds, which sat motionless in a row on the bough, with their heads all one way, and, by their weight, pressing down the twig so that it nearly touched the water. In this situation he watched them till he could see no longer. Repeated accounts of this sort, spring and fall, induce us greatly to suspect that *house-swallows* have some strong

attachment

attachment to water, independent of the matter of food; and, though they may not retire into that e ement, yet they may conceal themfelves in the banks of pools and rivers during the uncomfortable months of winter.

One of the keepers of *Woolmer-foreft* fent me a *peregrine-falcon*, which he fhot on the verge of that diftrict as it was devouring a wood-pigeon. The *falco peregrinus*, or *haggard falcon*, is a noble fpecies of hawk feldom feen in the fouthern counties. In winter 1767 one was killed in the neighbouring parifh of *Faringdon*, and fent by me to Mr. *Pennant* into *North-Wales* [b]. Since that time I have met with none till now. The fpecimen mentioned above was in fine prefervation, and not injured by the fhot: it meafured forty-two inches from wing to wing, and twenty-one from beak to tail, and weighed two pounds and an half ftanding weight. This fpecies is very robuft, and wonderfully formed for rapine: it's breaft was plump and mufcular; it's thighs long, thick, and brawny; and it's legs remarkably fhort and well fet: the feet were armed with moft formidable, fharp, long talons: the eyelids and cere of the bill were yellow; but the irides of the eyes dufky; the beak was thick and hooked, and of a dark colour, and had a jagged procefs near the end of the upper mandible on each fide: it's tail, or train, was fhort in proportion to the bulk of it's body: yet the wings, when clofed, did not extend to the end of the train. From it's large and fair proportions it might be fuppofed to have been a female; but I was not permitted to cut open the fpecimen. For one of the birds of prey, which are ufually lean, this was in high cafe: in it's craw were many barley-corns, which probably came from the crop of the wood-pigeon, on

---

[b] See my tenth and eleventh letter to that gentleman.

which

which it was feeding when fhot: for voracious birds do not eat grain; but, when devouring their quarry, with undiftinguifhing vehemence fwallow bones and feathers, and all matters, indifcriminately. This falcon was probably driven from the mountains of *North Wales* or *Scotland*, where they are known to breed, by rigorous weather and deep fnows that had lately fallen.

I am, &c.

LETTER LVIII.

TO THE SAME.

My near neighbour, a young gentleman in the fervice of the *Eaft-India* Company, has brought home a dog and a bitch of the *Chinefe* breed from *Canton*; fuch as are fattened in that country for the purpofe of being eaten: they are about the fize of a moderate fpaniel; of a pale yellow colour, with coarfe briftling hairs on their backs; fharp upright ears, and peaked heads, which give them a very fox-like appearance. Their hind legs are unufually ftraight, without any bend at the hock or ham, to fuch a degree as to give them an aukward gait when they trot. When they are in motion their tails are curved high over their backs like thofe of fome hounds, and have a bare place each on the outfide from the tip midway, that does not feem to be matter of

accident,

accident, but somewhat singular. Their eyes are jet-black, small, and piercing; the insides of their lips and mouths of the same colour, and their tongues blue. The bitch has a dew-claw on each hind leg; the dog has none. When taken out into a field the bitch showed some disposition for hunting, and dwelt on the scent of a covey of partridges till she sprung them, giving her tongue all the time. The dogs in *South America* are dumb; but these bark much in a short thick manner, like foxes; and have a surly, savage demeanour like their ancestors, which are not domesticated, but bred up in sties, where they are fed for the table with rice-meal and other farinaceous food. These dogs, having been taken on board as soon as weaned, could not learn much from their dam; yet they did not relish flesh when they came to *England*. In the islands of the *pacific* ocean the dogs are bred up on vegetables, and would not eat flesh when offered them by our circumnavigators.

We believe that all dogs, in a state of nature, have sharp, upright fox-like ears; and that hanging ears, which are esteemed so graceful, are the effect of choice breeding and cultivation. Thus, in the Travels of *Yhrandt Ides* from *Muscovy* to *China*, the dogs which draw the *Tartars* on snow-sledges near the river *Oby* are engraved with prick-ears, like those from *Canton*. The *Kamschat-dales* also train the same sort of sharp-eared peak-nosed dogs to draw their sledges; as may be seen in an elegant print engraved for Captain *Cook*'s last voyage round the world.

Now we are upon the subject of dogs, it may not be impertinent to add, that spaniels, as all sportsmen know, though they hunt partridges and pheasants as it were by instinct, and with much delight and alacrity, yet will hardly touch their bones when offered as food; nor will a mongrel dog of my own, though he

is

is remarkable for finding that fort of game. But, when we came to offer the bones of partridges to the two *Chinese* dogs, they devoured them with much greediness, and licked the platter clean.

No sporting dogs will flush woodcocks till inured to the scent and trained to the sport, which they then pursue with vehemence and transport; but then they will not touch their bones, but turn from them with abhorrence, even when they are hungry.

Now, that dogs should not be fond of the bones of such birds as they are not disposed to hunt is no wonder; but why they reject and do not care to eat their natural game is not so easily accounted for, since the end of hunting seems to be, that the chase pursued should be eaten. Dogs again will not devour the more rancid water-fowls, nor indeed the bones of any wild-fowls; nor will they touch the fœtid bodies of birds that feed on offal and garbage: and indeed there may be somewhat of providential instinct in this circumstance of dislike; for vultures [c], and kites, and ravens, and crows, &c. were intended to be messmates with dogs [d] over their carrion; and seem to be appointed by Nature as fellow-scavengers to remove all cadaverous nuisances from the face of the earth.

I am, &c.

[c] *Hasselquist*, in his Travels to the *Levant*, observes that the dogs and vultures at *Grand Cairo* maintain such a friendly intercourse as to bring up their young together in the same place.

[d] The *Chinese* word for a dog to an *European* ear sounds like *quibloh*.

LETTER LIX.

TO THE SAME.

THE fossil wood buried in the bogs of *Wolmer-forest* is not yet all exhausted; for the peat-cutters now and then stumble upon a log. I have just seen a piece which was sent by a labourer of *Oakhanger* to a carpenter of this village; this was the but-end of a small oak, about five feet long, and about five inches in diameter. It had apparently been severed from the ground by an axe, was very ponderous, and as black as ebony. Upon asking the carpenter for what purpose he had procured it; he told me that it was to be sent to his brother, a joiner at *Farnham*, who was to make use of it in cabinet work, by inlaying it along with whiter woods.

Those that are much abroad on evenings after it is dark, in spring and summer, frequently hear a nocturnal bird passing by on the wing, and repeating often a short quick note. This bird I have remarked myself, but never could make out till lately. I am assured now that it is the *Stone-curlew, (charadrius oedicnemus)*. Some of them pass over or near my house almost every evening after it is dark, from the uplands of the hill and *North field*, away down towards *Dorton*; where, among the streams and meadows, they find a greater plenty of food. Birds that fly by night are obliged to be noisy; their notes often repeated become signals or watch-words to keep them together, that they may not stray or lose each the other in the dark.

The

The evening proceedings and manœuvres of the rooks are curious and amufing in the autumn. Juſt before duſk they return in long ſtrings from the foraging of the day, and rendezvous by thouſands over *Selborne-down*, .where they wheel round in the air, and ſport and dive in a playful manner, all the while exerting their voices, and making a loud cawing, which, being blended and ſoftened by the diſtance that we at the village are below them, becomes a confuſed noiſe or chiding; or rather a pleaſing murmur, very engaging to the imagination, and not unlike the cry of a pack of hounds in hollow, echoing woods, or the ruſhing of the wind in tall trees, or the tumbling of the tide upon a pebbly ſhore. When this ceremony is over, with the laſt gleam of day, they retire for the night to the deep beechen woods of *Tiſted* and *Ropley*. We remember a little girl who, as ſhe was going to bed, uſed to remark on ſuch an occurrence, in the true ſpirit of *phyſico-theology*, that the rooks were ſaying their prayers; and yet this child was much too young to be aware that the ſcriptures have ſaid of the Deity—that " he feedeth the ravens who call upon him."

I am, &c.

LETTER

LETTER LX.

TO THE SAME.

In reading Dr. *Huxham's Obfervationes de Aëre*, &c. written at *Plymouth*, I find by thofe curious and accurate remarks, which contain an account of the weather from the year 1727 to the year 1748, inclufive, that though there is frequent rain in that diftrict of *Devonfhire*, yet the quantity falling is not great; and that fome years it has been very fmall: for in 1731 the rain meafured only 17^{inch}.—266^{thou}. and in 1741, 20—354; and again, in 1743 only 20—908. Places near the fea have frequent fcuds, that keep the atmofphere moift, yet do not reach far up into the country; making thus the maritime fituations appear wet, when the rain is not confiderable. In the wetteft years at *Plymouth* the Doctor meafured only once 36; and again once, viz. 1734, 37—114: a quantity of rain that has twice been exceeded at *Selborne* in the fhort period of my obfervations. Dr. *Huxham* remarks that frequent fmall rains keep the air moift; while heavy ones render it more dry, by beating down the vapours. He is alfo of opinion that the *dingy, fmoky appearance* in the fky, in very dry feafons, arifes from the want of moifture fufficient to let the light through, and render the atmofphere tranfparent; becaufe he had obferved feveral bodies more diaphanous when wet than dry; and did never recollect that the air had that look in rainy feafons.

My

My friend, who lives juft beyond the top of the down, brought his three fwivel guns to try them in my outlet, with their muzzles towards the *Hanger*, fuppofing that the report would have had a great effect; but the experiment did not anfwer his expectation. He then removed them to the *Alcove* on the *Hanger*; when the found, rufhing along the *Lythe* and *Comb-wood*, was very grand: but it was at the *Hermitage* that the echoes and repercuffions de-lighted the hearers; not only filling the *Lythe* with the roar, as if all the beeches were tearing up by the roots; but, turning to the left, they pervaded the vale above *Combwood-ponds*; and after a paufe feemed to take up the crafh again, and to extend round *Harteley-hangers*, and to die away at laft among the coppices and coverts of *Ward-le-ham*. It has been remarked before that this diftrict is an *anathoth*, a place of refponfes or echoes, and there-fore proper for fuch experiments: we may farther add that the paufes in echoes, when they ceafe and yet are taken up again, like the paufes in mufic, furprife the hearers, and have a fine effect on the imagination.

The gentleman abovementioned has juft fixed a barometer in his parlour at *Newton Valence*. The tube was firft filled here (at *Selborne*) twice with care, when the mercury agreed and ftood exactly with my own; but, being filled again twice at *Newton*, the mercury ftood, on account of the great elevation of that houfe, three-tenths of an inch lower than the barometers at this village, and fo continues to do, be the weight of the atmofphere what it may. The plate of the barometer at *Newton* is figured as low as 27; becaufe in ftormy weather the mercury there will fome-times defcend below 28. We have fuppofed *Newton-houfe* to ftand two hundred feet higher than this houfe: but if the rule holds good, which fays that mercury in a barometer finks one-tenth

of

of an inch for every hundred feet elevation, then the *Newton* barometer, by standing three-tenths lower than that of *Selborne*, proves that *Newton-house* muſt be three hundred feet higher than that in which I am writing, inſtead of two hundred.

It may not be impertinent to add, that the barometers at *Selborne* ſtand three-tenths of an inch lower than the barometers at *South Lambeth*: whence we may conclude that the former place is about three hundred feet higher than the latter; and with good, reaſon becauſe the ſtreams that riſe with us run into the *Thames* at *Weybridge*, and ſo to *London*. Of courſe therefore there muſt be lower ground all the way from *Selborne* to *South Lambeth*; the diſtance between which, all the windings and indentings of the ſtreams conſidered, cannot be leſs than an hundred miles.

I am, &c.

LETTER

LETTER LXI.

TO THE SAME.

SINCE the weather of a diftrict is undoubtedly part of it's natural hiftory, I fhall make no further apology for the four following letters, which will contain many particulars concerning fome of the great frofts and a few refpecting fome very hot fummers, that have diftinguifhed themfelves from the reft during the courfe of my obfervations.

As the froft in *January* 1768 was, for the fmall time it lafted, the moft fevere that we had then known for many years, and was remarkably injurious to ever-g eens, fome account of it's rigour, and reafon of it's ravages, may be ufeful, and not unacceptable to perfons that delight in planting and ornamenting; and may particularly become a work that profeffes never to lofe fight of utility.

For the laft two or three days of the former year there were confiderable falls of fnow, which lay deep and uniform on the ground without any drifting, wrapping up the more humble vegetation in perfect fecurity. From the firft day to the fifth of the new year more fnow fucceeded; but from that day the air became entirely clear; and the heat of the fun about noon had a confiderable influence in fheltered fituations.

It

It was in fuch an afpect that the fnow on the author's ever-greens was melted every day, and frozen intenfely every night; fo that the lauruftines, bays, laurels, and arbutufes looked, in three or four days, as if they had been burnt in the fire ; while a neigh-bour's plantation of the fame kind, in a high cold fituation, where the fnow was never melted at all, remained uninjured.

From hence I would infer that it is the repeated melting and freezing of the fnow that is fo fatal to vegetation, rather than the feverity of the cold. Therefore it highly behoves every planter, who wifhes to efcape the cruel mortification of lofing in a few days the labour and hopes of years, to beftir himfelf on fuch emer-gencies ; and, if his plantations are fmall, to avail himfelf of mats, cloths, peafe-haum, ftraw, reeds, or any fuch covering, for a fhort time ; or, if his fhrubberies are extenfive, to fee that his people go about with prongs and forks, and carefully diflodge the fnow from the boughs : fince the naked foliage will fhift much better for itfelf, than where the fnow is partly melted and frozen again.

It may perhaps appear at firft like a paradox ; but doubtlefs the more tender trees and fhrubs fhould never be planted in hot afpects ; not only for the reafon affigned above, but alfo becaufe, thus circumftanced, they are difpofed to fhoot earlier in the fpring, and to grow on later in the autumn, than they would otherwife do, and fo are fufferers by lagging or early frofts. For this reafon alfo plants from *Siberia* will hardly endure our climate ; becaufe, on the very firft advances of fpring, they fhoot away, and fo are cut off by the fevere nights of *March* or *April*.

Dr. *Fothergill* and others have experienced the fame inconvenience with refpect to the more tender fhrubs from *North-America* ; which they therefore plant under north-walls. There fhould alfo perhaps
be

be a wall to the eaft to defend them from the piercing blafts from that quarter.

This obfervation might without any impropriety be carried into animal life; for difcerning bee-mafters now find that their hives fhould not in the winter be expofed to the hot fun, becaufe fuch unfeafonable warmth awakens the inhabitants too early from their flumbers; and, by putting their juices into motion too foon, fubjects them afterwards to inconveniencies when rigorous weather returns.

The coincidents attending this fhort but intenfe froft were, that the horfes fell fick with an epidemic diftemper, which injured the winds of many, and killed fome; that colds and coughs were general among the human fpecies; that it froze under people's beds for feveral nights; that meat was fo hard frozen that it could not be fpitted, and could not be fecured but in cellars; that feveral redwings and thrufhes were killed by the froft; and that the large titmoufe continued to pull ftraws lengthwife from the eaves of thatched houfes and barns in a moft adroit manner, for a purpofe that has been explained already[d].

On the 3d of *January Benjamin Martin*'s thermometer within doors, in a clofe parlour where there was no fire, fell in the night to 20, and on the 4th to 18, and on the 7th to $17\frac{1}{2}$, a degree of cold which the owner never fince faw in the fame fituation; and he regrets much that he was not able at that juncture to attend his inftrument abroad. All this time the wind continued north and north-eaft; and yet on the 8th rooft-cocks, which had been filent, began to found their clarions, and crows to clamour, as prognoftic of milder weather; and, moreover, moles began to

[d] See Letter xli. to Mr. *Pennant.*

P p

heave

heave and work, and a manifeft thaw took place. From the latter circumftance we may conclude that thaws often originate under ground from warm vapours which arife; elfe how fhould fubterraneous animals receive fuch early intimations of their approach. Moreover, we have often obferved that cold feems to defcend from above; for, when a thermometer hangs abroad in a frofty night, the intervention of a cloud fhall immediately raife the mercury ten degrees; and a clear fky fhall again compel it to defcend to it's former gage.

And here it may be proper to obferve, on what has been faid above, that though frofts advance to their utmoft feverity by fomewhat of a regular gradation, yet thaws do not ufually come on by as regular a declenfion of cold; but often take place immediately from intenfe freezing; as men in ficknefs often mend at once from a paroxyfm.

To the great credit of *Portugal* laurels and *American* junipers, be it remembered that they remained untouched amidft the general havock: hence men fhould learn to ornament chiefly with fuch trees as are able to withftand accidental feverities, and not fubject themfelves to the vexation of a lofs which may befall them once perhaps in ten years, yet may hardly be recovered through the whole courfe of their lives.

As it appeared afterwards the ilexes were much injured, the cypreffes were half deftroyed, the arbutufes lingered on, but never recovered; and the bays, lauruftines, and laurels, were killed to the ground; and the very wild hollies, in hot afpects, were fo much affected that they caft all their leaves.

By the 14th of *January* the fnow was entirely gone; the turnips emerged not damaged at all, fave in funny places; the wheat looked delicately, and the garden plants were well preferved;

for

for fnow is the moft kindly mantle that infant vegetation can be wrapped in : were it not for that friendly meteor no vegetable life could exift at all in northerly regions. Yet in *Sweden* the earth in *April* is not divefted of fnow for more than a fortnight before the face of the country is covered with flowers.

LETTER LXI.

TO THE SAME.

THERE were fome circumftances attending the remarkable froft in *January* 1776 fo fingular and ftriking, that a fhort detail of them may not be unacceptable.

The moft certain way to be exact will be to copy the paffages from my journal, which were taken from time to time as things occurred. But it may be proper previoufly to remark that the firft week in *January* was uncommonly wet, and drowned with vaft rains from every quarter : from whence may be inferred, as there is great reafon to believe is the cafe, that intenfe frofts feldom take place till the earth is perfectly glutted and chilled with water[f]; and hence dry autumns are feldom followed by rigorous winters.

[f] The autumn preceding *January* 1768 was very wet, and particularly the month of *September*, during which there fell at *Lyndon*, in the county of *Rutland*, *fix inches and an half* of rain. And the terrible long froft in 1739-40 fet in after a rainy feafon, and when the fprings were very high.

January 7th.—Snow driving all the day; which was followed by frost, sleet, and some snow, till the 12th, when a prodigious mass overwhelmed all the works of men, drifting over the tops of the gates and filling the hollow lanes.

On the 14th the writer was obliged to be much abroad; and thinks he never before or since has encountered such rugged *Siberian* weather. Many of the narrow roads were now filled above the tops of the hedges; through which the snow was driven into most romantic and grotesque shapes, so striking to the imagination as not to be seen without wonder and pleasure. The poultry dared not to stir out of their roosting places; for cocks and hens are so dazzled and confounded by the glare of snow that they would soon perish without assistance. The hares also lay sullenly in their seats, and would not move till compelled by hunger; being conscious, poor animals, that the drifts and heaps treacherously betray their footsteps, and prove fatal to numbers of them.

From the 14th the snow continued to increase, and began to stop the road waggons and coaches, which could no longer keep on their regular stages; and especially on the western roads, where the fall appears to have been deeper than in the south. The company at *Bath*, that wanted to attend the *Queen's birth-day*, were strangely incommoded: many carriages of persons, who got in their way to town from *Bath* as far as *Marlborough*, after strange embarrassments, here met with a *ne plus ultra*. The ladies fretted, and offered large rewards to labourers if they would shovel them a track to *London*: but the relentless heaps of snow were too bulky to be removed; and so the 18th passed over, leaving the company in very uncomfortable circumstances at the *Castle* and other inns.

On

On the 20th the fun fhone out for the firft time fince the froft began; a circumftance that has been remarked before much in favour of vegetation. All this time the cold was not very intenfe, for the thermometer ftood at 29, 28, 25, and thereabout; but on the 21ft it defcended to 20. The birds now began to be in a very pitiable and ftarving condition. Tamed by the feafon, fky-larks fettled in the ftreets of towns, becaufe they faw the ground was bare; rooks frequented dunghills clofe to houfes; and crows watched horfes as they paffed, and greedily devoured what dropped from them; hares now came into men's gardens, and, fcraping away the fnow, devoured fuch plants as they could find.

On the 22d the author had occafion to go to *London* through a fort of *Laplandian-fcene,* very wild and grotefque indeed. But the metropolis itfelf exhibited a ftill more fingular appearance than the country; for, being bedded deep in fnow, the pavement of the ftreets could not be touched by the wheels or the horfes' feet, fo that the carriages ran about without the leaft noife. Such an exemption from din and clatter was ftrange, but not pleafant; it feemed to convey an uncomfortable idea of defolation:

"— — — — — — — — *ipfa filentia terrent.*"

On the 27th much fnow fell all day, and in the evening the froft became very intenfe. At *South Lambeth,* for the four following nights, the thermometer fell to 11, 7, 6, 6; and at *Selborne* to 7, 6, 10; and on the 31ft of *January,* juft before fun-rife, with rime on the trees and on the tube of the glafs, the quickfilver funk exactly to zero, being 32 degrees below the freezing point: but by eleven in the morning, though in the

shade,

shade, it sprung up to 16½ ℊ. — a most unusual degree of cold this for the south of *England!* During these four nights the cold was so penetrating that it occasioned ice in warm chambers and under beds; and in the day the wind was so keen that persons of robust constitutions could scarcely endure to face it. The *Thames* was at once so frozen over both above and below bridge that crowds ran about on the ice. The streets were now strangely encumbered with snow, which crumbled and trod dusty; and, turning grey, resembled bay-salt: what had fallen on the roofs was so perfectly dry that, from first to last, it lay twenty-six days on the houses in the city; a longer time than had been remembered by the oldest housekeepers living. According to all appearances we might now have expected the continuance of this rigorous weather for weeks to come, since every night increased in severity; but behold, without any apparent cause, on the 1st of *February* a thaw took place, and some rain followed before night; making good the observation above, that frosts often go off as it were at once, without any gradual declension of cold. On the 2d of *February* the thaw persisted; and on the 3d swarms of little insects were frisking and sporting in a court-yard at *South Lambeth*, as if they had felt no frost. Why the juices in the small bodies and smaller limbs of such minute beings are not frozen is a matter of curious inquiry.

Severe frosts seem to be partial, or to run in currents; for, at the same juncture, as the author was informed by accurate corre-

---

ℊ At *Selborne* the cold was greater than at any other place that the author could hear of with certainty: though some reported at the time that at a village in *Kent* the thermometer fell two degrees below zero, *viz.* 34 degrees below the freezing point.

The thermometer used at *Selborne* was graduated by *Benjamin Martin.*

spondents,

spondents, at *Lyndon*, in the county of *Rutland*, the thermometer stood at 19; at *Blackburn*, in *Lancashire*, at 19; and at *Manchester* at 21, 20, and 18. Thus does some unknown circumstance strangely overbalance latitude, and render the cold sometimes much greater in the southern than the northern parts of this kingdom.

The consequences of this severity were, that in *Hampshire*, at the melting of the snow, the wheat looked well, and the turnips came forth little injured. The laurels and laurustines were somewhat damaged, but only in *hot aspects*. No evergreens were quite destroyed; and not half the damage sustained that befell in *January* 1768. Those laurels that were a little scorched on the south-sides were perfectly untouched on their north-sides. The care taken to shake the snow day by day from the branches seemed greatly to avail the author's evergreens. A neighbour's laurel-hedge, in a high situation, and facing to the north, was perfectly green and vigorous; and the *Portugal laurels* remained unhurt.

As to the birds, the thrushes and blackbirds were mostly destroyed; and the partridges, by the weather and poachers, were so thinned that few remained to breed the following year.

LETTER

LETTER LXII.

As the froft in *December* 1784 was very extraordinary, you, I truft, will not be difpleafed to hear the particulars; and efpecially when I promife to fay no more about the feverities of winter after I have finifhed this letter.

The firft week in *December* was very wet, with the barometer very low. On the 7th, with the barometer at 28—five tenths, came on a vaft fnow, which continued all that day and the next, and moft part of the following night; fo that by the morning of the 9th the works of men were quite overwhelmed, the lanes filled fo as to be impaffable, and the ground covered twelve or fifteen inches without any drifting. In the evening of the 9th the air began to be fo very fharp that we thought it would be curious to attend to the motions of a thermometer: we therefore hung out two; one made by *Martin* and one by *Dollond*, which foon began to fhew us what we were to expect; for, by ten o'clock, they fell to 21, and at eleven to 4, when we went to bed. On the 10th, in the morning, the quickfilver of *Dollond*'s glafs was down to *half a degree below zero*; and that of *Martin*'s, which was abfurdly graduated only to four degrees *above zero*, funk quite into the brafs guard of the ball; fo that when the weather became moft interefting this was ufelefs. On the 10th, at eleven at night, though the air was perfectly ftill, *Dollond*'s glafs went down

to

to *one degree below zero!* This ſtrange ſeverity of the weather made me very deſirous to know what degree of cold there might be in ſuch an exalted and near ſituation as *Newton.* We had therefore, on the morning of the 10th, written to Mr. ———, and entreated him to hang out his thermometer, made by *Adams*; and to pay ſome attention to it morning and evening; expecting wonderful phænomena, in ſo elevated a region, at two hundred feet or more above my houſe. But, behold! on the 10th, at eleven at night, it was down only to 17, and the next morning at 22, when mine was at ten! We were ſo diſturbed at this unexpected reverſe of comparative local cold, that we ſent one of my glaſſes up, thinking that of Mr. ——— muſt, ſome how, be wrongly conſtructed. But, when the inſtruments came to be confronted, they went exactly together: ſo that, for one night at leaſt, the cold at *Newton* was 18 degrees leſs than at *Selborne*; and, through the whole froſt, 10 or 12 degrees; and indeed, when we came to obſerve conſequences, we could readily credit this; for all my lauruſtines, bays, ilexes, arbutuſes, cypreſſes, and even my *Portugal laurels* [h], and (which occaſions more regret) my fine ſloping laurel-hedge, were ſcorched up; while, at *Newton,* the ſame trees have not loſt a leaf!

We had ſteady froſt on to the 25th, when the thermometer in the morning was down to 10 with us, and at *Newton* only to 21. Strong froſt continued till the 31ſt, when ſome tendency to thaw was obſerved; and, by *January* the 3d, 1785, the thaw was confirmed, and ſome rain fell.

[h] Mr. *Miller,* in his Gardener's Dictionary, ſays poſitively that the *Portugal laurels* remained untouched in the remarkable froſt of 1739-40. So that either that accurate obſerver was much miſtaken, or elſe the froſt of *December* 1784 was much more ſevere and deſtructive than that in the year above-mentioned.

A circumſtance

A circumstance that I muſt not omit, becauſe it was new to us, is, that on *Friday, December* the 10th, being bright ſun-ſhine, the air was full of icy *ſpiculæ*, floating in all directions, like atoms in a ſun-beam let into a dark room. We thought them at firſt particles of the rime falling from my tall hedges; but were ſoon convinced to the contrary, by making our obſervations in open places where no rime could reach us. Were they watery particles of the air frozen as they floated; or were they evaporations from the ſnow frozen as they mounted?

We were much obliged to the thermometers for the early information they gave us; and hurried our apples, pears, onions, potatoes, &c. into the cellar, and warm cloſets; while thoſe who had not, or neglected ſuch warnings, loſt all their ſtore of roots and fruits, and had their very bread and cheeſe frozen.

I muſt not omit to tell you that, during thoſe two *Siberian* days, my parlour-cat was ſo electric, that had a perſon ſtroked her, and been properly *inſulated*, the ſhock might have been given to a whole circle of people.

I forgot to mention before, that, during the two ſevere days, two men, who were tracing hares in the ſnow, had their feet frozen; and two men, who were much better employed, had their fingers ſo affected by the froſt, while they were thraſhing in a barn, that a mortification followed, from which they did not recover for many weeks.

This froſt killed all the furze and moſt of the ivy, and in many places ſtripped the hollies of all their leaves. It came at a very early time of the year, before old *November* ended; and yet may be allowed from it's effects to have exceeded any ſince 1739-40.

———————

LETTER

LETTER LXIII.

TO THE SAME.

As the effects of heat are seldom very remarkable in the northerly climate of *England*, where the summers are often so defective in warmth and sun-shine as not to ripen the fruits of the earth so well as might be wished, I shall be more concise in my account of the severity of a summer season, and so make a little amends for the prolix account of the degrees of cold, and the inconveniences that we suffered from some late rigorous winters.

The summers of 1781 and 1783 were unusually hot and dry; to them therefore I shall turn back in my journals, without recurring to any more distant period. In the former of these years my peach and nectarine-trees suffered so much from the heat that the rind on the bodies was scalded and came off; since which the trees have been in a decaying state. This may prove a hint to assiduous gardeners to fence and shelter their wall-trees with mats or boards, as they may easily do, because such annoyance is seldom of long continuance. During that summer also, I observed that my apples were coddled, as it were, on the trees; so that they had no quickness of flavour, and would not keep in the winter. This circumstance put me in mind of what I have heard travellers assert, that they never ate a good apple or apricot in the south of *Europe*,

Q q 2

where

where the heats were fo great as to render the juices vapid and infipid.

The great pefts of a garden are wafps, which deftroy all the finer fruits juft as they are coming into perfection. In 1781 we had none; in 1783 there were myriads; which would have devoured all the produce of my garden, had not we fet the boys to take the nefts, and caught thoufands with hazel-twigs tipped with bird-lime: we have fince employed the boys to take and deftroy the large breeding wafps in the fpring. Such expedients have a great effect on thefe marauders, and will keep them under. Though wafps do not abound but in hot fummers, yet they do not prevail in every hot fummer, as I have inftanced in the two years abovementioned.

In the fultry feafon of 1783 honey-dews were fo frequent as to deface and deftroy the beauties of my garden. My honeyfuckles, which were one week the moft fweet and lovely objects that the eye could behold, became the next the moft loathfome; being enveloped in a vifcous fubftance, and loaded with black aphides, or fmother-flies. The occafion of this clammy appearance feems to be this, that in hot weather the effluvia of flowers in fields and meadows and gardens are drawn up in the day by a brifk evaporation, and then in the night fall down again with the dews, in which they are entangled; that the air is ftrongly fcented, and therefore impregnated with the particles of flowers in fummer weather, our fenfes will inform us; and that this clammy fweet fubftance is of the vegetable kind we may learn from bees, to whom it is very grateful: and we may be affured that it falls in the night, becaufe it is always firft feen in warm ftill mornings.

On

On chalky and fandy foils, and in the hot villages about *London*, the thermometer has been often obferved to mount as high as 83 or 84; but with us, in this hilly and woody diftrict, I have hardly ever feen it exceed 80; nor does it often arrive at that pitch. The reafon, I conclude, is, that our denfe clayey foil, fo much fhaded by trees, is not fo eafily heated through as thofe above-mentioned : and, befides, our mountains caufe currents of air and breezes ; and the vaft effluvia from our woodlands temper and moderate our heats.

LETTER LXIV.

TO THE SAME.

THE fummer of the year 1783 was an amazing and portentous one, and full of horrible phænomena ; for, befides the alarming meteors and tremendous thunder-ftorms that affrighted and dif-treffed the different counties of this kingdom, the peculiar *haze*, or fmokey fog, that prevailed for many weeks in this ifland, and in every part of *Europe*, and even beyond it's limits, was a moft extraordinary appearance, unlike any thing known within the memory of man. By my journal I find that I had noticed this ftrange occurrence from *June* 23 to *July* 20 inclufive, during which period the wind varied to every quarter without making any alter-ation

ation in the air. The sun, at noon, looked as blank as a clouded moon, and shed a rust-coloured ferruginous light on the ground, and floors of rooms; but was particularly lurid and blood-coloured at rising and setting. All the time the heat was so intense that butchers' meat could hardly be eaten on the day after it was killed; and the flies swarmed so in the lanes and hedges that they rendered the horses half frantic, and riding irksome. The country people began to look with a superstitious awe at the red, louring aspect of the sun; and indeed there was reason for the most enlightened person to be apprehensive; for, all the while, *Calabria* and part of the isle of *Sicily*, were torn and convulsed with earth-quakes; and about that juncture a *volcano* sprung out of the sea on the coast of *Norway*. On this occasion *Milton*'s noble simile of the sun, in his first book of *Paradise Lost*, frequently occurred to my mind; and it is indeed particularly applicable, because, towards the end, it alludes to a superstitious kind of dread, with which the minds of men are always impressed by such strange and unusual phænomena.

> " — — — As when the *sun,* new risen,
> " Looks through the horizontal, *misty* air,
> " *Shorn* of his *beams*; or from behind the moon,
> " In *dim* eclipse, *disastrous twilight sheds*
> " On half the nations, and with *fear* of *change*
> " *Perplexes* monarchs — — — — — —"

LETTER

LETTER LXV.

TO THE SAME.

WE are very feldom annoyed with thunder-ftorms : and it is no lefs remarkable than true, that thofe which arife in the fouth have hardly been known to reach this village ; for, before they get over us, they take a direction to the eaft or to the weft, or fometimes divide into two, and go in part to one of thofe quarters, and in part to the other ; as was truly the cafe in fummer 1783, when, though the country round was continually haraffed with tempefts, and often from the fouth, yet we efcaped them all ; as appears by my journal of that fummer. The only way that I can at all account for this fact—for fuch it is—is that, on that quarter, between us and the fea, there are continual mountains, hill behind hill, fuch as *Nore-hill*, the *Barnet*, *Butfer-hill*, and *Ports-down*, which fome how divert the ftorms, and give them a different direction. High promontories, and elevated grounds, have always been obferved to attract clouds and difarm them of their mifchievous contents, which are difcharged into the trees and fummits as foon as they come in contact with thofe turbulent meteors ; while the humble vales efcape, becaufe they are fo far beneath them.

But, when I fay I do not remember a thunder-ftorm from the fouth, I do not mean that we never have fuffered from thunder-

ftorms

ftorms at all; for on *June* 5th, 1784, the thermometer in the morning being at 64, and at noon at 70, the barometer at 29—fix tenths one-half, and the wind north, I obferved a blue mift, fmelling ftrongly of fulphur, hanging along our floping woods, and feeming to indicate that thunder was at hand. I was called in about two in the afternoon, and fo miffed feeing the gathering of the clouds in the north; which they who were abroad affured me had fomething uncommon in it's appearance. At about a quarter after two the ftorm began in the parifh of *Hartley*, moving flowly from north to fouth; and from thence it came over *Norton-farm*, and fo to *Grange-farm*, both in this parifh. It began with vaft drops of rain, which were foon fucceeded by round hail, and then by convex pieces of ice, which meafured three inches in girth. Had it been as extenfive as it was violent, and of any continuance (for it was very fhort), it muft have ravaged all the neighbourhood. In the parifh of *Hartley* it did fome damage to one farm; but *Norton*, which lay in the center of the ftorm, was greatly injured; as was *Grange*, which lay next to it. It did but juft reach to the middle of the village, where the hail broke my north windows, and all my garden-lights and hand-glaffes, and many of my neighbours' windows. The extent of the ftorm was about two miles in length and one in breadth. We were juft fitting down to dinner; but were foon diverted from our repaft by the clattering of tiles and the jingling of glafs. There fell at the fame time prodigious torrents of rain on the farms above-mentioned, which occafioned a flood as violent as it was fudden; doing great damage to the meadows and fallows, by deluging the one and wafhing away the foil of the other. The hollow lane towards *Alton* was fo torn and difordered as not to be paffable till mended, rocks being removed that weighed 200 weight. Thofe that faw the effect which the

great

great hail had on ponds and pools say that the dashing of the water made an extraordinary appearance, the froth and spray standing up in the air three feet above the surface. The rushing and roaring of the hail, as it approached, was truly tremendous.

Though the clouds at *South Lambeth*, near *London*, were at that juncture thin and light, and no storm was in sight, nor within hearing, yet the air was strongly electric; for the bells of an electric machine at that place rang repeatedly, and fierce sparks were discharged.

When I first took the present work in hand I proposed to have added an *Annus Historico-naturalis*, or The Natural History of the Twelve Months of the Year; which would have comprised many incidents and occurrences that have not fallen in my way to be mentioned in my series of letters;—but, as Mr. *Aikin* of *Warrington* has lately published somewhat of this sort, and as the length of my correspondence has sufficiently put your patience to the test, I shall here take a respectful leave of you and natural history together; And am,

With all due deference and regard,

Your most obliged,

And most humble servant,

GIL. WHITE.

SELBORNE,
June 25, 1787.

R r

THE

ANTIQUITIES

OF

SELBORNE,

IN THE

COUNTY OF SOUTHAMPTON.

– – – – – JUVAT IRE – – – –
DESERTOSQUE VIDERE LOCOS – – – – – VIRGIL.

THE

ANTIQUITIES

OF

SELBORNE.

LETTER I.

IT is reasonable to suppose that in remote ages this woody and mountainous district was inhabited only by bears and wolves. Whether the *Britons* ever thought it worthy their attention, is not in our power to determine; but we may safely conclude, from circumstances, that it was not unknown to the *Romans*. Old people remember to have heard their fathers and grandfathers say that, in dry summers and in windy weather, pieces of money were sometimes found round the verge of *Woolmer-pond*; and tradition had inspired the foresters with a notion that the bottom of that lake contained great stores of treasure. During the spring and summer

of

of 1740 there was little rain; and the following fummer alfo, 1741, was fo uncommonly dry, that many fprings and ponds failed, and this lake in particular, whofe bed became as dufty as the furrounding heaths and waftes. This favourable juncture induced fome of the foreft-cottagers to begin a fearch, which was attended with fuch fuccefs, that all the labourers in the neighbourhood flocked to the fpot, and with fpades and hoes turned up great part of that large area. Inftead of pots of coins, as they expected, they found great heaps, the one lying on the other, as if fhot out of a bag; many of which were in good prefervation. Silver and gold thefe inquirers expected to find; but their difcoveries confifted folely of many hundreds of *Roman* copper-coins, and fome medallions, all of the lower empire. There was not much *virtù* ftirring at that time in this neighbourhood; however, fome of the gentry and clergy around bought what pleafed them beft; and fome dozens fell to the fhare of the author.

The owners at firft held their commodity at an high price; but, finding that they were not likely to meet with dealers at fuch a rate, they foon lowered their terms, and fold the faireft as they could. The coins that were rejected became current, and paffed for farthings at the petty fhops. Of thofe that we faw, the greater part were of *Marcus Aurelius*, and the Emprefs *Fauftina*, his wife, the father and mother of *Commodus*. Some of *Fauftina* were in high relief, and exhibited a very agreeable fet of features, which probably refembled that lady, who was more celebrated for her beauty than for her virtues. The medallions in general were of a paler colour than the coins. To pretend to account for the means of their coming to this place would be fpending time in conjecture. The fpot, I think, could not be a *Roman* camp, becaufe it is commanded by hills on two fides; nor does it fhew the leaft traces

of

of entrenchments; nor can I suppose that it was a *Roman* town, because I have too good an opinion of the taste and judgment of those polished conquerors to imagine that they would settle on so barren and dreary a waste.

———————

LETTER II.

THAT *Selborne* was a place of some distinction and note in the time of the *Saxons* we can give most undoubted proofs. But, as there are few if any accounts of villages before *Domesday*, it will be best to begin with that venerable record. " Ipse rex tenet " *Selesburne*. *Eddid* regina tenuit, et nunquam geldavit. De isto " manerio dono dedit rex *Radfredo* presbytero dimidiam hidam " cum ecclesia. Tempore regis *Edwardi* et post, valuit duodecim " solidos et sex denarios; modo octo solidos et quatuor denarios." Here we see that *Selborne* was a royal manor; and that *Editha*, the queen of *Edward* the Confessor, had been lady of that manor; and was succeeded in it by the Conqueror; and that it had a church. Beside these, many circumstances concur to prove it to have been a *Saxon* village; such as the name of the place itself[1],

———————

[1] *Selesburne, Seleburne, Selburn, Selbourn, Selborne*, and *Selborn*, as it has been variously spelt at different periods, is of *Saxon* derivation; for *Sel* signifies *great*, and *burn* torrens, a *brook* or *rivulet*: so that the name seems to be derived from the great perennial stream that breaks out at the upper end of the village.—*Sel* also signifies *bonus*, item, *fœcundus, fertilis.* " Sel ᵹæꞃꞃ-ꞇon : *fœcunda graminis clausura*; *fertile pascuum*: a meadow in the parish of *Godelming* is still called *Sal-gars-ton*."

Lye's Saxon Dictionary, in the Supplement, by Mr. *Manning*.

the

the names of many fields, and some families[k], with a variety of words in husbandry and common life, still subsisting among the country people.

What probably first drew the attention of the *Saxons* to this spot was the beautiful spring or fountain called *Well-head*[l], which induced them to build by the banks of that perennial current; for ancient settlers loved to reside by brooks and rivulets, where they could dip for their water without the trouble and expense of digging wells and of drawing.

It remains still unsettled among the antiquaries at what time tracts of land were first appropriated to the chase alone for the amusement of the sovereign. Whether our *Saxon* monarchs had any royal forests does not, I believe, appear on record; but the *Constitutiones de Foresta* of *Canute*, the Dane, are come down to us. We shall not therefore pretend to say whether *Woolmer-forest* existed as a royal domain before the conquest. If it did not, we may suppose it

[k] Thus the name of *Aldred* signifies *all-reverend*, and that of *Kemp* means a *soldier*. Thus we have a *church-litton*, or enclosure for dead bodies, and not a *church-yard*: there is also a *Culver-croft* near the *Grange-farm*, being the enclosure where the priory *pigeon-house* stood, from *culver* a pigeon. Again there are three steep pastures in this parish called the *Lithe*, from *Hlithe*, *clivus*. The wicker-work that binds and fastens down a hedge on the top is called *ether*, from *ether* an hedge. When the good women call their hogs they cry *sic, sic* *, not knowing that *sic* is Saxon, or rather Celtic, for a hog. Coppice or brush wood our countrymen call *rise*, from *hris*, *frondes*; and talk of a load of *rise*. Within the author s memory the *Saxon* plurals, *housen* and *peason*, were in common use. But it would be endless to instance in every circumstance: he that wishes for more specimens must frequent a farmer's kitchen. I have therefore selected some words to shew how familiar the *Saxon* dialect was to this district, since in more than seven hundred years it is far from being obliterated.

[l] *Well-head* signifies *spring-head*, and not a deep pit from whence we draw water. For particulars about which see Letter I. to Mr. *Pennant*.

* Σικα, porcus, apud *Lacones*; un Porceau chez les *Lacedemoniens*: ce mot a sans doute esté pris des *Celtes*, qui disoent *sic*, pour marquer un porceau. Encore aujourd'huy quand les *Bretons* chassent ces animaux, ils ne disent point autrement, que *sic, sic*.

Antiquité de la Nation, et de la Langue des Celtes, par Pezron.

was laid out by some of our earliest *Norman* kings, who were exceedingly attached to the pleasures of the chase, and resided much at *Winchester*, which lies at a moderate distance from this district. The *Plantagenet* princes seem to have been pleased with *Woolmer*; for tradition says that king *John* resided just upon the verge, at *Ward-le-ham*, on a regular and remarkable mount, still called *King John's Hill* and *Lodge Hill*; and *Edward* III. had a chapel in his park, or enclosure, at *Kingsley*[l]. *Humphrey*, duke of *Gloucester*, and *Richard*, duke of *York*, say my evidences, were both, in their turns, *wardens of Woolmer-forest*; which seems to have served for an appointment for the younger princes of the royal family, as it may again.

I have intentionally mentioned *Edward* III. and the dukes *Humphrey* and *Richard*, before king *Edward* II. because I have reserved, for the entertainment of my readers, a pleasant anecdote respecting that prince, with which I shall close this letter.

As *Edward* II. was hunting on *Woolmer-forest*, *Morris Ken*, of the kitchen, fell from his horse several times; at which accidents the king laughed immoderately: and, when the chase was over, ordered him twenty shillings[m]; an enormous sum for those days! Proper allowances ought to be made for the youth of this monarch, whose spirits also, we may suppose, were much exhilarated by the sport of the day: but, at the same time, it is reasonable to remark

---

[l] The parish of *Kingsley* lies between, and divides *Woolmer-forest* from *Ayles Holt-forest*.
See Letter IX. to Mr. *Pennant*.

[m] " Item, paid at the lodge at *Woolmer*, when the king was stag-hunting there, to *Morris Ken*, of the kitchen, because he rode before the king and often fell from his horse, at which the king laughed exceedingly — a gift, by command, of twenty shillings."

A MSS. in possession of *Thomas Astle*, esq. containing the private expenses of Edward II.

that,

that, whatever might be the occafion of *Ken*'s firſt fall, the ſubſe-
quent ones ſeem to have been deſigned. The ſcullion appears
to have been an artful fellow, and to have ſeen the king's foible;
which furniſhes an early ſpecimen of that his eaſy ſoftneſs and
facility of temper, of which the infamous *Gaveſton* took ſuch advantages, as brought innumerable calamities on the nation, and
involved the prince at laſt in miſfortunes and ſufferings too de-
plorable to be mentioned without horror and amazement.

LETTER III.

Fʀᴏᴍ the ſilence of *Domeſday* reſpecting churches, it has been
ſuppoſed that few villages had any at the time when that record
was taken; but *Selborne*, we ſee, enjoyed the benefit of one:
hence we may conclude, that this place was in no abject ſtate even
at that very diſtant period. How many fabrics have ſucceeded
each other ſince the days of *Radfredrus* the *preſbyter*, we cannot
pretend to ſay; our buſineſs leads us to a deſcription of the preſent edifice, in which we ſhall be circumſtantial.

Our church, which was dedicated to the *Virgin Mary*, conſiſts
of three ailes, and meaſures fifty-four feet in length by forty-
ſeven in breadth, being almoſt as broad as it is long. The preſent building has no pretenſions to antiquity; and is, as I ſuppoſe,
of no earlier date than the beginning of the reign of *Henry* VII.
It

South View of SELBORNE CHURCH.

Published Nov. 1. 1790. as the Act directs by B. White, Fleet.

J. Mazell Sculp.

S. H. Grimm del.

It is perfectly plain and unadorned, without painted glafs, carved work, fculpture, or tracery. But when I fay it has no claim to antiquity, I would mean to be underftood of the fabric in general; for the pillars which fupport the roof are undoubtedly old, being of that low, fquat, thick order, ufually called *Saxon*. Thefe, I fhould imagine, upheld the roof of a former church, which, falling into decay, was rebuilt on thofe maffy props, becaufe their ftrength had preferved them from the injuries of time[n]. Upon thefe reft blunt *gothic* arches, fuch as prevailed in the reign above-mentioned, and by which, as a criterion, we would prove the date of the building.

At the bottom of the fouth aile, between the weft and fouth doors, ftands the font, which is deep and capacious, and confifts of three maffy round ftones, piled one on another, without the leaft ornament or fculpture: the cavity at the top is lined with lead, and has a pipe at bottom to convey off the water after the facred ceremony is performed.

The eaft end of the fouth aile is called the *South Chancel*, and, till within thefe thirty years, was divided off by old carved *gothic* frame-work of timber, having been a private chantry. In this opinion we are more confirmed by obferving two *gothic* niches within the fpace, the one in the eaft wall and the other in the fouth, near which there probably ftood images and altars.

In the middle aile there is nothing remarkable; but I remember when it's beams were hung with garlands in honour of

---

[n] In the fame manner, to compare great things with fmall, did *Wykeham*, when he new-built the cathedral at *Winchefter*, from the tower weftward, apply to his purpofe the old piers or pillars of Bifhop *Walkelin*'s church, by blending *Saxon* and *Gothic* architecture together. See *Lowth*'s Life of Wykeham.

young

young women of the parish, reputed to have died virgins; and recollect to have seen the clerk's wife cutting, in white paper, the resemblances of gloves, and ribbons to be twisted into knots and roses, to decorate these memorials of chastity. In the church of *Faringdon*, which is the next parish, many garlands of this sort still remain.

The north aile is narrow and low, with a sloping ceiling, reaching within eight or nine feet of the floor. It had originally a flat roof covered with lead, till, within a century past, a churchwarden stripping off the lead, in order, as he said, to have it mended, sold it to a plumber, and ran away with the money. This aile has no door, for an obvious reason; because the north-side of the church-yard, being surrounded by the vicarage garden, affords no path to that side of the church. Nothing can be more irregular than the pews of this church, which are of all dimensions and heights, being patched up according to the fancy of the owners: but whoever nicely examines them will find that the middle aile had, on each side, a regular row of benches of solid oak, all alike, with a low back-board to each. These we should not hesitate to say are coeval with the present church : and especially as it is to be observed that, at their ends, they are ornamented with carved blunt *gothic* niches, exactly correspondent to the arches of the church, and to a niche in the south wall. The south aile also has a row of these benches; but some are decayed through age, and the rest much disguised by modern alterations.

At the upper end of this aile, and running out to the north, stands a transept, known by the name of the *North Chancel*, measuring twenty-one feet from south to north, and nineteen feet from east to west: this was intended, no doubt, as a private chantry; and was also, till of late, divided off by a *gothic* frame-

work

work of timber. In its north wall, under a very blunt *gothic* arch, lies perhaps the founder of this edifice, which, from the shape of its arch, may be deemed no older than the latter end of the reign of *Henry* VII. The tomb was examined some years ago, but contained nothing except the scull and thigh-bones of a large tall man, and the bones of a youth or woman, lying in a very irregular manner, without any escutcheon or other token to ascertain the names or rank of the deceased. The grave was very shallow, and lined with stone at the bottom and on the sides.

From the east wall project four stone brackets, which I conclude supported images and crucifixes. In the great thick pilaster, jutting out between this transept and the chancel, there is a very sharp *gothic* niche, of older date than the present chantry or church. But the chief pieces of antiquity are two narrow stone coffin-lids, which compose part of the floor, and lie from west to east, with the very narrow ends eastward : these belong to remote times; and, if originally placed here, which I doubt, must have been part of the pavement of an older transept. At present there are no coffins under them, whence I conclude they have been removed to this place from some part of a former church. One of these lids is so eaten by time, that no sculpture can be discovered upon it; or, perhaps, it may be the wrong side uppermost; but on the other, which seems to be of stone of a closer and harder texture, is to be discerned a *discus*, with a cross on it, at the end of a staff or rod, the well-known symbol of a *Knight-Templar* [o].

This order was distinguished by a red cross on the left shoulder of their cloak, and by this attribute in their hand. Now, if these

[o] See *Dugdale, Monasticon Anglicanum*, Vol. II. where there is a fine engraving of a *Knight-Templar*, by *Hollar*.

stones

ftones belonged to *Knights Templars*, they muſt have lain here many
centuries; for this order came into England early in the reign
of king *Stephen* in 1113; and was diſſolved in the time of
Edward II. in 1312, having ſubſiſted only one hundred and ninety-
nine years. Why I ſhould ſuppoſe that *Knights Templars* were
occaſionally buried at this church, will appear in ſome future
letter, when we come to treat more particularly concerning the
property they poſſeſſed here, and the intercourſe that ſubſiſted between them and the *priors* of *Selborne*.

We muſt now proceed to the chancel, properly ſo called, which
ſeems to be coeval with the church, and is in the ſame plain unadorned ſtyle, though neatly kept. This room meaſures thirty-
one feet in length, and ſixteen feet and an half in breadth, and
is wainſcoted all round, as high as to the bottom of the windows.
The ſpace for the communion table is raiſed two ſteps above the
reſt of the floor, and railed in with oaken baluſters. Here I ſhall
ſay ſomewhat of the windows of the chancel in particular, and
of the whole fabric in general. They are moſtly of that ſimple
and unadorned ſort called *Lancet*, ſome ſingle, ſome double, and
ſome in triplets. At the eaſt end of the chancel are two of a
moderate ſize, near each other; and in the north wall two very
diſtant ſmall ones, unequal in length and height: and in the
ſouth wall are two, one on each ſide of the chancel door, that are
broad and ſquat, and of a different order. At the eaſt end of
the ſouth aile of the church there is a large lancet-window in a
triplet; and two very ſmall, narrow, ſingle ones in the ſouth
wall, and a broad ſquat window beſide, and a double lancet one
in the weſt end; ſo that the appearance is very irregular. In the
north aile are two windows, made ſhorter when the roof was
ſloped; and in the north tranſept a large triple window, ſhortened

at

at the time of a repair in 1721; when over it was opened a round one of confiderable fize, which affords an agreeable light, and renders that chantry the moft cheerful part of the edifice.

The church and chancels have all coved roofs, ceiled about the year 1683; before which they were open to the tiles and fhingles, fhowing the naked rafters, and threatening the congregation with the fall of a fpar, or a blow from a piece of loofe mortar.

On the north wall of the chancel is fixed a large oval white marble monument, with the following infcription; and at the foot of the wall, over the deceafed, and infcribed with his name, age, arms, and time of death, lies a large flab of black marble:

Prope hunc parietem fepelitur
GILBERTUS WHITE, SAMSONIS WHITE, de
Oxon. militis filius tertius, Collegii Magdale-
-nenfis ibidem alumnus, & focius. Tandem faven-
-te collegio ad hanc ecclefiam promotus; ubi primæ-
-vâ morum fimplicitate et diffusâ erga omnes bene-
-volentiâ feliciter confenuit.
Paftor fidelis, comis, affabilis,
Maritus, et pater amantiffimus,
A conjuge invicem, et liberis, atque
A parochianis impenfé dilectus.
Pauperibus ita beneficus
ut decimam partem censûs
moribundus
piis ufibus confecravit.
Meritis demum juxta et annis plenus
ex hac vitâ migravit Feb. 13°.
anno falutis 172⅞
Ætatis fuæ 77.
Hoc pofuit Rebecca
Conjux illius mæftiffima,
mox fecutura.

On

On the fame wall is newly fixed a fmall fquare table-monument of white marble, infcribed in the following manner.

Sacred to the memory
of the Revᵈ. ANDREW ETTY, B. D.
23 Years Vicar of this parifh:
In whofe character
The conjugal, the parental, and the facerdotal virtues
were fo happily combined
as to deferve the imitation of mankind.
And if in any particular he followed more invariably
the fteps of his bleffed Mafter,
It was in his humility.
His parifhioners,
efpecially the fick and neceffitous,
as long as any traces of his memory fhall remain,
muft lament his death.
To perpetuate fuch an example, this ftone is erected;
as while living he was a preacher of righteoufnefs,
fo, by it, he being dead yet fpeaketh.
He died April 8ᵗʰ. 1784. aged 66 years.

LETTER

LETTER IV.

WE have now taken leave of the infide of the church, and fhall pafs by a door at the weft end of the middle aile into the belfry. This room is part of a handfome fquare embattled tower of forty-five feet in height, and of much more modern date than the church; but old enough to have needed a thorough repair in 1781, when it was neatly ftuccoed at a confiderable expenfe, by a fet of workmen who were employed on it for the greateft part of the fummer. The old bells, three in number, loud and out of tune, were taken down in 1735, and caft into four; to which *Sir Simeon Stuart*, the grandfather of the prefent baronet, added a fifth at his own expenfe: and, beftowing it in the name of his favourite daughter Mrs. *Mary Stuart*, caufed it to be caft with the following motto round it:

" Clara puella dedit, dixitque mihi efto Maria:
" Illius et laudes nomen ad aftra fono."

The day of the arrival of this tuneable peal was obferved as an high feftival by the village, and rendered more joyous, by an order from the donor, that the treble-bell fhould be fixed bottom upward in the ground, and filled with punch, of which all prefent were permitted to partake.

The porch of the church, to the fouth, is modern, and would not be worthy attention did it not fhelter a fine fharp *gothic* door-way. This is undoubtedly much older than the prefent fabric; and,

T t

being

being found in good prefervation, was worked into the wall, and is the grand entrance into the church : nor are the folding-doors to be paffed over in filence; fince, from their thick and clumfy ftructure, and the rude flourifhed-work of their hinges, they may poffibly be as ancient as the door-way itfelf.

The whole roof of the fouth aile, and the fouth-fide of the roof of the middle aile, is covered with oaken fhingles inftead of tiles, on account of their lightnefs, which favours the ancient and crazy timber-frame. And, indeed, the confideration of accidents by fire excepted, this fort of roofing is much more eligible than tiles. For fhingles well feafoned, and cleft from quartered timber, never warp, nor let in drifting fnow; nor do they fhiver with froft; nor are they liable to be blown off, like tiles; but, when well nailed down, laft for a long period, as experience has fhown us in this place, where thofe that face to the north are known to have endured, untouched, by undoubted tradition for more than a century.

Confidering the fize of the church, and the extent of the parifh, the church-yard is very fcanty; and efpecially as all wifh to be buried on the fouth-fide, which is become fuch a mafs of mortality that no perfon can be there interred without difturbing or difplacing the bones of his anceftors. There is reafon to fuppofe that it once was larger, and extended to what is now the vicarage court and garden; becaufe many human bones have been dug up in thofe parts feveral yards without the prefent limits. At the eaft end are a few graves; yet none till very lately on the north-fide but, as two or three families of beft repute have begun to bury in that quarter, prejudice may wear out by degrees, and their example be followed by the reft of the neighbourhood.

In

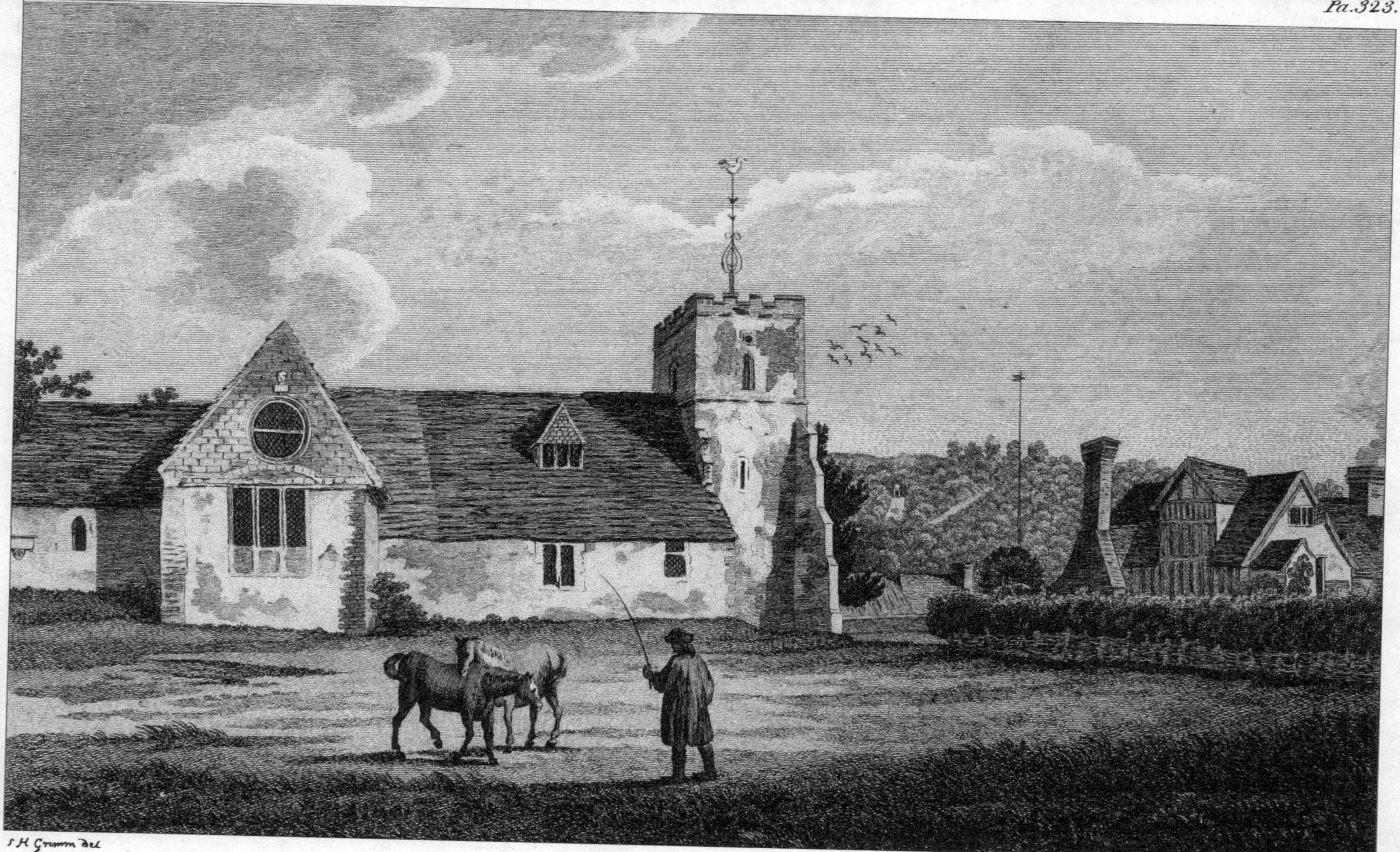

S H Grimm del

P Mazell Sculp

North View of SELBORNE CHURCH.

In speaking of the church, I have all along talked of the east and west-end, as if the chancel stood exactly true to those points of the compass; but this is by no means the case, for the fabric bears so much to the north of the east that the four corners of the tower, and not the four sides, stand to the four cardinal points. The best method of accounting for this deviation seems to be, that the workmen, who probably were employed in the longest days, endeavoured to set the chancels to the rising of the sun.

Close by the church, at the west end, stands the vicarage-house; an old, but roomy and convenient edifice. It faces very agreeably to the morning sun, and is divided from the village by a neat and cheerful court. According to the manner of old times, the hall was open to the roof; and so continued, probably, till the vicars became family-men, and began to want more conveniencies; when they flung a floor across, and, by partitions, divided the space into chambers. In this hall we remember a date, some time in the reign of *Elizabeth*; it was over the door that leads to the stairs.

Behind the house is a garden of an irregular shape, but well laid out; whose terrace commands so romantic and picturesque a prospect, that the first master in landscape might contemplate it with pleasure, and deem it an object well worthy of his pencil.

LETTER

LETTER V.

IN the church-yard of this village is a *yew-tree*, whofe afpect befpeaks it to be of a great age : it feems to have feen feveral centuries, and is probably coeval with the church, and therefore may be deemed an antiquity : the body is fquat, fhort, and thick, and meafures twenty-three feet in the girth, fupporting an head of fuitable extent to it's bulk. This is a male tree, which in the fpring fheds clouds of duft, and fills the atmofphere around with it's farina.

As far as we have been able to obferve, the males of this fpecies become much larger than the females ; and it has fo fallen out that moft of the yew-trees in the church-yards of this neighbourhood are males : but this muft have been matter of mere accident, fince men, when they firft planted yews, little dreamed that there were fexes in trees.

In a yard, in the midft of the ftreet, till very lately grew a middle-fized female tree of the fame fpecies, which commonly bore great crops of berries. By the high winds ufually prevailing about the autumnal equinox, thefe berries, then ripe, were blown down into the road, where the hogs ate them. And it was very remarkable, that, though barrow-hogs and young fows found no inconvenience from this food, yet milch-fows often died after fuch a repaft : a circumftance that can be accounted for only by fuppofing that the latter, being much exhaufted and hungry, devoured a larger quantity.

While

While mention is making of the bad effects of yew-berries, it may be proper to remind the unwary that the twigs and leaves of yew, though eaten in a very small quantity, are certain death to horses and cows, and that in a few minutes. An horse tied to a yew-hedge, or to a faggot-stack of dead yew, shall be found dead before the owner can be aware that any danger is at hand : and the writer has been several times a sorrowful witness to losses of this kind among his friends ; and in the island of *Ely* had once the mortification to see nine young steers or bullocks of his own all lying dead in an heap from browzing a little on an hedge of yew in an old garden, into which they had broken in snowy weather. Even the clippings of a yew-hedge have destroyed a whole dairy of cows when thrown inadvertently into a yard. And yet sheep and turkies, and, as park-keepers say, deer, will crop these trees with impunity.

Some intelligent persons assert that the branches of yew, while *green*, are not noxious ; and that they will kill only when *dead* and *withered*, by lacerating the stomach : but to this assertion we cannot by any means assent, because, among the number of cattle that we have known fall victims to this deadly food, not one has been found, when it was opened, but had a lump of *green* yew in it's paunch. True it is, that yew-trees stand for twenty years or more in a field, and no bad consequences ensue : but at some time or other cattle, either from wantonness when full, or from hunger when empty, (from both which circumstances we have seen them perish) will be meddling, to their certain destruction ; the yew seems to be a very improper tree for a pasture-field.

Antiquaries seem much at a loss to determine at what period this tree first obtained a place in church-yards. A statute passed A. D. 1307 and 35 *Edward* I. the title of which is " Ne rector
" arbores

" arbores in cemeterio prosternat." Now if it is recollected that we seldom see any other very large or ancient tree in a church-yard but yews, this statute must have principally related to this species of tree; and consequently their being planted in church-yards is of much more ancient date than the year 1307.

As to the use of these trees, possibly the more respectable parishioners were buried under their shade before the improper custom was introduced of burying within the body of the church, where the living are to assemble. *Deborah, Rebekah*'s nurse q, was buried under an oak; the most honourable place of interment probably next to the cave of *Machpelah* r, which seems to have been appropriated to the remains of the patriarchal family alone.

The farther use of yew-trees might be as a screen to churches, by their thick foliage, from the violence of winds; perhaps also for the purpose of archery, the best long bows being made of that material: and we do not hear that they are planted in the church-yards of other parts of *Europe*, where long bows were not so much in use. They might also be placed as a shelter to the congregation assembling before the church-doors were opened, and as an emblem of mortality by their funereal appearance. In the south of *England* every church-yard almost has it's tree, and some two; but in the north, we understand, few are to be found.

The idea of R. C. that the *yew-tree* afforded it's branches instead of palms for the processions on *Palm-Sunday*, is a good one, and deserves attention. See Gent. Mag. Vol. L. p. 128.

q Gen. xxxv, 8.   r Gen. xxiii, 9.

LETTER

LETTER VI.

THE living of *Selborne* was a very small vicarage; but, being in the patronage of *Magdalen-college*, in the university of *Oxford*, that society endowed it with the great tithes of *Selborne*, more than a century ago: and since the year 1758 again with the great tithes of *Oakhanger*, called *Bene's parsonage*: so that, together, it is become a respectable piece of preferment, to which one of the fellows is always presented. The vicar holds the great tithes, by lease, under the college. The great disadvantage of this living is, that it has not one foot of glebe near home[s].

ITS PAYMENTS ARE,

	£.	s.	d.
King's books — — — —	8	2	1
Yearly tenths — — — —	0	16	$2\frac{1}{2}$
Yearly procurations for *Blackmore* and *Oakhanger* Chap: with acquit: — —	0	1	7
Selborne procurations and acquit: — —	0	9	0

I am unable to give a complete list of the vicars of this parish till towards the end of the reign of queen *Elizabeth*; from which period the registers furnish a regular series.

In *Domesday* we find thus — " De isto manerio dono dedit Rex " *Radfredo* presbytero dimidiam hidam cum ecclesia." So that before *Domesday*, which was compiled between the years 1081 and 1086, here was an officiating minister at this place.

---

[s] At *Bene's*, or *Bin's*, *parsonage* there is a house and stout barn, and seven acres of glebe: *Bene's* parsonage is three miles from the church.

After

After this, among my documents, I find occasional mention of a vicar here and there: the first is

Roger, instituted in 1254.

In 1410 *John Lynne* was vicar of *Selborne*.

In 1411 *Hugo Tybbe* was vicar.

The presentations to the vicarage of *Selborne* generally ran in the name of the prior and the convent; but *Tybbe* was presented by prior *John Wynechestre* only.

June 29, 1528, *William Fisher*, vicar of *Selborne*, resigned to *Miles Peyrson*.

1594, *William White* appears to have been vicar to this time. Of this person there is nothing remarkable, but that he hath made a regular entry twice in the register of *Selborne* of the funeral of *Thomas Cowper*, bishop of *Winchester*, as if he had been buried at *Selborne*; yet this learned prelate, who died 1594, was buried at *Winchester*, in the cathedral, near the episcopal throne [t].

1595, *Richard Boughton*, vicar.

1596, *William Inkforbye*, vicar.

May 1606, *Thomas Phippes*, vicar.

June 1631, *Ralph Austine*, vicar.

July 1632, *John Longworth*. This unfortunate gentleman, living in the time of *Cromwell's* usurpation, was deprived of his preferment for many years, probably because he would not take the league and covenant: for I observe that his father-in-law, the Reverend *Jethro Beal*, rector of *Faringdon*, which is the next parish, enjoyed his benefice during the whole of that unhappy period. *Longworth*, after he was dispossessed, retired to a little tenement about one hundred and fifty yards from the church, where he earned a

[t] See *Godwin* de præsulibus, folio *Cant.* 1743, page 239.

small

small pittance by the practice of physic. During those dismal times it was not uncommon for the deposed clergy to take up a medical character; as was the case in particular, I know, with the Reverend Mr. *Yalden*, rector of *Compton*, near *Guildford*, in the county of *Surrey*. Vicar *Longworth* used frequently to mention to his sons, who told it to my relations, that, the *Sunday* after his deprivation, his puritanical successor stepped into the pulpit with no small petulance and exultation; and began his sermon from *Psalm* xx. 8. " *They are brought down and fallen; but we are risen* " *and stand upright.*" This person lived to be restored in 1660, and continued vicar for eighteen years; but was so impoverished by his misfortunes, that he left the vicarage-house and premises in a very abject and dilapidated state.

July 1678. *Richard Byfield*, who left eighty pounds by will, the interest to be applied to apprentice out poor children: but this money, lent on private security, was in danger of being lost, and the bequest remained in an unsettled state for near twenty years, till 1700; so that little or no advantage was derived from it. About the year 1759 it was again in the utmost danger by the failure of a borrower; but, by prudent management, has since been raised to one hundred pounds stock in the three per cents reduced. The trustees are the vicar and the renters or owners of *Temple, Priory, Grange, Blackmore,* and *Oakhanger-house,* for the time being. This gentleman seemed inclined to have put the vicarial premises in a comfortable state; and began, by building a solid stone wall round the front-court, and another in the lower yard, between that and the neighbouring garden; but was interrupted by death from fulfilling his laudable intentions.

April, 1680, *Barnabas Long* became vicar.

June

June 1681. This living was now in such low estimation in *Magdalen-college* that it descended to a junior fellow, *Gilbert White*, M. A. who was instituted to it in the thirty-first year of his age. At his first coming he ceiled the chancel, and also floored and wainscoted the parlour and hall, which before were paved with stone, and had naked walls; he enlarged the kitchen and brewhouse, and dug a cellar and well: he also built a large new barn in the lower yard, removed the hovels in the front court, which he laid out in walks and borders; and entirely planned the back garden, before a rude field with a stone-pit in the midst of it. By his will he gave and bequeathed " the sum of forty pounds to be laid " out in the most necessary repairs of the church; that is, in " strengthening and securing such parts as seem decaying and " dangerous." With this sum two large buttresses were erected to support the east end of the south wall of the church; and the gable-end wall of the west end of the south aile was new built from the ground.

By his will also he gave " One hundred pounds to be laid " out on lands; the yearly rents whereof shall be employed in " teaching the poor children of *Selbourn* parish to read and write, " and say their prayers and catechism, and to sew and knit:— " and be under the direction of his executrix as long as she " lives; and, after her, under the direction of such of his " children and their issue, as shall live in or within five miles " of the said parish: and on failure of any such, then under the " direction of the vicar of *Selbourn* for the time being; but still " to the uses above-named." With this sum was purchased, of *Thomas Turville*, of *Hawkeley*, in the county of *Southampton*, yeoman, and *Hannah* his wife, *two closes* of freehold land, commonly called *Collier's*, containing, by estimation, *eleven* acres, lying in *Hawkeley* aforesaid.

aforesaid. These closes are let at this time, 1785, on lease, at the rate of three pounds by the year.

This vicar also gave by will *two hundred pounds* towards the repairs of the highways" in the parish of *Selborne*. That sum was carefully and judiciously laid out in the summer of the year 1730 by his son *John White*, who made a solid and firm causey from *Rood-green*, all down *Honey-lane*, to a farm called *Oak-woods*, where the sandy soil begins. This miry and gulfy lane was chosen as worthy of repair, because it leads to the forest, and thence through the *Holt* to the town of *Farnham* in *Surrey*, the only market in those days for men who had wheat to sell in this neighbourhood. This causey was so deeply bedded with stone, so properly raised above the level of the soil, and so well drained, that it has, in some degree, withstood fifty-four years of neglect and abuse; and might, with moderate attention, be rendered a solid and comfortable road. The space from *Rood-green* to *Oak-woods* measures about three quarters of a mile.

In 1727, *William Henry Cane*, B. D. became vicar; and, among several alterations and repairs, new-built the back front of the vicarage-house.

On *February* 1, 1740, *Duncombe Bristowe*, D. D. was instituted to this living. What benefactions this vicar bestowed on the parish will be best explained by the following passages from his will :—
" *Item*, I hereby give and bequeath to the minister and church-
" wardens of the parish of *Selbourn*, in the county of *Southampton*,
" a mahogany table, which I have ordered to be made for the
" celebration of the Holy Communion; and also the sum of

" Such legacies were very common in former times, before any effectual laws were " made for the repairs of highways," Sir John Cullum's Hawsted. p. 15.

U u 2 " thirty

" thirty pounds, in truſt, to be applied in manner following; that
" is, ten pounds towards the charge of erecting a gallery at the
" weſt end of the church; and ten pounds to be laid out for
" cloathing, and ſuch like neceſſaries, among the poor (and
" eſpecially among the ancient and infirm) of the ſaid pariſh:
" and the remaining ten pounds to be diſtributed in bread, at
" twenty ſhillings a week, at the diſcretion of *John White*, eſq.
" or any of his family, who ſhall be reſident in the ſaid pariſh."

On *November* 12, 1758, *Andrew Etty*, B. D. became vicar.
Among many uſeful repairs he new-roofed the body of the vicar-
age-houſe; and wainſcoted, up to the bottom of the windows,
the whole of the chancel; to the neatneſs and decency of which
he always paid the moſt exact attention.

On *September* 25, 1784, *Chriſtopher Taylor*, B. D. was inducted
into the vicarage of *Selborne*.

LETTER

LETTER VII.

I SHALL now proceed to the *Priory*, which is undoubtedly the most interesting part of our history.

The *Priory* of *Selborne* was founded by *Peter de la Roche*, or *de Rupibus* [x], one of those accomplished foreigners that resorted to the court of king *John*, where they were usually caressed, and met with a more favourable reception than ought, in prudence, to have been shown by any monarch to strangers. This adventurer was a *Poictevin* by birth, had been bred to arms in his youth, and distinguished by knighthood. Historians all agree not to speak very favourably of this remarkable man; they allow that he was possessed of courage and fine abilities, but then they charge him with arbitrary principles, and violent conduct. By his insinuating manners he soon rose high in the favour of *John*; and in 1205, early in the reign of that prince, was appointed bishop of *Winchester*. In 1214 he became lord chief justiciary of *England*, the first magistrate in the state, and a kind of viceroy, on whom depended all the civil affairs in the kingdom. After the death of *John*, and during the minority of his son *Henry*, this prelate took upon him the entire management of the realm, and was soon appointed protector of the king and kingdom.

x See *Godwin* de Præsulibus Anglia. Folio. London. 1743. p. 217.

The

The barons faw with indignation a ftranger poffeffed of all the power and influence, to part of which they thought they had a claim; they therefore entered into an affociation againft him, and determined to wreft fome of that authority from him which he had fo unreafonably ufurped. The bifhop difcerned the ftorm at a diftance; and, prudently refolving to give way to that torrent of envy which he knew not how to withftand, withdrew quietly to the Holy Land, where he refided fome time.

At this juncture a very fmall part of *Paleftine* remained in the hands of the Chriftians: they had been by *Saladine* difpoffeffed of *Jerufalem*, and all the internal parts, near forty years before; and with difficulty maintained fome maritime towns and garrifons: yet the bufy and enterprifing fpirit of *de Rupibus* could not be at reft; he diftinguifhed himfelf by the fplendour and magnificence of his expenfes, and amufed his mind by ftrengthening fortreffes and caftles, and by removing and endowing of churches. Before his expedition to the eaft he had fignalized himfelf as a founder of convents, and as a benefactor to hofpitals and monafteries.

In the year 1231 he returned again to *England*; and the very next year, in 1232, began to build and endow the PRIORY of SELBORNE. As this great work followed fo clofe upon his return, it is not improbable that it was the refult of a vow made during his voyage; and efpecially as it was dedicated to the *Virgin Mary*. Why the bifhop made choice of *Selborne* for the fcene of his munificence can never be determined now: it can only be faid that the parifh was in his diocefe, and lay almoft midway between *Winchefter* and *Farnham*, or *South Waltham* and *Farnham*; from either of which places he could without much trouble overlook his workmen, and obferve what progrefs they made; and that the fituation was retired, with a ftream running by it, and

fequeftered

sequestered from the world, amidst woods and meadows, and so far proper for the site of a religious house [y].

The first person with whom the founder treated about the purchase of land was *Jacobus de Achangre*, or *Ochangre*, a gentleman of property who resided at that hamlet; and, as appears, at the house now called *Oakhanger-house*. With him he agreed for a croft, or little close of land, known by the name of *La liega*, or *La lyge*, which was to be the immediate site of the *Priory*.

De Achangre also accommodated the bishop at the same instant with three more adjoining crofts, which for a time was all the footing that this institution obtained in the parish. The seller in the conveyance says "Warantizabimus, defendemus, et æquietabimus " *contra omnes gentes;*" viz. "We will warrant the thing sold " against all claims from any quarter." In modern conveyancing this would be termed a covenant for *further assurance*. Afterwards is added — " Pro hac autem donacione, &c. dedit mihi pred. " Episcopus sexdecim marcas argenti in *Gersumam:*" i. e. " the " bishop gave me sixteen silver marks as a consideration for the " thing purchased."

y The institution at *Selborne* was a priory of *Black-Canons* of the order of St. *Agustine*, called also *Canons-Regular.* *Regular-Canons* were such as lived in a conventual manner, under one roof, had a common refectory and dormitary, and were bound by vows to observe the rules and statutes of their order: in fine, they were a kind of religious, whose discipline was less rigid than the *monks.* The chief rule of these *canons* was that of St. *Augustine,* who was constituted bishop of *Hippo,* A. D. 395: but they were not brought into *England* till after the conquest; and seem not to have obtained the appellation of *Augustine canons* till some years after. Their habit was a long black *cassock,* with a white *rocket* over it; and over that a black *cloak and hood.* The monks were always shaved: but these canons wore their hair and beards, and caps on their heads. There were of these *canons,* and women of the same order called *Canonesses,* about 175 houses.

As

As the grant from *Jac. de Achangre* was without date [z], and the next is circumstanced in the same manner, we cannot say exactly what interval there was between the two purchases; but we find that *Jacobus de Nortun*, a neighbouring gentleman, also soon sold to the bishop of *Winchester* some adjoining grounds, through which our stream passes, that the priory might be accommodated with a *mill*, which was a common necessary appendage to every manor: he also allowed access to these lands by a road for carts and waggons.—" *Jacobus de Nortun* concedit *Petro Winton* episcopo totum " cursum aque que descèndit de *Molendino de Durton* usq; ad " boscum *Will. Mauduit*, et croftam terre vocat: *Edriche croft*, " cum extensione ejusdem et abuttamentis; ad fundandam domum " religiosam de ordine *Sti. Augustini*. Concedit etiam viam ad " carros, et caretas," &c. This vale, down which runs the brook, is now called the *Long Lithe*, or *Lythe*. Bating the following particular expression, this grant runs much in the style of the former; " Dedit mihi episcopus predictus triginta quinque " marcas argenti *ad me acquietandum versus Judæos*."—that is, " the " bishop advanced me thirty-five marks of silver to pay my debts " to the jews, who were then the only lenders of money."

Finding himself still streightened for room, the founder applied to his royal master, *Henry*, who was graciously pleased to bestow certain lands in the manor at *Selborne* on the new priory of his favourite minister. These grounds had been the property of *Stephen de Lucy*; and, abutting upon the narrow limits of the convent, became a very commodious and agreeable acquisition. This grant, I find, was made on *March* the 9th, in the eighteenth year of *Henry*, viz. 1234, being two years after the foundation of the

---

[z] The custom of affixing dates to deeds was not become general in the reign of *Henry* III.

monastery.

monaftery. The royal donor beftowed his favour with a good grace, by adding to it almoft every immunity and privilege that could have been fpecified in the law-language of the times.—
" Quare volumus prior, &c. habeant totam terram, &c. cum " omnibus libertatibus in bofco et plano, in viis et femitis, pratis " et pafcuis ; aquis et pifcariis ; infra burgum, et extra burgum, " cum foka et faca, Thol et Them, Infangenethef et Utfangene-" thef, et hamfocne et blodwite, et pecunia que dari folet pro " murdro et forftal, et flemeneftrick, et cum quietancia de omni " fcotto et geldo, et de omnibus auxiliis regum, vicecomitum, " et omn : miniftralium fuorum ; et hidagio et exercitibus, et " fcutagiis, et tallagiis, et fhiris et hundredis, et placitis et " querelis, et warda et wardpeny, et opibus caftellorum et pon-" tium, et claufuris parcorum, et omni carcio et fumagio, et " domor : regal : edificatione, et omnimoda reparatione, et cum " omnibus aliis libertatibus." This grant was made out by *Richard* bifhop of *Chichefter*, then chancellor, at the town of *Northampton*, before the lord chief jufticiary, who was the founder himfelf.

The *charter* of foundation of the *Priory*, dated 1233, comes next in order to be confidered ; but being of fome length, I fhall not interrupt my narrative by placing it here ; and therefore refer the reader to the Appendix, N° I. This my copy, taken from the original, I have compared with *Dugdale*'s copy, and find that they perfectly agree ; except that in the latter the preamble and the names of the witneffes are omitted. Yet I think it proper to quote a paffage from this charter—" Et ipfa domus religiofa *a cujuflibet* " *alterius domûs religiofæ fubjectione libera* permaneat, et in omnibus " *abfoluta*" — to fhew how much *Dugdale* was miftaken when he inferted *Selborne* among the *alien priories* ; forgetting that this difpofition of the convent contradicted the grant that he had

X x   publifhed.

publiſhed. In the *Monaſticon Anglicanum*, in Engliſh, p. 119, is part of his catalogue of *alien priories*, ſuppreſſed 2 *Henry* V. *viz.* 1414, where may be ſeen as follows,

S.

Sele, Suſſex.

SELEBURN.

Shirburn.

This appeared to me from the firſt to have been an overſight, before I had ſeen my authentic evidences. For *priories alien*, a few *conventual* ones excepted, were little better than *granges* to foreign abbies; and their priors little more than bailiffs, removeable at will: whereas the *priory* of *Selborne* poſſeſſed the valuable eſtates and manors of *Selborne, Achangre, Norton, Brompden, Baſſinges, Baſingſtoke,* and *Natele*; and the prior challenged the right of *Pillory, Thurcet,* and *Furcas,* and every manerial privilege.

I find next a grant from *Jo. de Venur,* or *Venuz,* to the prior of *Selborne*—" de tota mora [a moor or bog] ubi *Beme* oritur, uſque " ad campum vivarii, et de prato voc. *Sydenmeade* cum abutt: et " de curſu aque molendini." And alſo a grant in reverſion " unius virgate terre," [a yard land] in *Achangre* at the death of *Richard Actedene* his ſiſter's huſband, who had no child. He was to preſent a pair of gloves of one penny value to the prior and canons, to be given annually by the ſaid *Richard*; and to quit all claim to the ſaid lands in reverſion, provided the prior and canons would engage annually to pay to the king, through the hands of his bailiffs of *Aulton*, ten ſhillings at four quarterly payments, " pro omnibus ſerviciis, conſuetudinibus, exactionibus, et " demandis."

This

This *Jo. de Venur* was a man of property at *Oakhanger*, and lived probably at the ſpot now called *Chapel-farm*. The grant bears date the 17th year of the reign of *Henry* III. [*viz.* 1233.]

It would be tedious to enumerate every little grant for lands or tenements that might be produced from my vouchers. I ſhall therefore paſs over all ſuch for the preſent, and conclude this letter with a remark that muſt ſtrike every thinking perſon with ſome degree of wonder. No ſooner had a monaſtic inſtitution got a footing, but the neighbourhood began to be touched with a ſecret and religious awe. Every perſon round was deſirous to promote ſo good a work; and either by ſale, by grant, or by gift in reverſion, was ambitious of appearing a benefactor. They who had not lands to ſpare gave roads to accommodate the infant foundation. The religious were not backward in keeping up this pious propenſity, which they obſerved ſo readily influenced the breaſts of men. Thus did the more opulent monaſteries add houſe to houſe, and field to field; and by degrees manor to manor: till at laſt " there was no place left;" but every diſtrict around became appropriated to the purpoſes of their founders, and every precinct was drawn into the vortex.

LETTER

LETTER VIII.

OUR forefathers in this village were no doubt as bufy and buftling, and as important, as ourfelves: yet have their names and tranfactions been forgotten from century to century, and have funk into oblivion; nor has this happened only to the vulgar, but even to men remarkable and famous in their generation. I was led into this train of thinking by finding in my vouchers that Sir *Adam Gurdon* was an inhabitant of *Selborne*, and a man of the firft rank and property in the parifh. By Sir *Adam Gurdon* I would be underftood to mean that leading and accomplifhed malecontent in the *Mountfort* faction, who diftinguifhed himfelf by his daring conduct in the reign of *Henry* III. The firft that we hear of this perfon in my papers is, that with two others he was bailiff of *Alton* before the fixteenth of *Henry* III. *viz.* about 1231, and then not knighted. Who *Gurdon* was, and whence he came, does not appear: yet there is reafon to fufpect that he was ori-ginally a mere foldier of fortune, who had raifed himfelf by mar-rying women of property. The name of *Gurdon* does not feem to be known in the fouth; but there is a name fo like it in an adjoining kingdom, and which belongs to two or three noble families, that it is probable this remarkable perfon was a *North Briton*; and the more fo, fince the Chriftian name of *Adam* is a diftinguifhed one to this day among the family of the *Gordons*.—But, be this as it may, Sir *Adam Gurdon* has been noticed by all the writers of *Englifh* hiftory for his bold difpofition and difaffected

fpirit,

ſpirit, in that he not only figured during the ſucceſsful rebellion of *Leiceſter*, but kept up the war after the defeat and death of that baron's entrenching himſelf in the woods of *Hampſhire*, towards the town of *Farnham*. After the battle of *Eveſham*, in which *Mountfort* fell, in the year 1265, *Gurdon* might not think it ſafe to return to his houſe for fear of a ſurpriſe; but cautiouſly fortified himſelf amidſt the foreſts and woodlands with which he was ſo well acquainted. Prince *Edward*, deſirous of putting an end to the troubles which had ſo long haraſſed the kingdom, purſued the arch-rebel into his faſtneſſes; attacked his camp; leaped over the entrenchments; and, ſingling out *Gurdon*, ran him down, wounded him, and took him priſoner[a].

There is not perhaps in all hiſtory a more remarkable inſtance of command of temper, and magnanimity, than this before us: that a young prince, in the moment of victory, when he had the fell adverſary of the crown and royal family at his mercy, ſhould be able to withhold his hand from that vengeance which the vanquiſhed ſo well deſerved. A cowardly diſpoſition would have been blinded by reſentment: but this gallant heir-apparent ſaw at once a method of converting a moſt deſperate foe into a laſting friend. He raiſed the fallen veteran from the ground, he pardoned him, he admitted him into his confidence, and introduced him to the queen, then lying at *Guildford*, that very evening[a]. This unmerited and unexpected lenity melted the heart of the rugged *Gurdon* at once; he became in an inſtant a loyal and uſeful ſubject, truſted and employed in matters of moment by *Edward* when king, and confided in till the day of his death.

[a] M. Paris, p. 675. & Triveti Annale.

LETTER

LETTER IX.

IT has been hinted in a former letter that Sir *Adam Gurdon* had availed himself by marrying women of property. By my evidences it appears that he had three wives, and probably in the following order: *Constantia*, *Ameria*, and *Agnes*. The first of these ladies, who was the companion of his middle life, seems to have been a person of considerable fortune, which she inherited from *Thomas Makerel*, a gentleman of *Selborne*, who was either her father or uncle. The second, *Ameria*, calls herself the quondam wife of Sir *Adam*, " quæ fui uxor," &c. and talks of her sons under age. Now *Gurdon* had no son: and beside *Agnes* in another document says, " Ego *Agnes* quondam uxor Domini *Adæ Gurdon* in pura et " ligea viduitate mea:" but *Gurdon* could not leave two widows; and therefore it seems probable that he had been divorced from *Ameria*, who afterwards married, and had sons. By *Agnes* Sir *Adam* had a daughter *Johanna*, who was his heiress, to whom *Agnes* in her life-time surrendered part of her jointure:—he had also a bastard son.

Sir *Adam* seems to have inhabited the house now called *Temple*, lying about two miles east of the church, which had been the property of *Thomas Makerel*.

In the year 1262 he petitioned the prior of *Selborne* in his own name, and that of his wife *Constantia* only, for leave to build him an *oratory* in his manor-house, " in curia sua." Licenses of this sort were frequently obtained by men of fortune and rank from the bishop of the diocese, the archbishop, and sometimes, as I have

seen

S. H. Grimm del.

TEMPLE, in the Parish of SELBORNE.

Published Nov.r 1. 1790. as the Act directs, by B. White & Son.

D. Lerpinière sculp.

seen inftances, from the pope; not only for convenience-fake, and on account of diftance, and the badnefs of the roads, but as a matter of ftate and diftinction. Why the owner fhould apply to the *prior*, in preference to the *bifhop* of the *diocefe*, and how the former became competent to fuch a grant, I cannot fay; but that the priors of *Selborne* did take that privilege is plain, becaufe fome years afterward, in 1280, Prior *Richard* granted to *Henry Waterford* and his wife *Nicholaa* a licenfe to build an *oratory* in their court-houfe, " curia fua *de Waterford*," in which they might celebrate divine fervice, faving the rights of the mother church of *Bafynges*. Yet all the while the prior of *Selborne* grants with fuch referve and caution, as if in doubt of his power, and leaves *Gurdon* and his lady anfwerable in future to the bifhop, or his ordinary, or to the vicar for the time being, in cafe they fhould infringe the rights of the mother church of *Selborne*.

The manor-houfe called *Temple* is at prefent a fingle building, running in length from fouth to north, and has been occupied as a common farm houfe from time immemorial. The fouth end is modern, and confifts of a brew-houfe, and then a kitchen. The middle part is an hall twenty-feven feet in length, and nineteen feet in breadth; and has been formerly open to the top; but there is now a floor above it, and alfo a chimney in the weftern wall. The roofing confifts of ftrong maffive rafter-work ornamented with carved rofes. I have often looked for the *lamb* and *flag*, the arms of the *knights templars*, without fuccefs; but in one corner found a a fox with a goofe on his back, fo coarfely executed, that it re-quired fome attention to make out the device.

Beyond the hall to the north is a fmall parlour with a vaft heavy ftone chimney-piece; and, at the end of all, the *chapel* or *oratory*, whofe maffive thick walls and narrow windows at once befpeak

great

great antiquity. This room is only fixteen feet by fixteen feet eight inches; and full feventeen feet nine inches in height. The ceiling is formed of vaft joifts, placed only five or fix inches apart. Modern delicacy would not much approve of fuch a place of worfhip: for it has at prefent much more the appearance of a dungeon than of a room fit for the reception of people of condition. For the outfide I refer the reader to the plate, in which Mr. *Grimm* has reprefented it with his ufual accuracy. The field on which this oratory abuts is ftill called Chapel-field. The fituation of this houfe is very particular, for it ftands upon the immediate verge of a fteep abrupt hill.

Not many years fince this place was ufed for an hop-kiln, and was divided into two ftories by a loft, part of which remains at prefent, and makes it convenient for peat and turf, with which it is ftowed.

LETTER

Pl. IX.

J. H. Grimm del.

The Picture called Pitminster

P. Mazell Sculp

LETTER X.

THE Priory at times was much obliged to *Gurdon* and his family. As *Sir Adam* began to advance in years he found his mind influenced by the prevailing opinion of the reasonableness and efficacy of prayers for the dead; and, therefore, in conjunction with his wife *Constantia*, in the year 1271, granted to the prior and convent of *Selborne* all his right and claim to a certain place, *placea*, called *La Pleystow*, in the village aforesaid, " in *liberam*, *puram*, et *per-* " *petuam elemosinam.*" This *Pleystow* [b], *locus ludorum*, or play-place, is a level area near the church of about forty-four yards by thirty-six, and is known now by the name of the *Plestor* [c].

It continues still, as it was in old times, to be the scene of recreation for the youths and children of the neighbourhood; and impresses an idea on the mind that this village, even in *Saxon* times, could not be the most abject of places, when the inhabitants thought proper to asign so spacious a spot for the sports and amusements of it's young people [d].

[b] In *Saxon* Pleȝeȝtop, or Pleȝȝtop; viz. *Plegestow*, or *Plegstow*.

[c] At this juncture probably the vast oak, mentioned p. 5, was planted by the prior, as an ornament to his new acquired market place. According to this supposition the oak was aged 432 years when blown down.

[d] For more circumstances respecting the *Plestor*, see Letter II. to Mr. *Pennant*.

Y y

As

As soon as the prior became possessed of this piece of ground, he procured a charter for a *market* e from king *Henry* III. and began to erect houses and stalls, " *seldas*," around it. From this period *Selborne* became a market town : but how long it enjoyed that privilege does not appear. At the same time *Gurdon* reserved to himself, and his heirs, a way through the said *Plestor* to a tenement and some crofts at the upper end, abutting on the south corner of the church-yard. This was, in old days, the manerial house of the street manor, though now a poor cottage ; and is known at present by the modern name of *Elliot's*. *Sir Adam* also did, for the health of his own soul, and that of his wife *Constantia*, their predecessors and successors, grant to the prior and canons quiet possession of all the tenements and gardens, " *curtillagia*," which they had built and laid out on the lands in *Selborne*, on which he and his vassals, " *homines*," had undoubted right of common : and moreover did grant to the convent the full privilege of that right of common ; and empowered the religious to build tenements and make gardens along the king's highway in the village of *Selborne*.

From circumstances put together it appears that the above were the first grants obtained by the Priory in the village of *Selborne*, after it had subsisted about thirty-nine years : moreover they explain the nature of the mixed manor still remaining in and about the village, where one field or tenement shall belong to *Magdalen-*

e Bishop *Tanner*, in his *Notitia Monastica*, has made a mistake respecting the *market* and *fair* at *Selborne* : for in his references to *Dodsworth*, cart. 54 *Hen.* III. m. 3. he says, " *De mercatu, et feria de Seleburn*." But this reference is wrong ; for, instead of *Seleburn*, it proves that the place there meant was *Lekeborne*, or *Legeborne*, in the county of *Lincoln*. This error was copied from the index of the Cat. MSS. Angl. It does not appear that there ever was a *chartered* fair at *Selborne*. For several particulars respecting the present fair at *Selborne* see Letter XXVI. of these Antiquities.

college

college in the univerſity of *Oxford*, and the next to *Norton Powlet*, eſq. of *Rotherfield* houſe ; and ſo down the whole ſtreet. The caſe was, that the whole was once the property of *Gurdon*, till he made his grants to the convent; ſince which ſome belongs to the ſucceſſors of *Gurdon* in the manor, and ſome to the college; and this is the occaſion of the ſtrange jumble of property. It is remarkable that the tenement and crofts which Sir *Adam* reſerved at the time of granting the Pleſtor ſhould ſtill remain a part of the *Gurdon-manor*, though ſo deſirable an addition to the vicarage that is not as yet poſſeſſed of one inch of glebe at home : but of late, *viz.* in *January* 1785, *Magdalen-college* purchaſed that little eſtate, which is life-holding, in reverſion, for the generous purpoſe of beſtowing it, and it's lands, being twelve acres (three of which abut on the church-yard and vicarage-garden) as an improvement hereafter to the living, and an eligible advantage to future incumbents.

The year after *Gurdon* had beſtowed the *Pleſtor* on the Priory, *viz.* in 1272, *Henry* III. king of *England* died, and was ſucceeded by his ſon *Edward*. This magnanimous prince continued his regard for Sir *Adam*, whom he eſteemed as a brave man, and made him *warden*, " *cuſtos*," of the foreſt of *Wolmer*[f]. Though little emolument

f Since the letters reſpecting *Wolmer-foreſt* and *Ayles-holt*, from p. 14 to 26, were printed, the author has been favoured with the following extracts :

In the " Act of Reſumption, 1 Hen. VII." it was provided, that it be not prejudicial to " *Harry at Lode*, ranger of our foreſt of *Wolmere*, to him by oure letters patents " before tyme gevyn." Rolls of Parl. Vol. VI. p. 370.

In the 11 Hen. VII. 1495—" *Warlham* [Wardleham] and the office of foreſt [foreſter] of *Wolmere*" were held by *Edmund* duke of *Suffolk*.—Rolls,. ib. 474.

Act of general pardon, 14 *Hen.* VIII. 1523, not to extend to " *Rich.* Bp. of *Wynton* " [biſhop *Fox*] for any ſeizure or forfeiture of liberties, &c. within the foreſt of *Wolmer*, " *Alyſholt*, lument might hang to this appointment, yet are there reasons why it might be highly acceptable; and, in a few reigns after, it was given to princes of the blood [g]. In old days gentry resided more at home on their estates, and, having fewer resources of elegant in-door amusement, spent most of their leisure hours in the field

" *Alysholt*, and *Newe Forest*; nor to any person for waste, &c. within the manor of *Ward-* " *lam*, or parish of *Wardlam* [*Wardleham*]; nor to abusing, &c. of any office or fee, " within the said forests of *Wolmer* or *Alysholt*, or the said park of *Wardlam*."—County Suth't.   Rolls prefixt to 1st Vol. of Journals of the Lords, p. xciii. b.

To these may be added some other particulars, taken from a book lately published, entitled " An Account of all the Manors, Messuages, Lands, &c. in the different " Counties of *England* and *Wales*, held by Lease from the Crown; as contained in the " Report of the Commissioners appointed to inquire into the State and Condition of the " Royal Forests," &c.——London, 1787.

" Southampton."

P. 64, " A fee-farm rent of $\begin{array}{ccc} \pounds. & s. & d. \\ 31 & 2 & 11 \end{array}$ out of the manors of East and West " *Wardleham*; and also the office of *lieutenant* or *keeper* of the forest or chase of *Aliceholt* " and *Wolmer*, with all offices, fees, commodities, and privileges thereto belonging.
" Names of lessees, *William* earl of *Dartmouth* and others (in trust.)
" Date of the last lease, March 23, 1780; granted for such term as would fill up the " subsisting term to 31 years.
" Expiration March 23, 1811."

" Appendix, N°. III."
" Southampton."
" Hundreds—Selborne and Finchdeane."
" Honours and manors," &c.

" *Aliceholt* forest, three parks there.
" *Bensted* and *Kingsley*; a petition of the parishioners concerning the three parks in " *Aliceholt* forest."
William, first earl of *Dartmouth*, and paternal grandfather to the present lord *Stawel*, was a lessee of the forests of *Aliceholt* and *Wolmer* before brigadier-general *Emanuel Scroope Howe*.

[g] See Letter II. of these Antiquities.

and

and the pleasures of the chase. A large domain, therefore, at little more than a mile distance, and well stocked with game, must have been a very eligible acquisition, affording him influence as well as entertainment; and especially as the manerial house of *Temple*, by its exalted situation, could command a view of near two-thirds of the forest.

That *Gurdon*, who had lived some years the life of an outlaw, and at the head of an army of insurgents, was, for a considerable time, in high rebellion against his sovereign, should have been guilty of some outrages, and should have committed some depredations, is by no means matter of wonder. Accordingly we find a *distringas* against him, ordering him to restore to the bishop of *Winchester* some of the temporalities of that see, which he had taken by violence and detained; viz. some lands in *Hocheleye*, and a mill [h]. By a *breve*, or writ, from the king he is also enjoined to readmit the bishop of *Winchester*, and his tenants of the parish and town of *Farnham*, to pasture their horses, and other *larger* cattle, " *averia*," in the forest of *Wolmer*, as had been the usage from time immemorial. This writ is dated in the tenth year of the reign of *Edward*, viz. 1282.

All the king's writs directed to *Gurdon* are addressed in the following manner: " *Edwardus*, Dei gratia, &c. dilecto et fideli " suo *Ade Gurdon* salutem;" and again, " Custodi foreste sue de " *Wolvemere*."

In the year 1293 a quarrel between the crews of an *English* and a *Norman* ship, about some trifle, brought on by degrees such serious consequences, that in 1295 a war broke out between the

[h] *Hocheleye*, now spelt *Hawkley*, is in the hundred of *Selborne*, and has a mill at this day.

two nations. The *French* king, *Philip the Hardy*, gained some advantages in *Gascony*; and, not content with those, threatened *England* with an invasion, and, by a sudden attempt, took and burnt *Dover*.

Upon this emergency *Edward* sent a writ to *Gurdon*, ordering him and four others to enlist three thousand soldiers in the counties of *Surrey*, *Dorset*, and *Wiltshire*, able-bodied men, " tam sagittare " quam balistare potentes ;" and to see that they were marched, by the feast of *All Saints*, to *Winchelsea*, there to be embarked aboard the king's transports.

The occasion of this armament appears also from a summons to the bishop of *Winchester* to parliament, part of which I shall transcribe on account of the insolent menace which is said therein to have been denounced against the *English* language :—" qualiter " rex *Franciæ* de terra nostra *Gascon* nos fraudulenter et cautelose " decepit, eam nobis nequiter detinendo . . . vero predictis " fraude et nequitia non contentus, ad expugnationem regni " nostri classe maxima et bellatorum copiosa multitudine congre- " gatis, cum quibus regnum nostrum et regni ejusdem incolas " hostiliter jam invasurus, *linguam Anglicam*, si concepte iniquitatis " proposito detestabili potestas correspondeat, quod Deus avertat, " *omnino de terra delere proponit.*" Dated 30th September, in the year of king *Edward*'s reign xxiii [1].

The above are the last traces that I can discover of *Gurdon*'s appearing and acting in public. The first notice that my evidences give of him is, that, in 1232, being the 16th of *Henry* III. he was the king's bailiff, with others, for the town of *Alton*. Now,

[1] Reg. *Wynton, Stratford,* but query *Stratford*; for *Stratford* was not bishop of *Winton* till 1323, near thirty years afterwards.

from

from 1232 to 1295 is a space of sixty-three years; a long period for one man to be employed in active life! Should any one doubt whether all these particulars can relate to one and the same person, I should wish him to attend to the following reasons why they might. In the first place, the documents from the priory mention but one Sir *Adam Gurdon*, who had no son lawfully begotten: and in the next, we are to recollect that he must have probably been a man of uncommon vigour both of mind and body; since no one, unsupported by such accomplishments, could have engaged in such adventures, or could have borne up against the difficulties which he sometimes must have encountered: and, moreover, we have modern instances of persons that have maintained their abilities for near that period.

Were we to suppose *Gurdon* to be only twenty years of age in 1232, in 1295 he would be eighty-three; after which advanced period it could not be expected that he should live long. From the silence, therefore, of my evidences it seems probable that this extraordinary person finished his life in peace, not long after, at his mansion of *Temple*. *Gurdon's* seal had for its device —a man, with an helmet on his head, drawing a cross-bow; the legend, " *Sigillum Ade de Gurdon* ;" his arms were, " Goulis, " iii floures argent issant de testes de leopards [k]."

If the stout and unsubmitting spirit of *Gurdon* could be so much influenced by the belief and superstition of the times, much more might the hearts of his ladies and daughter. And accordingly we find that *Ameria*, by the consent and advice of her sons, though said to be all under age, makes a grant for ever of some lands

[k] From the collection of *Thomas Martin*, Esq. in the Antiquarian Repertory, p. 109, No. XXXI.

down

down by the ſtream at *Durton*; and alſo of her right of the common
of *Durton* itſelf [1]. *Johanna*, the daughter and heireſs of Sir *Adam*,
was married, I find, to *Richard Achard*; ſhe alſo grants to the prior
and convent lands and tenements in the village of *Selborne*, which
her father obtained from *Thomas Makerel*; and alſo all her goods
and chattels in *Selborne* for the conſideration of two hundred
pounds ſterling. This laſt buſineſs was tranſacted in the firſt year
of *Edward* II. viz. 1307. It has been obſerved before that *Gurdon*
had a natural ſon: this perſon was called by the name of *John
Daſtard*, alias *Waſtard*, but more probably *Baſtard*; ſince baſtardy
in thoſe days was not eſteemed any diſgrace, though daſtardy
was eſteemed the greateſt. He was married to *Gunnorie Duncun*;
and had a tenement and ſome land granted him in *Selborne* by his
ſiſter *Johanna*.

---

[1] *Durton*, now called *Dorton*, is ſtill a common for the copyholders of *Selborne*
manor.

LETTER

LETTER XI.

THE *Knights Templars* [m], who have been mentioned in a former letter, had confiderable property in *Selborne*; and alfo a *preceptory* at *Sudington*, now called *Southington*, a hamlet lying one mile to the

The MILITARY ORDERS of the RELIGIOUS.

[m] The *Knights Hofpitalars* of *St. John of Jerufalem*, afterwards called *Knights of Rhodes*, now of *Malta*, came into *England* about the year 1100. 1 *Hen.* I.

The *Knights Templars* came into *England* pretty early in *Stephen*'s reign, which commenced 1135. The order was diffolved in 1312, and their eftates given by act of Parliament to the *Hofpitalars* in 1323. (all in *Edw.* II.) though many of their eftates were never actually enjoyed by the faid *Hofpitalars*. *Vid. Tanner*, p. xxiv. x.

The *commandries* of the *Hofpitalars*, and *preceptories* of *Templars*, were each fubordinate to the principal houfe of their refpective religion in *London*. Although thefe are the different denominations, which *Tanner* at p. xxviii. affigns to the cells of thefe different orders, yet throughout the work very frequent inftances occur of *preceptories* attributed to the *Hofpitalars*; and if in fome paffages of *Notitia Monaft. commandries* are attributed to the *Templars*, it is only where the place afterwards became the property of the *Hofpitalars*, and fo is there indifferently ftyled *preceptory* or *commandry*; fee p. 243, 263, 276, 577, 678. But, to account for the firft obferved inaccuracy, it is probable the *preceptories* of the *Templars*, when given to the *Hofpitalars*, were ftill vulgarly, however, called by their old name of *preceptories*; whereas in propriety the focieties of the *Hofpitalars* were indeed (as has been faid) *commandries*. And fuch deviation from the ftrictnefs of expreffion in this cafe might occafion thofe focieties of *Hofpitalars* alfo to be indifferently called *preceptories*, which had *originally* been vefted in *them*, having never belonged to the *Templars* at all.—See in *Archer*, p. 609. *Tanner*, p. 300. col. 1. 720. note *e*.

Z z It

the east of the village. Bishop *Tanner* mentions only two such houses of the *Templars* in all the county of *Southampton*, viz. *Godesfield*, founded by *Henry de Blois*, bishop of *Winchester*, and *South Badeisley*, a preceptory of the *Knights Templars*, and afterwards of *St. John of Jerusalem*, valued at one hundred and eighteen pounds sixteen shillings and seven pence *per annum*. Here then was a *preceptory* unnoticed by antiquaries, between the village and *Temple*. Whatever the edifice of the *preceptory* might have been, it has long since been dilapidated; and the whole hamlet contains now only one mean farm-house, though there were two in the memory of man.

It has been usual for the religious of different orders to fall into great dissensions, and especially when they were near neighbours. Instances of this sort we have heard of between the monks of

It is observable that the very statute for the dissolution of the *Hospitalars* holds the same language; for there, in the enumeration of particulars, occur " *commandries, pre-* " *ceptories.*" *Codex*, p. 1190. Now this intercommunity of names, and that in an act of parliament too, made some of our ablest antiquaries look upon a *preceptory* and *commandry* as strictly synonymous; accordingly we find *Camden*, in his *Britannia*, explaining *praceptoria* in the text by a *commandry* in the margin, p. 356. 510.

J. L.

Commandry, a manor or chief messuage with lands, &c. belonging to the priory of *St. John of Jerusalem*; and he who had the government of such house was called the *commander*, who could not dispose of it but to the use of the priory, only taking thence his own sustenance, according to his degree, who was usually a brother of the same priory. *Cowell*. He adds (confounding these with *preceptories*) they are in many places termed *Temples*, as *Temple Bruere* in *Lincolnshire*, &c. *Preceptories* were possessed by the more eminent sort of *Templars*, whom the chief master created and called *Praceptores Templi*. *Cowell*, who refers to *Stephens* de Jurisd. lib. 4. c. 10. num. 27.

Placita de juratis et assis coram Salom. de Roff et sociis suis justic. Itiner. apud Wynton. &c. anno regni R. Edwardi fil. Reg. Hen. octavo.—"et Magr. Milicie Tem-" pli in Angl. ht emendasse panis, & suis [cerevisiæ] in *Sodington*, & nescint q°. war. et "—et magist. Milicie Templi nōn vēn iō distr." *Chapter-house, Westminster.*

Canterbury;

Canterbury; and again between the old abbey of St. *Swythun*, and the comparatively new minfter of *Hyde* in the city of *Winchefter*ⁿ. These feuds arofe probably from different orders being crowded within the narrow limits of a city, or garrifon-town, where every inch of ground was precious, and an object of contention. But with us, as far as my evidences extend, and while *Robert Saunford* was *mafter*°, and *Richard Carpenter* was *preceptor*, the *Templars* and the *Priors* lived in an intercourfe of mutual good offices.

My papers mention three tranfactions, the exact time of which cannot be afcertained, becaufe they fell out before dates were

Notitia Monaftica, p. 155.

ⁿ " *Winchefter, Newminfter.* King *Alfred* founded here firft only a houfe and chapel " for the learned monk *Grimbald*, whom he had brought out of *Flanders*: but after- " wards projected, and by his will ordered, a noble church or religious houfe to be built " in the cemetery on the north fide of the old minfter or cathedral; and defigned that *Grim-* " *bald* fhould prefide over it.   This was begun A. D. 901, and finifhed to the honour " of the *Holy Trinity, Virgin Mary,* and *St. Peter,* by his fon king *Edward,* who placed " therein fecular canons : but A. D. 963 they were expelled, and an abbot and monks " put in poffeffion by bifhop *Ethelwold.*

" Now the churches and habitations of thefe two focieties being fo very near together, " the differences which were occafioned by their finging, bells, and other matters, arofe " to fo great a height, that the religious of the new monaftery thought fit, about A. D. " 1119, to remove to a better and more quiet fituation without the walls, on the north " part of the city called HYDE, where king *Henry* I. at the inftance of *Will. Gifford,* " bifhop of *Winton,* founded a ftately abbey for them. St. *Peter* was generally accounted " patron; though it is fometimes called the monaftery of St. *Grimbald,* and fometimes " of St. *Barnabas,*" &c.

Note.   A few years fince a county bridewell, or houfe of correction, has been built on the immediate fite of *Hide Abbey*.   In digging up the old foundations the workmen found the head of a crofier in good prefervation.

° *Robert Saunforde* was *mafter* of the *Temple* in 1241; *Guido de Forefta* was the next in 1292.   The former is fifth in a lift of the *mafters* in a MS. *Bib. Cotton. Nero.* E. VI.

ufually

usually inserted; though probably they happened about the middle of the thirteenth century, not long after *Saunford* became *master*. The first of these is that the *Templars* shall pay to the priory of *Selborne*, annually, the sum of ten shillings at two half yearly payments from their chamber, " *camera*," at *Sudington*, " per manum " *preceptoris*, vel *ballivi* nostri, qui pro tempore fuerit ibidem, till they can provide the *prior* and canons with an equivalent in lands or rents within four or five miles of the said convent. It is also further agreed that, if the *Templars* shall be in arrears for one year, that then the *prior* shall be empowered to distrain upon their live stock in *Bradeseth*. The next matter was a grant from *Robert de Saunford* to the *priory* for ever, of a good and sufficient road, " *cheminum*," capable of admitting carriages, and proper for the drift of their larger cattle, from the way which extends from *Sudington* towards *Blakemere*, on to the lands which the convent possesses in *Bradeseth*.

The third transaction (though for want of dates we cannot say which happened first and which last) was a grant from *Robert Samford* to the priory of a tenement and its appurtenances in the village of *Selborne*, given to the *Templars* by *Americus de Vasci* [p]. This property, by the manner of describing it,—" totum tene- " mentum cum omnibus pertinentiis suis, scilicet in terris, & " *hominibus*, in pratis & pascuis, & nemoribus," &c. seems to have been no inconsiderable purchase, and was sold for two hundred marks sterling, to be applied for the buying of more land for the support of the holy war.

---

[p] *Americus Vasci*, by his name, must have been an *Italian*, and had been probably a soldier of fortune, and one of *Gurdon*'s captains. *Americus Vespucio*, the person who gave name to the new world, was a *Florentine*.

Prior

Prior *John* is mentioned as the person to whom *Vasci's* land is conveyed. But in *Willis's* list there is no prior *John* till 1339, several years after the dissolution of the order of the *Templars* in 1312; so that unless *Willis* is wrong, and has omitted a prior *John* since 1262, (that being the date of his first prior) these transactions must have fallen out before that date.

I find not the least traces of any concerns between *Gurdon* and the *Knights Templars*; but probably after his death his daughter *Johanna* might have, and might bestow, *Temple* on that order in support of the holy land: and, moreover, she seems to have been moving from *Selborne* when she sold her goods and chattels to the priory, as mentioned above.

Temple no doubt did belong to the knights, as may be asserted, not only from it's name, but also from another corroborating circumstance of it's being still a manor *tithe-free*; " for, by virtue " of their order," says Dr. *Blackstone*, " the lands of the *Knights* " *Templars* were privileged by the pope with a discharge from " tithes."

Antiquaries have been much puzzled about the terms *preceptores* and *preceptorium*, not being able to determine what officer or edifice was meant. But perhaps all the while the passage quoted above from one of my papers " per manum *preceptoris* vel *ballivi* " nostri, qui pro tempore fuerit ibidem," may help to explain the difficulty. For if it be allowed here that *preceptor* and *ballivus* are synonymous words, then the brother who took on him that office resided in the house of the *Templars* at *Sudington*, a *preceptory*; where he was their *preceptor*, superintended their affairs, received their money; and, as in the instance there mentioned, paid from their chamber, " *camera*," as directed: so that, according to this
explanation,

explanation, a *preceptor* was no other than a steward, and a *preceptorium* was his residence. I am well aware that, according to strict Latin, the *vel* should have been *seu* or *five*, and the order of the words "*preceptoris nostri*, vel ballivi, qui" — et "*ibidem*" should have been *ibi*; *ibidem* necessarily having reference to *two* or more persons: but it will hardly be thought fair to apply the niceties of classic rules to the Latinity of the thirteenth century, the writers of which seem to have aimed at nothing farther than to render themselves intelligible.

There is another remark that we have made, which, I think, corroborates what has been advanced; and that is, that *Richard Carpenter*, *preceptor* of *Sudington*, at the time of the transactions between the *Templars* and *Selborne* Priory, did always sign *last* as a witness in the three deeds: he calls himself *frater*, it is true, among many other brothers, but subscribes with a kind of deference, as if, for the time being, his *office* rendered him an inferior in the community q.

---

p In two or three ancient records relating to St. *Oswald's* hospital in the city of *Worcester*, printed by Dr. *Nash*, p. 227 and 228, of his collections for the history of *Worcestershire*, the words *preceptorium* and *preceptoria* signify the *mastership* of the said hospital: " ad *preceptorium* five *magisterium* presentavit—*preceptorii* five *magisterii* patro- " nus. Vacavit dicta *preceptoria* seu *magisterium*—ad *preceptoriam* et regimen dicti " hospitalis---Te *preceptorem* five magistrum prefecimus."

Where *preceptorium* denotes a building or apartment it may probably mean the master's lodgings, or at least the *preceptor's* apartment, whatsoever may have been the office or employment of the said preceptor.

A *preceptor* is mentioned in *Thoresby's Ducatus Leodiensis*, or History of *Leeds*, p. 225, and a deed witnessed by the *preceptor* and chaplain before dates were inserted. --- *Du Fresne's* Supplement : " *Preceptoriae*, prædia *preceptoribus* assignata."---*Cowel*, in his Law Dictionary, enumerates sixteen *preceptoriae*, or *preceptories*, in *England*; but *Sudington* is not among them.—It is remarkable that *Gurtlerus*, in his *Historia Templariorum* Amstel. 1691, never once mentions the words *preceptor* or *preceptorium*.

LETTER

LETTER XII.

THE ladies and daughter of Sir *Adam Gurdon* were not the only benefactreffes to the Priory of *Selborne*; for, in the year 1281, *Ela Longfpee* obtained maffes to be performed for her foul's health; and the prior entered into an engagement that one of the convent fhould every day fay a fpecial mafs for ever for the faid bene- factrefs, whether living or dead. She alfo engaged within five years to pay to the faid convent one hundred marks of filver for the fupport of a *chantry* and *chantry-chaplain*, who fhould perform his maffes daily in the parifh church of *Selborne* r. In the eaft end of the fouth aile there are two fharp-pointed gothic niches; one of thefe probably was the place under which thefe maffes were per- formed; and there is the more reafon to fuppofe as much, becaufe, till within thefe thirty years, this fpace was fenced off with *gothic* wooden railing, and was known by the name of the fouth chancel s.

The folicitude expreffed by the donor plainly fhews her piety and firm perfuafion of the efficacy of prayers for the dead; for

r A *chantry* was a chapel joined to fome cathedral or parifh church, and endowed with annual revenues for the maintenance of one or more priefts to fing mafs daily for the foul of the founder, and others.

s For what is faid more refpecting this chantry fee Letter III. of thefe Antiquities.---- Mention is made of a *Nicholas Langrifh*, capellanus de *Selborne*, in the time of *Henry* VIII. Was he chantry-chaplain to *Ela Longfpee*, whofe maffes were probably continued to the time of the reformation? More will be faid of this perfon hereafter.

fhe

she seems to have made every provision for the payment of the sum stipulated within the appointed time; and to have felt much anxiety lest her death, or the neglect of her executors or assigns, might frustrate her intentions.—" Et si contingat me in solucione
" predicte pecunie annis predictis in parte aut in toto deficere,
" quod absit; concedo et obligo pro me et assignatis meis, quod
" *Vice-Comes - - - Oxon* et - - - - - qui pro tempore fuerint, per omnes
" terras et tenementa, et omnia bona mea mobilia et immobilia
" ubicunque in balliva sua fuerint inventa ad solucionem predictam
" faciendam possent nos compellere."   And again—" Et si con-
" tingat dictos religiosos labores seu expensas facere circa pre-
" dictam pecuniam, seu circa partem dicte pecunie; volo quod
" dictorum religiosorum impense et labores levantur ita quod pre-
" dicto priori vel uni canonicorum suorum superhiis simplici
" verbo credatur sine alterius honere probacionis; et quod utrique
" predictorum virorum in unam marcam argenti pro cujuslibet
" distrincione super me facienda tenear. — Dat. apud *Wareborn die*
" *sabati* proxima ante festum *St. Marci* evangeliste, anno regni regis
" *Edwardi* tertio decimo [t]."

But the reader perhaps would wish to be better informed respect-ing this benefactress, of whom as yet he has heard no particulars.

The *Ela Longspee* therefore above-mentioned was a lady of high birth and rank, and became countess to *Thomas de Newburgh*, the sixth earl of *Warwick*: she was the second daughter of the famous *Ela Longspee* countess of *Salisbury*, by *William Longspee*, natural son of king *Henry* II. by *Rosamond*.

[t] Ancient deeds are often dated on a *Sunday*, having been executed in churches and church-yards for the sake of notoriety, and for the conveniency of procuring several witnesses to attest.

Our

Our lady, following the steps of her illustrious mother[u], " was
" a great benefactress to the university of *Oxford*, to the canons
" of *Oseney*, the nuns of *Godstow*, and other religious houses in
" *Oxfordshire*. She died very aged in the year 1300[x], and was
" buried before the high altar in the abbey church of *Oseney*,
" at the head of the tomb of *Henry D'Oily*, under a flat marble,
" on which was inlaid her portraiture, in the habit of a vowess,
" engraved on a copper-plate." —— *Edmondson's History and
Genealogical Account of the Grevilles*, p. 23.

[u] *Ela Longspee*, countess of *Salisbury*, in 1232 founded a monastery at *Lacock*, in
the county of *Wilts*, and also another at *Hendon*, in the county of *Somerset*, in her widow-
hood, to the honour of the Blessed Virgin and St. *Bernard*. CAMDEN.

[x] Thus she survived the foundation of her chantry at *Selborne* fifteen years. About
this lady and her mother consult *Dugdale's* Baronage, I. 72, 175, 177.——*Dugdale's* War-
wickshire, I. 383,---*Leland's* Itin. II. 45.

LETTER

LETTER XIII.

THE reader is here prefented with five forms refpecting the chufing of a prior; but as they are of fome length they muft be referved for the Appendix; their titles are N°. 108. " Charta " petens licentiam elegendi *prelatum* a Domino epifcopo *Wintoni-* " *enfi*: — " Forma licentie conceffe :" — " Forma deereti poft " electionem conficiendi :"—108. " Modus procedendi ad elec- " tionem per formam fcrutinii :" — et " Forma ricte prefentandi " electum." Such evidences are rare and curious, and throw great light upon the general *monaftico-ecclefiaftical* hiftory of this kingdom, not yet fufficiently underftood.

In the year 1324 there was an election for a prior at *Selborne*; when fome difficulties occurring, and a devolution taking place, application was made to *Stratford*, who was bifhop of *Winchefter* at that time, and of courfe the vifitor and patron of the convent at the fpot above-mentioned y.

An Extract from REG. STRATFORD. *Winton.*

P. 4. " Commiffio facta fub-priori de *Selebourne*" by the bifhop enjoining him to preferve the difcipline of the order in the con- vent during the vacancy made by the late death of the prior,

---

y *Stratford* was bifhop of *Winchefter* from 1323 to 1333, when he was tranflated to *Canterbury*.

(" nuper

("" nuper paſtoris ſolatio deſtituta,"") dated 4th. kal. Maii. ann. 2do ſc. of his conſecration. [ſc. 1324.]

P. 6. "" Cuſtodia Prioratus de *Seleburne* vacantis,"" committed by the biſhop to *Nicholas de la* - - -, a layman, it belonging to the biſhop "" ratione vacationis ejuſdem,"" in *July* 1324, ibid. "" nego-"" tium electionis de *Selebourne*. Acta coram *Johanne* Epiſcopo, &c. "" 1324 in negotio electionis de fratre *Waltero de Inſula* concanonico "" prioratus de *Selebourne*,"" lately elected by the ſub-prior and convent, by way of ſcrutiny: that it appeared to the biſhop, by certificate from the dean of *Alton*, that ſolemn citation and procla-mation had been made in the church of the convent where the election was held that any who oppoſed the ſaid election or elected ſhould appear.—Some difficulties were ſtarted, which the biſhop over-ruled, and confirmed the election, and admitted the new prior *ſub hac forma* :—

"" In Dei nomine Amen. Ego *Johannes* permiſſione divina, &c. "" te *Walterum de Inſula* eccleſie de *Selebourne* noſtre dioceſeos "" noſtrique patronatus vacantis, canonicum et cantorem, virum "" utique providum, et diſcretum, literarum ſcientia preditum, "" vita moribus et converſatione merito commendatum, in ordine "" ſacerdotali et etate legitima conſtitutum, de legitimo matrimonio "" procreatum, in ordine et religione *Sancti Auguſtini de Selebourne* "" expreſſe profeſſum, in ſpiritualibus et temporalibus circumſpec-"" tum, *jure* nobis hac vice *devoluto* in hac parte, in dicte eccleſie "" de *Selebourne* perfectum priorem; curam et adminiſtrationem "" ejuſdem tibi in ſpiritualibus et temporalibus committentes. Dat. "" apud *Selebourne* XIII kalend. Auguſti anno ſupradicto.""

There

There follows an order to the sub-prior and convent pro obedientia :

A mandate to *Nicholas* above-named to release the Priory to the new prior :

A mandate for the induction of the new prior.

LETTER XIV.

" In the year 1373 *Wykeham*, bishop of *Winchester*, held a visitation
" of his whole diocese ; not only of the secular clergy through
" the several deaneries, but also of the monasteries, and religious
" houses of all sorts, which he visited in person. The next year
" he sent his commissioners with power to correct and reform the
" several irregularities and abuses which he had discovered in the
" course of his visitation.

" Some years afterward, the bishop having visited *three several*
" *times* all the religious houses throughout his diocese, and being
" well informed of the state and condition of each, and of the
" particular abuses which required correction and reformation,
" besides the orders which he had already given, and the remedies
" which he had occasionally applied by his commissioners, now
" issued his injunctions to each of them. They were accommo-
" dated to their several exigencies, and intended to correct the
" abuses introduced, and to recall them all to a strict observation
" of

" of the rules of their respective orders. Many of these injunc-
" tions are still extant, and are evident monuments of the care
" and attention with which he discharged this part of his episcopal
" duty [z]."

Some of these injunctions I shall here produce; and they are
such as will not fail, I think, to give satisfaction to the antiquary,
both as never having been published before, and as they are a
curious picture of monastic irregularities at that time.

The documents that I allude to are contained in the *Notabilis
Visitatio de Seleburne*, held at the Priory of that place, by *Wykeham*
in person, in the year 1387.

This evidence, in the original, is written on two skins of parchment; the one large, and the other smaller, and consists of a *preamble*, 36 *items*, and a *conclusion*, which altogether evince the
patient investigation of the visitor, for which he had always been
so remarkable in all matters of moment, and how much he had
at heart the regularity of those institutions, of whose efficacy in
their prayers for the dead he was so firmly persuaded. As the
bishop was so much in earnest, we may be assured that he had
nothing in view but to correct and reform what he found amiss;
and was under no bias to blacken, or misrepresent, as the commissioners of *Thomas* Lord *Cromwell* seem in part to have done at
the time of the reformation [a]. We may therefore with reason
suppose that the bishop gives us an exact delineation of the
morals and manners of the canons of *Selborne* at that juncture; and
that what he found they had omitted he enjoins them; and for
what they have done amiss, and contrary to their rules and

[z] See *Lowth's* Life of *Wykeham*.

[a] Letters of this sort from Dr. *Layton* to *Thomas* Lord *Cromwell* are still extant.

statutes,

statutes, he reproves them; and threatens them with punishment suitable to their irregularities.

This *visitatio* is of confiderable length, and cannot be introduced into the body of this work; we fhall therefore refer the reader to the *Appendix*, where he will find every particular, while we fhall take fome notice, and make fome remarks, on the moft fingular *items* as they occur.

In the preamble the vifitor fays—" Confidering the charge " lying upon us, that your blood may not be required at our " hands, we came down to vifit your Priory, as our office re- " quired : and every time we repeated our vifitation we found " fomething ftill not only contrary to regular rules but alfo re- " pugnant to religion and good reputation."

In the firft article after the preamble—" he commands them " on their obedience, and on pain of the greater excommunication, " to fee that the canonical hours by night and by day be fung in " their choir, and the maffes of the Bleffed *Mary*, and other " accuftomed maffes, be celebrated at the proper hours with devo- " tion, and at moderate paufes ; and that it be not allowed to " any to abfent themfelves from the hours and maffes, or to " withdraw before they are finifhed."

Item 2d. He enjoins them to obferve that filence to which they are fo ftrictly bound by the rule of Saint *Auguftine* at ftated times, and wholly to abftain from frivolous converfation.

Item 4th. " Not to permit fuch frequent paffing of fecular " people of both fexes through their convent, as if a thorough- " fare, from whence many diforders may and have arifen."

Item 5th. " To take care that the doors of their church and " Priory be fo attended to that no fufpected and diforderly " females, ' fufpectæ et aliæ inhoneftæ,' pafs through their choir

" and

" and cloister in the dark;" and to see that the doors of their church between the nave and the choir, and the gates of their cloister opening into the fields, be constantly kept shut until their first choir-service is over in the morning, at dinner time, and when they meet at their evening collation [b].

Item 6th mentions that several of the canons are found to be very ignorant and illiterate, and enjoins the prior to see that they be better instructed by a proper master.

Item 8th. The canons are here accused of refusing to accept of their statutable clothing year by year, and of demanding a certain specified sum of money, as if it were their annual rent and due. This the bishop forbids, and orders that the canons shall be clothed out of the revenue of the Priory, and the old garments be laid by in a chamber and given to the poor, according to the rule of Saint *Augustine.*

In *Item* 9th is a complaint that some of the canons are given to wander out of the precincts of the convent without leave; and that others ride to their manors and farms, under pretence of inspecting the concerns of the society, when they please, and stay as long as they please. But they are enjoined never to stir either about their own private concerns or the business of the convent without leave from the prior: and no canon is to go alone, but to have a grave brother to accompany him.

The injunction in *Item* 10th, at this distance of time, appears rather ludicrous; but the visitor seems to be very serious on the occasion, and says that it has been evidently proved to him that some of the canons, living dissolutely after the flesh, and not after the spirit, sleep naked in their beds without their breeches

---

[b] A collation was a meal or repast on a fast day in lieu of a supper.

and

and shirts, " absque femoralibus et camisiis[c]." He enjoins that these culprits shall be punished by severe fasting, especially if they shall be found to be faulty a third time; and threatens the prior and sub-prior with suspension if they do not correct this enormity.

In *Item* 11th the good bishop is very wroth with some of the canons, whom he finds to be professed hunters and sportsmen, keeping hounds, and publicly attending hunting-matches. These pursuits, he says, occasion much dissipation, danger to the soul and body, and frequent expense; he, therefore, wishing to extirpate this vice wholly from the convent, " *radicibus extirpare*," does absolutely enjoin the canons never intentionally to be present at any public noisy tumultuous huntings; or to keep any hounds, by themselves or by others, openly or by stealth, within the convent, or without[d].

In *Item* 12th he forbids the canons in office to make their business a plea for not attending the service of the choir; since by these means either divine worship is neglected or their brother-canons are over-burdened.

By *Item* 14th we are informed that the original number of canons at the Priory of *Selborne* was *fourteen*; but that at this visitation they were found to be let down to *eleven*. The visitor therefore strongly

[c] The rule alluded to in *Item* 10th, of not sleeping naked, was enjoined the *Knights Templars*, who also were subject to the rules of St. *Augustine*.

See *Gurtleri Hist. Templariorum.*

[d] Considering the strong propensity in human nature towards the pleasures of the chase, it is not to be wondered that the canons of *Selborne* should languish after hunting, when, from their situation so near the precincts of *Wolmer-forest*, the king's hounds must have been often in hearing, and sometimes in sight from their windows.----If the bishop was so offended at these sporting canons, what would he have said to our modern fox-hunting divines?

and

and earneftly enjoins them that, with all due fpeed and diligence, they fhould proceed to the election of proper perfons to fill up the vacancies, under pain of the greater excommunication.

In *Item* 17th. the prior and canons are accufed of fuffering, through neglect, notorious dilapidations to take place among their manerial houfes and tenements, and in the walls and enclofures of the convent itfelf, to the fhame and fcandal of the inftitution; they are therefore enjoined, under pain of fufpenfion, to repair all defects within the fpace of fix months.

Item 18th. charges them with grievoufly burthening the faid Priory by means of fales, and grants of *liveries* [e] and *corrodies* [f].

The bifhop, in *item* 19th, accufes the canons of neglect and omiffion with refpect to their perpetual *chantry-fervices*.

Item 20th. The vifitor here conjures the prior and canons not to withhold their original *alms*, " *eleemofynas*;" nor thofe that they were enjoined to diftribute for the good of the fouls of founders and benefactors: he alfo ftrictly orders that the fragments and broken victuals, both from the hall of their prior and their common refectory fhould be carefully collected together by their *eleemofynarius*, and given to the poor without any diminution; the officer to be fufpended for neglect or omiffion.

---

[e] " *Liberationes,* or *liberaturæ,* allowances of corn, &c. to fervants, *delivered* at " certain times, and in certain quantities, as *clothes* were among the allowances from " religious houfes to their dependants. See the *corrodies* granted by *Croyland abbey.*

Hift. of Croyland, Appendix, Nº XXXIV.

" It is not improbable that the word in after-ages came to be confined to the uniform " of the retainers or fervants of the great, who were hence called *livery fervants.*"

Sir John Cullum's Hift. of Hawfted.

[f] A *corrody* is an allowance to a fervant living in an abbey or priory.

Item 23d. He bids them diftribute their *pittances*, "*pitancias* [g]," regularly on obits, anniverfaries, feftivals, &c.

Item 25th. All and every one of the canons are hereby inhibited from ftanding godfather to *any boy* for the future, " ne com-" *patres alicujus pueri* de cetero fieri prefumatis," unlefs by exprefs licenfe from the bifhop obtained; becaufe from fuch relationfhip favour and affection, nepotifm, and undue influence, arife, to the injury and detriment of religious inftitutions [h].

Item 26th. The vifitor herein feverely reprimands the canons for appearing publicly in what would be called in the univerfities an *unftatutable manner*, and for wearing of boots, " caligæ de " *Burneto*, et *fotularium*————in ocrearum loco, ad modum fotu-" larium [i]."

[g] " *Pitancia*, an allowance of bread and beer, or other provifion to any pious ufe, " efpecially to the religious in a monaftery, &c. for augmentation of their commons."
Glofſ. to Kennett's Par. Antiq.

[h] " The relationfhip between fponfors and their god-children, who were called " *fpiritual fons* and *daughters*, was formerly efteemed much more facred than at pre-" fent. The prefents at chriftenings were fometimes very confiderable: the connexion " lafted through life, and was clofed with a legacy. This laft mark of attention feems " to have been thought almoft indifpenfable: for, in a will, from whence no extracts " have been given, the teftator left every one of his god-children a bufhel of barley."
Sir John Cullum's Hiſt. of Hawſted.
" D. *Margaretæ* filiæ Regis primogenitæ, quam *filiolam*, quia ejus in baptifmo " *compater* fuit, appellat, cyphum aureum et quadraginta libras, legavit."—Archbifhop *Parker* de Antiquitate Ecclef. Brit. fpeaking of Archbifhop *Morton*.

[i] *Du Frefne* is copious on *caligæ* of feveral forts. " Hcc item de Clericis, prefertim " beneficiatis: *caligis* fcacatis (chequered) rubeis, et viridibus publice utentibus dici-" mus effe cenfendum." *Statut. Ecclef. Tutel.* The chequered boots feem to be the highland plaid ftockings.---" *Burnetum*, i. e. *Brunetum*, pannus non ex lanâ nativi coloris " confectus."---" *Sotularium*, i. e. fubtalaris, quia fub talo eft. Peculium genus, quibus " maxime Monachi nocte utebantur in æftate; in hyeme vero Soccis."

This writer gives many quotations concerning *Sotularia*, which were not to be made too fhapely; nor were the *caligæ* to be laced on too nicely.

It

It is remarkable that the bifhop expreffes more warmth againſt this than any other irregularity; and ſtrictly enjoins them, under pain of eccleſiaſtical cenſures, and even impriſonment if neceſſary (a threat not made uſe of before), for the future to wear boots, " ocreis ſeu botis," according to the regular uſage of their ancient order.

Item 29th. He here again, but with leſs earneſtneſs, forbids them foppiſh ornaments, and the affectation of appearing like beaux with garments edged with coſtly furs, with fringed gloves, and ſilken girdles trimmed with gold and ſilver. It is remarkable that no puniſhment is annexed to this injunction.

Item 31ſt. He here ſingly and ſeverally forbids each canon not admitted to a cure of ſouls to adminiſter extreme unction, or the ſacrament, to clergy or laity; or to perform the ſervice of matrimony, till he has taken out the licenſe of the pariſh prieſt.

Item 32d. The biſhop ſays in this *item* that he had obſerved and found, in his ſeveral viſitations, that the ſacramental plate and cloths of the altar, ſurplices, &c. were ſometimes left in ſuch an uncleanly and difguſting condition as to make the beholders ſhudder with horror ;—" quod aliquibus ſunt horrori [k] :" he therefore enjoins them for the future to ſee that the plate, cloths, and veſtments, be kept bright, clean, and in decent order : and, what

[k] " Men abhorred the offering of the Lord." 1 *Sam. chap.* ii. *v.* 17. Strange as this account may appear to modern delicacy, the author, when firſt in orders, twice met with ſimilar circumſtances attending the ſacrament at two churches belonging to two obſcure villages. In the firſt he found the inſide of the chalice covered with birds' dung; and in the other the communion-cloth ſoiled with cabbage and the greaſy drippings of a gammon of bacon. The good dame at the great farm-houſe, who was to furniſh the cloth, being a notable woman, thought it beſt to ſave her clean linen, and ſo ſent a foul cloth that had covered her own table for two or three *Sundays* before.

muſt

muſt ſurpriſe the reader, adds—that he expects for the future that
the ſacriſt ſhould provide for the ſacrament good wine, pure and
unadulterated; and not, as had often been the practice, that which
was ſour, and tending to decay:—he ſays farther, that it ſeems
quite prepoſterous to omit in ſacred matters that attention to decent cleanlineſs, the neglect of which would diſgrace a common
convivial meeting [1].

Item 33d ſays that, though the relics of ſaints, the plate, holy
veſtments, and books of religious houſes, are forbidden by canonical inſtitutes to be pledged or lent out upon pawn; yet, as the
viſitor finds this to be the caſe in his ſeveral viſitations, he therefore ſtrictly enjoins the prior forthwith to recall thoſe pledges, and
to reſtore them to the convent; and orders that all the papers and
title deeds thereto belonging ſhould be ſafely depoſited, and kept
under three locks and keys.

In the courſe of the *Viſitatio Notabilis* the *conſtitutions* of *Legate
Ottobonus* are frequently referred to. *Ottobonus* was afterwards
Pope Adrian V. and died in 1276. His conſtitutions are in *Lyndewood's
Provinciale*, and were drawn up in the 52d of *Henry* III.

In the *Viſitatio Notabilis* the uſual puniſhment is faſting on bread
and beer; and in caſes of repeated delinquency on bread and water.
On theſe occaſions *quarta feria*, et *ſexta feria*, are mentioned often,
and are to be underſtood of the days of the *week* numerically on
which ſuch puniſhment is to be inflicted.

---

[1] "—— —— —— ne turpe toral, ne ſordida mappa
" Corruget nares; ne non et cantharus, et lanx
" Oſtendat tibi te —— —— ——"

LETTER

LETTER XV.

THOUGH bifhop *Wykeham* appears fomewhat ftern and rigid in his vifitatorial character towards the Priory of *Selborne*, yet he was on the whole a liberal friend and benefactor to that convent, which, like every fociety or individual that fell in his way, partook of the generofity and benevolence of that munificent prelate.

" In the year 1377 *William* of *Wykeham*, out of his mere good
" will and liberality, difcharged the whole debts of the prior
" and convent of *Selborne*, to the amount of one hundred and ten
" marks eleven fhillings and fixpence[m]; and, a few years before
" he died, he made a free gift of one hundred marks to the fame
" Priory: on which account the prior and convent voluntarily
" engaged for the celebration of two maffes a day by two canons
" of the convent for ten years, for the bifhop's welfare, if he
" fhould live fo long; and for his foul if he fhould die before
" the expiration of this term[n]."

At this diftance of time it feems matter of great wonder to us how thefe focieties, fo nobly endowed, and whofe members were exempt by their very inftitution from every means of perfonal and

[m] Yet in ten years time we find, by the *Notabilis Vifitatio,* that all their relics, plate, veftments, title-deeds, &c. were in pawn.

[n] *Lowth's* Life of *Wykeham.*

family

family expenfe, could poffibly run in debt without fquandering their revenues in a manner incompatible with their function.

Religious houfes might fometimes be diftreffed in their revenues by fires among their buildings, or large dilapidations from ftorms, &c.; but no fuch accident appears to have befallen the Priory at *Selborne.* Thofe fituate on public roads, or in great towns where there were fhrines of faints, were liable to be intruded on by travellers, devotees, and pilgrims; and were fubject to the importunity of the poor, who fwarmed at their gates to partake of doles and broken victuals. Of thefe difadvantages fome convents ufed to complain, and efpecially thofe at *Canterbury*; but this Priory, from it's fequeftered fituation, could feldom be fubject to either of thefe inconveniencies, and therefore we muft attribute it's frequent debts and embarraffments, well endowed as it was, to the bad conduct of it's members, and a general inattention to the interefts of the inftitution.

LETTER

LETTER XVI.

BEAUFORT was bifhop of *Winchefter* from 1405 to 1447; and yet, notwithftanding this long epifcopate, only tom. I. of *Beaufort's* Regifter is to be found. This lofs is much to be regretted, as it muft unavoidably make a gap in the hiftory of *Selborne* Priory, and perhaps in the lift of it's priors.

In 1410 there was an election for a prior, and again in 1411.

In vol. I. p. 24, of *Beaufort's* Regifter, is the inftrument of the election of *John Wyncheftre* to be prior—the fubftance as follows :

Richard Elftede, fenior canon, fignifies to the bifhop that brother *Thomas Wefton,* the late prior, died *October* 18th, 1410, and was buried *November* 11th.—That the bifhop's licenfe to elect having been obtained he and the whole convent met in the chapter-houfe, on the fame day, about the hour of vefpers, to confider of the election :—that brother *John Wyncheftre,* then fub-prior, with the general confent, appointed the 12th of *November, ad horam ejufdem diei capitularem,* for the bufinefs :—when they met in the chapter-houfe, *poft miffam de fancto Spiritu,* folemnly celebrated in the church ;—to wit, *Richard Elftede; Thomas Halyborne; John Lemyngton,* facrifta ; *John Stepe,* cantor ; *Walter Ffarnham ; Richard Putworth,* celerarius ; *Hugh London, Henry Brampton,* alias *Brompton; John Wyncheftre,* fenior ; *John Wyncheftre,* junior ;—then " Propofito " primitus verbo Dei," and then ympno " Veni Creator Spiritus"
being

being folemnly fung, cum " verficulo et oratione," as ufual, and his letter of licenfe, with the appointment of the hour and place of election being read, *alta voce*, in valvis of the chapter-houfe; — *John Wynchestre*, fenior, the fub-prior, in his own behalf and that of all the canons, and by their mandate, " quafdam monicionem " et proteftacionem in fcriptis redactas fecit, legit, et interpofuit"— that all perfons difqualified, or not having right to be prefent, fhould immediately withdraw; and protefting againft their voting, &c.—that then having read the conftitution of the general council " Quia propter," and explained the modes of proceeding to elec-tion, they agreed unanimoufly to proceed " per viam feu formam "*fimplicis compromiffi*;" when *John Wynchestre*, fub-prior, and all the others (the commiffaries undernamed excepted) named and chofe brothers *Richard Elftede, Thomas Halyborne, John Lemyngton* the facrift, *John Stepe*, chantor, and *Richard Putworth*, canons, to be commiffaries, who were fworn each to nominate and elect a fit perfon to be prior : and empowered by letters patent under the common feal, to be in force only until the darknefs of the night of the fame day; — that they, or the greater part of them, fhould elect for the whole convent, within the limited time, from their own number, or from the reft of the convent ;—that one of them fhould publifh their confent in common before the clergy and people :—they then all promifed to receive as prior the perfon thefe five canons fhould fix on. Thefe commiffaries feceded from the chapter-houfe to the refectory of the Priory, and were fhut in with mafter *John Penkefter*, bachelor of laws; and *John Couke* and *John Lynne*, perpetual vicars of the parifh churches of *Newton* and *Selborne* ; and with *Sampfon Maycock*, a public notary; where they treated of the election ; when they unanimoufly agreed on *John Wynchestre*, and appointed *Thomas Halyborne* to chufe him in com-

mon,

mon for all, and to publish the election, as customary; and returned long before it was dark to the chapter-house, where *Thomas Halyborne* read publicly the instrument of election; when all the brothers, the new prior excepted, singing solemnly the hymn " Te Deum laudamus," *fecerunt deportari novum electum*, by some of the brothers, from the chapter-house to the high altar of the church°; and the hymn being sung, *dictisque versiculo et oratione consuetis in hac parte*, Thomas Halyborne, *mox tunc ibidem*, before the clergy and people of both sexes solemnly published the election *in vulgari*.   Then *Richard Elstede*, and the whole convent by their proctors and nuncios appointed for the purposes, *Thomas Halyborne* and *John Stepe*, required several times the assent of the elected; " et tandem post diutinas interpellationes, et deliberationem " providam penes se habitam, in hac parte divine nolens, ut " asseruit, resistere voluntati," within the limited time he signified his acceptance in the usual written form of words.   The bishop is then supplicated to confirm their election, and do the needful, under common seal, in the chapter-house. *November* 14, 1410.

The bishop, *January* 6, 1410, *apud* Esher *in camera inferiori*, declared the election duly made, and ordered the new prior to be inducted—for this the archdeacon of *Winchester* was written to; " stallumque in choro, et locum in capitulo juxta morem preteriti " temporis," to be assigned him; and every thing beside necessary to be done.

° It seems here as if the canons used to *chair* their new elected prior from the chapter-house to the high altar of their convent church.   In letter XXI, on the same occasion, it is said---" et sic canentes dictum electum ad majus altare ecclesie *deduximus*, ut apud nos " moris est."

BEAUFORT'S REGISTER, Vol. I.

P. 2. Taxatio fpiritualis Decanatus de *Aulton*, Ecclefia de *Selebourn*, cum Capella,—xxx marc. decima x lib. iii fol. Vicaria de *Selebourn* non taxatur propter exilitatem.

P. 9. Taxatio bonorum temporalium religioforum in Archidiac. *Wynton*.

Prior de *Sclebourn* habet meneria de

Bromdene taxat. ad - - - - - - - -	xxx s. iid.
Apud *Schete* ad - - - - - - - - - - -	xvii s.
P. *Selebourne* ad - - - - - - - - - -	vi lib.
In civitate *Wynton* de reddit. - - - - - - -	vi lib. viii ob.
Tannaria fua taxat. ad - - - - - - -	x lib. s.

Summa tax. xxxviii lib. xiiii d. ob. Inde decima vi lib. s. q. ob.

———

LETTER

LETTER XVII.

INFORMATION being sent to *Rome* respecting the havock and spoil that was carrying on among the revenues and lands of the Priory of *Selborne*, as we may suppose by the bishop of *Winchester*, it's visitor, *Pope Martin* P, as soon as the news of these proceedings came before him, issued forth a *bull*, in which he enjoins his commissary immediately to revoke all the property that had been alienated.

In this instrument his holiness accuses the prior and canons of having granted away (they themselves and their predecessors) to certain clerks and laymen their tithes, lands, rents, tenements, and possessions, to some of them for their lives, to others for an undue term of years, and to some again for a perpetuity, to the great and heavy detriment of the monastery: and these leases were granted, he continues to add, under their own hands, with the sanction of an oath and the renunciation of all right and claims, and under penalties, if the right was not made good.—— But it will be best to give an abstract from the *bull*.

N. 298. *Pope Martin's bull* touching the revoking of certaine things alienated from the Priory of *Seleburne*. Pontif. sui ann. 1.

P *Pope Martin* V. chosen about 1417. He attempted to reform the church, but died in 1431, just as he had summoned the council of *Basil.*

" *Martinus*

" *Martinus* Epf. fervus fervorum Dei. Dilecto filio Priori de
" *Suthvale* ¹ *Wyntonien.* dioc. Salutem & apoftolicam ben. Ad
" audientiam noftram pervenit quam tam dilecti filii prior et
" conventus monafterii de *Seleburn* per Priorem foliti gubernari
" ordinis Sᵗⁱ. *Auguftini Winton.* dioc. quam de predeceffores eorum
" decimas, terras, redditus, domos, poffeffiones, *vineas*ʳ, et
" quedam alia bona ad monafterium ipfum fpectantia, datis fuper
" hoc litteris, interpofitis juramentis, factis renuntiationibus, et
" penis adjectis, in gravem ipfius monafterii lefionem nonnullis
" clericis et laicis, aliquibus eorum ad vitam, quibufdam vero
" ad non modicum tempus, & aliis perpetuo ad firmam, vel
" fub cenfu annuo concefferunt; quorum aliqui dicunt fuper
" hiis a fede aplica in communi forma confirmationis litteras
" impetraffe. Ouia vero noftra intereft lefis monafteriis fub-
" venire — [He the Pope here commands] — ea ad jus et pro-
" prietatem monafterii ftudeas legitime revocare," &c.

The conduct of the religious had now for fome time been
generally bad. Many of the monaftic focieties, being very opulent, were become voluptuous and licentious, and had deviated
entirely from their original inftitutions. The laity faw with indignation the wealth and poffeffions of their pious anceftors perverted
to the fervice of fenfuality and indulgence; and fpent in gratifi-
cations highly unbecoming the purpofes for which they were

ⁱ Should have been no doubt *Southwick*, a priory under *Portfdown.*

ʳ Mr. *Barrington* is of opinion that anciently the Englifh *vinea* was in almoft every
inftance an orchard; not perhaps always of apples merely, but of other fruits; as cherries,
plums, and currants. We ftill fay a plum or cherry-orchard. See vol. III. of
Archæologia.

In the inftance above the pope's fecretary might infert *vineas* merely becaufe they
were a fpecies of cultivation familiar to him in *Italy.*

given. A total difregard to their refpective rules and difcipline drew on the monks and canons a heavy load of popular odium. Some good men there were who endeavoured to oppofe the general delinquency; but their efforts were too feeble to ftem the torrent of monaftic luxury. As far back as the year 1381 *Wickliffe*'s principles and doctrines had made fome progrefs, were well received by men who wifhed for a reformation, and were defended and maintained by them as long as they dared; till the bifhops and clergy began to be fo greatly alarmed, that they procured an act to be paffed by which the fecular arm was empowered to fupport the corrupt doctrines of the church; but the firft lollard was not burnt until the year 1401.

The wits alfo of thofe times did not fpare the grofs morals of the clergy, but boldly ridiculed their ignorance and profligacy. The moft remarkable of thefe were *Chaucer*, and his contemporary *Robert Langelande*, better known by the name of *Piers Plowman*. The laughable tales of the former are familiar to almoft every reader; while the *vifions* of the latter are but in few hands. With a quotation from the *Paffus Decimus* of this writer I fhall conclude my letter; not only on account of the remarkable prediction therein contained, which carries with it fomewhat of the air of a prophecy; but alfo as it feems to have been a ftriking picture of monaftic infolence and diffipation; and a fpecimen of one of the keeneft pieces of fatire now perhaps fubfifting in any language, ancient or modern.

" Now is religion a rider, a romer by ftreate;
" A leader of love-days, and a loud begger;
" A pricker on a palfrey from maner to maner,
" A heape of hounds at his arfe, as he a lord were.
" And

" And but if his knave kneel, that shall his cope bring,
" He loureth at him, and asketh him who taught him curtesie.
" Little had lords to done, to give lands from her heirs,
" To religious that have no ruth if it rain on her altars.
" In many places ther they persons be, by himself at ease:
" Of the poor have they no pity, and that is her charitie;
" And they letten hem as lords, her lands lie so broad.
" And *there shal come a king*[r], and confess you religious;
" And beate you, as the bible telleth, for breaking your rule,
" And amend monials, and monks, and chanons,
" And put hem to her penaunce *ad pristinum statum ire.*"

[r] F. l. a. " This prediction, although a probable conclusion concerning a king who
" after a time would suppress the religious houses, is remarkable. I imagined it might
" have been foisted into the copies in the reign of king *Henry* VLII. but it is to be found
" in MSS. of this poem, older than the year 1400." fol. l. a. b.
  " Again, where he, *Piers Plowman*, alludes to the *Knights Templars*, lately sup-
" pressed, he says

"——— ——— ——— Men of holie kirk
" Shall *turn as Templars* did; *the tyme approacheth nere.*"

" This, I suppose, was a favourite doctrine in *Wickliffe's* discourses."

Warton's Hist. of English Poetry, Vol. I. p. 282.

LETTER

LETTER XVIII.

WILLIAM of *Waynflete* became bifhop of *Winchefter* in the year 1447, and feems to have purfued the generous plan of *Wykeham* in endeavouring to reform the Priory of *Selborne.*

When *Waynflete* came to the fee he found prior *Stype,* alias *Stepe,* ftill living, who had been elected as long ago as the year 1411.

Among my documents I find a curious paper of the things put into the cuftody of *Peter Bernes* the facrift, and efpecially fome *relics :* the title of this evidence is " Nº. 50, Indentura " prioris de *Selborne* quorundam tradit. *Petro Bernes* facriftæ, " ibidem, ann. Hen. VI. - - - una cum confiff. ejufdem *Petri* " fcript." The occafion of this catalogue, or lift of effects, being drawn between the prior and facrift does not appear, nor the date when ; only that it happened in the reign of *Hen.* VI. This tranfaction probably took place when *Bernes* entered on his office ; and there is the more reafon to fuppofe that to be the cafe, becaufe the lift confifts of veftments and implements, and relics, fuch as belonged to the church of the Priory, and fell under the care of the facrift. For the numerous *items* I fhall refer the curious reader to the Appendix, and fhall juft mention the relics, although they are not all fpecified ; and the ftate of the live ftock of the monaftery at that juncture.

" Item

" Item 2. *osculatōr*. argent.

" Item 1. *osculatorium* cum *osse digiti auriculār.—St.* *Johannis*
        Baptistæ [s].

" Item 1. parvam *crucem* cum V. *reliquiis*.

" Item 1. *anulum* argent. et deauratum *St. Edmundi* [t].

" Item 2. *osculat*. de coper.

" Item 1. *junctorium St. Ricardi* [u].

" Item 1. *pecten St. Ricardi* [x]."

The *staurum*, or live stock, is quite ridiculous, consisting only
of " 2 vacce, 1 sus, 4 hoggett. et 4 porcell." viz. two cows,
one sow, four porkers, and four pigs.

---

[s] How the convent came by the bone of the little finger of *Saint John the Baptist*
does not appear: probably the founder, while in *Palestine*, purchased it among the
Asiatics, who were at that time great traders in relics. We know from the best authority that as soon as *Herod* had cruelly beheaded that holy man " his disciples came
" and took up the body and buried it, and went and told *Jesus*." Matt. iv. 12.—Farther would be difficult to say.

[t] *November* 20, in the calendar, *Edmund* king and martyr, in the 9th century. See
also a *Sanctus Edmundus* in *Godwin*, among the archbishops of *Canterbury*, in the 13th
century; his surname *Rich*, in 1234.

[u] *April* 3, ibid. *Richard* bishop of *Chichester*, in the 13th century; his surname *De la
Wich*, in 1245.

Junctorium, perhaps a *joint* or *limb* of St. *Richard*; but what particular joint the religious were not such osteologists as to specify. This barbarous word was not to be
found in any dictionary consulted by the author.

[x] " *Pecten* inter ministeria sacra recensetur, quo scil. sacerdotes ac clerici, antequam
" in ecclesiam procederent, crines pecterent. E quibus colligitur monachos, tunc
" temporis, non omnino tonsos fuisse."   *Du Fresne.*

The author remembers to have seen in great farm houses a family comb chained to a
post for the use of the hinds when they came in to their meals.

LETTER

LETTER XIX.

Stepe died towards the end of the year 1453, as we may suppose pretty far advanced in life, having been prior forty-four
years.

On the very day that the vacancy happened, viz. *January* 26,
1453-4, the sub-prior and convent petitioned the visitor—" vos
" unicum levamen nostrum, et spem unanimiter rogamus, qua-
" tinus eligendum ex nobis unum confratrem de gremio nostro,
" in nostra religione probatum et expertem, licenciam vestram
" paternalem cum plena libertate nobis concedere dignemini
" graciose."   *Reg. Waynflete*, tom. I.

Instead of the license requested we find next a commission
" custodie prioratus de *Selebourne* durante vacatione," addressed
to brother *Peter Berne*, canon-regular of the priory of *Selebourne*,
and of the order of St. *Augustine*, appointing him keeper of the said
priory, and empowering him to collect and receive the profits
and revenues, and " alia bona" of the said priory; and to exer-
cise in every respect the full power and authority of a prior; but
to be responsible to the visitor finally, and to maintain this supe-
riority during the bishop's pleasure only. This instrument is
dated from the bishop's manor-house in *Southwark*, *March* 1,
1453-4, and the seventh of his consecration.

D d d

After

After this transaction it does not appear that the chapter of the Priory proceeded to any election; on the contrary, we find that at six months end from the vacancy the visitor declared that a lapse had taken place; and that therefore he did confer the priorship on canon *Peter Berne*.—" Prioratum vacantem et ad nostram colla-
" tionem, seu provisionem jure ad nos in hac parte *per lapsum*
" temporis legitime devoluto spectantem, tibi (sc. *P. Berne)* de
" legitimo matrimonio procreato, &c.—conferimus," &c. This deed bears date July 28, 1454.

Reg. Waynflete, tom. I. p. 69.

On *February* 8, 1462, the visitor issued out a power of sequestra-tion against the Priory of *Selborne* on account of notorious dilapida-tions, which threatened manifest ruin to the roofs, walls, and edifices, of the said convent; and appointing *John Hammond*, B. D. rector of the parish church of *Hetlegh*, *John Hylling*, vicar of the parish church of *Newton Valence*, and *Walter Gorfin*, inhabitant of the parish of *Selborne*, his sequestrators, to exact, collect, levy, and receive, all the profits and revenues of the said convent : he adds " ac ea sub areto, et tuto custodiatis, custodirive faciatis;" as they would answer it to the bishop at their peril.

In consequence of these proceedings prior *Berne*, on the last day of *February*, and the next year, produced a state of the revenues of the Priory, Nᵒ. 381, called " A paper conteyning the value of " the manors and lands pertayning to the Priory of *Selborne*. " 4 *Edward* III. with a note of charges yssuing out of it."

This is a curious document, and will appear in the Appendix. From circumstances in this paper it is plain that the sequestration produced good effects; for in it are to be found bills of repairs to a considerable amount.

By

By this evidence also it appears that there were at that juncture only four canons at the Priory[u]; and that these, and their four household servants, during this sequestration for their clothing, wages, and diet, were allowed *per ann.* xxx lib.; and that the annual pension of the lord prior, reside where he would, was to be x lib.

In the year 1468, prior *Berne*, probably wearied out by the dissensions and want of order that prevailed in the convent, resigned his priorship into the hands of the bishop.

Reg. Waynflete, tom. I. pars 1ma, fol. 157.

March 28, A. D. 1468. " In quadam alta camera juxta mag-
" nam portam manerii of the bishop of *Wynton de Waltham* coram
" eodem rev. patre ibidem tunc sedente, *Peter Berne*, prior of
" *Selborne*, ipsum prioratum in sacras, et venerabiles manus of the
" bishop, viva voce libere resignavit: and his resignation was
" admitted before two witnesses and a notary-public. In conse-
" quence, *March* 29th, before the bishop, in capella manerii sui ante
" dicti pro tribunali sedente, comparuerunt fratres" *Peter Berne*,
Thomas London, *William Wyndesor*, and *William Paynell*, alias *Stret-ford*, canons regular of the Priory, " capitulum, et conventum
" ejusdem ecclesie facientes; ac jus et voces in electione futura
" prioris dicti prioratus solum et in solidum, ut asseruerunt,
" habentes;" and after the bishop had notified to them the vacancy of a prior, with his free license to elect, deliberated awhile, and then, by way of compromise, as they affirmed, unanimously transferred their right of election to the bishop before witnesses. In consequence of this the bishop, after full deliberation, proceeded,

---

[u] If bishop *Wykeham* was so disturbed (see *Notab. Visitatio*) to find the number of canons reduced from fourteen to eleven, what would he have said to have seen it diminished below one third of that number?

April

April 7th, " in capella manerii sui de *Waltham*," to the election of
a prior; " et fratrem *Johannem Morton*, priorem ecclesie conventualis
" de *Reygate* dicti ordinis S^{ti}. *Augustini Wynton.* dioc. in priorem
" vice et nomine omnium et singulorum canonicorum predictorum
" elegit, in ordine sacerdotali, et etate licita constitutum, &c."
And on the same day, in the same place, and before the same
witnesses, *John Morton* resigned to the bishop the priorship of *Reygate*
viva voce. The bishop then required his consent to his own election; " qui licet in parte renitens tanti reverendi patris se confirmans," obeyed, and signified his consent *oraculo vive vocis.* Then
was there a mandate citing any one who would gainsay the said
election to appear before the bishop or his commissary in his chapel
at *Farnham* on the second day of *May* next. The dean of the
deanery of *Aulton* then appeared before the chancellor, his commissary, and returned the citation or mandate dated *April* 22d, 1468,
with signification, in writing, of his having published it as required,
dated *Newton Valence*, *May* 1st, 1468. This certificate being read,
the four canons of *Selborne* appeared and required the election to
be confirmed; *et ex super abundanti* appointed *William Long* their
proctor to solicit in their name that he might be canonically confirmed. *John Morton* also appeared, and proclamation was made;
and no one appearing against him, the commissary pronounced all
absentees contumacious, and precluded them from objecting at
any other time; and, at the instance of *John Morton* and the proctor,
confirmed the election by his decree, and directed his mandate to
the rector of *Hedley* and the vicar of *Newton Valence* to install him
in the usual form.

Thus,

Thus, for the firſt time, was a perſon, a ſtranger to the convent of *Selborne*, and never canon of that monaſtery, elected prior; though the ſtyle of the petitions in former elections uſed to run thus, — " Vos - - - - rogamus quatinus eligendum ex *nobis* " unum *confratrem* de *gremio noſtro*, — licentiam veſtram — nobis " concedere dignemini."

LETTER XX.

Prior *Morton* dying in 1471, two canons, by themſelves, proceeded to election, and choſe a prior; but two more (one of them *Berne*) complaining of not being ſummoned, objected to the proceedings as informal; till at laſt the matter was compromiſed that the biſhop ſhould again, for that turn, nominate as he had before. But the circumſtances of this election will be beſt explained by the following extract:

Reg. Waynflete, tom. II, pars 1ᵐᵃ, fol. 7.

Memorandum. A.D. 1471. Auguſt 22.

William Wyndeſor, a canon-regular of the Priory of *Selburne*, having been elected prior on the death of brother *John*, appeared in perſon before the biſhop in his chapel at *South Waltham*. He

was

was attended on this occasion by *Thomas London* and *John Bromes-grove*, canons, who had elected him. *Peter Berne* and *William Stratfeld*, canons, also presented themselves at the same time, complaining that in this business they had been overlooked, and not summoned; and that therefore the validity of the election might with reason be called in question, and quarrels and dissensions might probably arise between the newly chosen prior and the parties thus neglected.

After some altercation and dispute they all came to an agreement with the new prior, that what had been done should be rejected and annulled; and that they would again, for this turn, transfer to the bishop their power to elect, order, and provide them another prior, whom they promised unanimously to admit.

The bishop accepted of this offer before witnesses; and on *September* 27, in an inner chamber near the chapel abovementioned, after full deliberation, chose brother *Thomas Fairwise*, vicar of *Somborne*, a canon-regular of Saint *Augustine* in the Priory of *Bruscough*, in the diocese of *Coventry* and *Litchfield*, to be prior of *Selborne*. The form is nearly as above in the last election. The canons are again enumerated; *W. Wyndesor*, sub-prior, *P. Berne*, *T. London*, *W. Stratfeld*, *J. Bromesgrove*, who had formed the chapter, and had requested and obtained license to elect, but had unanimously conferred their power on the bishop. In consequence of this proceeding, the bishop taking the business upon himself, that the Priory might not suffer detriment for want of a governor, appoints the aforesaid *T. Fairwise* to be prior. A citation was ordered as above for gainsayers to appear *October* 4th, before the bishop or his commissaries at *South Waltham*; but none appearing, the commissaries admitted the said *Thomas*, ordered him to be installed, ftalled, and fent the ufual letter to the convent to render him due obedience.

Thus did the bifhop of *Winchefter* a fecond time appoint a ftranger to be prior of *Selborne*, inftead of one chofen out of the chapter. For this feeming irregularity the vifitor had no doubt good and fufficient reafons, as probably may appear hereafter.

LETTER XXI.

WHATEVER might have been the abilities and difpofition of prior *Fairwife*, it could not have been in his power to have brought about any material reformation in the Priory of *Selborne*, becaufe he departed this life in the month of *Auguft* 1472, before he had prefided one twelvemonth.

As foon as their governor was buried the chapter applied to their vifitor for leave to chufe a new prior, which being granted, after deliberating for a time, they proceeded to an election by a *fcrutiny*. But as this mode of voting has not been defcribed but by the mere form in the *Appendix*, an extract from the bifhop's regifter, reprefenting the manner more fully, may not be difagreeable to feveral readers.

WAYNEFLETE REG. tom. II. pars 1ma, fol. 15.

" Reverendo &c. ac noftro patrono graciofiffimo veftri humiles, " et devote obedientie filii," &c.

To

To the right reverend Father in God, and our moſt gracious patron, we, your obedient and devoted ſons, *William Wyndeſor,* preſident of the chapter of the Priory of *Selborne,* and the convent of that place, do make known to your lordſhip, that our prior-ſhip being lately vacant by the death of *Thomas Fairwiſe,* our late prior, who died Auguſt 11th, 1472, having committed his body to decent ſepulture, and having requeſted, according to cuſtom, leave to elect another, and having obtained it under your ſeal, we, *William Wyndeſor,* preſident of the convent on the 29th of Auguſt, in our chapter-houſe aſſembled, and making a chapter, taking to us in this buſineſs *Richard ap Jenkyn,* and *Galfrid Bryan,* chaplains, that our ſaid Priory might not by means of this vacancy incur harm or loſs, unanimouſly agreed on *Auguſt* the laſt for the day of election; on which day, having firſt cele-brated maſs, " De ſancto ſpiritu," at the high altar, and hav-ing called a chapter by tolling a bell about ten o' the clock, we, *William Wyndeſor,* preſident, *Peter Berne, Thomas London,* and *William Stratfield,* canons, who alone had voices, being the only canons, about ten o' the clock, firſt ſung " Veni Creator," the letters and licenſe being read in the preſence of many perſons there. Then *William Wyndeſor,* in his own name, and that of all the canons, made ſolemn proclamation, enjoining all who had no right to vote to depart out of the chapter-houſe. When all were withdrawn except *Guyllery de Lacuna,* in decretis Baccalarius, and *Robert Peverell,* notary-public, and alſo the two chaplains, the firſt was requeſted to ſtay, that he might direct and inform us in the mode of election; the other, that he might record and atteſt the tranſactions; and the two laſt that they might be wit-neſſes to them.

Then

Then, having read the conftitution of the general council "Quia propter," and the forms of elections contained in it being fufficiently explained to them by *De Lacuna*, as well in Latin as the vulgar tongue, and having deliberated in what mode to proceed in this election, they refolved on that of *fcrutiny*. Three of the canons, *Wyndefor*, *Berne*, and *London*, were made *fcrutators*: *Berne*, *London*, and *Stratfeld*, chufing *Wyndefor*; *Wyndefor*, *London*, and *Stratfeld*, chufing *Berne*; *Wyndefor*, *Berne*, and *Stratfeld*, chufing *London*.

They were empowered to take each other's vote, and then that of *Stratfeld*; " et ad inferiorem partem angularem" of the chapter-houfe, "juxta oftium ejufdem declinentes," with the other perfons, (except *Stratfeld*, who ftaid behind) proceeded to voting, two fwearing, and taking the voice of the third, in fucceffion, privately. *Wyndefor* voted firft : " Ego credo *Petrum Berne* meliorem et utili-" orem ad regimen iftius ecclefie, et in ipfum confentio, ac eum " nomino," &c. *Berne* was next fworn, and in like manner nominated *Wyndefor*; *London* nominated *Berne*: *Stratfeld* was then called and fworn, and nominated *Berne*.

" Quibus in fcriptis redactis," by the notary-public, they re-turned to the upper part of the chapter-houfe, where by *Wyndefor* " fic purecta fecerunt in communi," and then folemnly, in form written, declared the election of *Berne*: when all, " antedicto " noftro electo excepto, approbantes et ratificantes, cepimus " decantare folemniter ' *Te Deum laudamus*,' et fic canentes dictum " electum ad majus altare ecclefie deduximus, ut apud nos eft " moris. Then *Wyndefor* electionem clero et populo infra chorum " dicte ecclefie congregatis publicavit, et perfonam electi publice " et perfonaliter oftendit." We then returned to the chapter-houfe, except our prior; and *Wyndefor* was appointed by the other two

E e e

their

their proctor, to defire the affent of the elected, and to notify what had been done to the bifhop; and to defire him to confirm the election, and do whatever elfe was neceffary. Then their proctor, before the witneffes, required *Berne*'s affent in the chapter-houfe: " qui quidem inftanciis et precibus multiplicatis devictus," confented, " licet indignus electus," in writing. They therefore requeft the bifhop's confirmation of their election " fic canonice " et folemniter celebrata," &c. &c. Sealed with their common feal, and fubfcribed and attefted by the notary. Dat. in the chapter-houfe *September* 5th. 1472.

In confequence, *September* 11th, 1472, in the bifhop's chapel at *Efher*, and before the bifhop's commiffary, appeared *W. Wyndefor*, and exhibited the above inftrument, and a mandate from the bifhop for the appearance of gainfayers of the election there on that day:—and no one appearing, the abfentees were declared contumacious, and the election confirmed; and the vicar of *Aulton* was directed to induct and inftall the prior in the ufual manner.

Thus did canon *Berne*, though advanced in years, reaffume his abdicated priorfhip for the fecond time, to the no fmall fatisfaction, as it may feem, of the bifhop of *Winchefter*, who profeffed, as will be fhown not long hence, an high opinion of his abilities and integrity.

LETTER

LETTER XXII.

As prior *Berne*, when chosen in 1454, held his priorship only to 1468, and then made a voluntary resignation, wearied and disgusted, as we may conclude, by the disorder that prevailed in his convent; it is no matter of wonder that, when re-chosen in 1472, he should not long maintain his station; as old age was then coming fast upon him, and the increasing anarchy and misrule of that declining institution required unusual vigour and resolution to stem that torrent of profligacy which was hurrying it on to it's dissolution. We find, accordingly, that in 1478 he resigned his dignity again into the hands of the bishop.

WAYNFLETE REG. fol. 55.

Resignatio Prioris de Seleborne.

May 14, 1478. *Peter Berne* resigned the priorship. *May* 16 the bishop admitted his resignation " in manerio suo *de Waltham*," and declared the priorship void; " et priorat. solacio destitutum " esse;" and granted his letters for proceeding to a new election: when all the religious, assembled in the chapter-house, did transfer their power under their seal to the bishop, by the following public instrument.

"In Dei nomine Amen," &c. A. D. 1478, Maii 19. In the chapter-houſe for the election of a prior for that day, on the free reſignation of *Peter Berne*, having celebrated in the firſt place maſs at the high altar " De ſpiritu ſancto," and having called a chapter by tolling a bell, *ut moris eſt*; in the preſence of a notary and witneſſes appeared perſonally *Peter Berne, Thomas Aſhford, Stephen Clydgrove*, and *John Aſhton*, preſbyters, and *Henry Canwood*ˣ, in chapter aſſembled; and after ſinging the hymn ' *Veni Creator Spiritus*,' " cum verſiculo et oratione ' *Deus qui corda*;' declarataque li- " centia Fundatoris et patroni; futurum priorem eligendi con- " ceſſa, et conſtitutione conſilii generalis que incipit ' *Quia* " *propter*' declaratis; viiſque per quas poſſent ad hanc electionem " procedere," by the *decretorum doctorem*, whom the canons had taken to direct them—they all and every one " dixerunt et affir- " marunt ſe nolle ad aliquam viam procedere :—but, for this turn only, renounced their right, and unanimouſly transferred their power to the biſhop, the ordinary of the place, promiſing to receive whom he ſhould provide; and appointed a proctor to preſent the inſtrument to the biſhop under their ſeal; and required their notary to draw it up in due form, &c. ſubſcribed by the notary.

After the viſitor had fully deliberated on the matter, he proceeded to the choice of a prior, and elected, by the following inſtrument, *John Sharp*, alias *Glaſtenbury*.

ˣ Here we ſee that all the canons were changed in ſix years; and that there was quite a new chapter, *Berne* excepted, between 1472 and 1478; for, inſtead of *Wyndeſor, London*, and *Stratfeld*, we find *Aſhford, Clydgrove, Aſhton*, and *Canwood*, all new men, who were ſoon gone in their turn off the ſtage, and are heard of no more. For, in ſix years after, there ſeem to have been no canons at all.

Fol.

Fol. 56. Provisio Prioris per Epm.

Willmus, &c. to our beloved brother in Christ *John Sharp*, alias *Glaſtenbury*, Eccleſie conventualis de *Bruton*, of the order of St. *Auſtin*, in the dioceſe of *Bath* and *Wells*, canon-regular—*ſalutem* &c. " De tue circumſpectionis induſtria plurimum confidentes, " te virum providum et diſcretum, literarum ſcientia, et moribus " merito commendandum," &c.—do appoint you prior—under our ſeal. " Dat. in manerio noſtro de *Suthwaltham, May* 20," 1478, " et noſtre Confec. 31.

Thus did the biſhop, three times out of the four that he was at liberty to nominate, appoint a prior from a diſtance, a ſtranger to the place, to govern the convent of *Selborne*, hoping by this method to have broken the cabal, and to have interrupted that habit of miſmanagement that had pervaded the ſociety: but he acknowledges, in an evidence lying before us, that he never did ſuceed to his wiſhes with reſpect to thoſe late governors,—" quos " tamen male ſe habuiſſe, et inutiliter adminiſtrare, et admini- " ſtraſſe uſque ad preſentia tempora poſt debitam inveſtigationem, " &c. invenit." The only time that he appointed from among the canons, he made choice of *Peter Berne*, for whom he had con- ceived the greateſt eſteem and regard.

When prior *Berne* firſt relinquiſhed his priorſhip, he returned again to his former condition of canon, in which he continued for ſome years: but when he was re-choſen, and had abdicated a ſecond time, we find him in a forlorn ſtate, and in danger of being re- duced to beggary, had not the biſhop of *Wincheſter* interpoſed in his favour, and with great humanity inſiſted on a proviſion for him.

him for life. The reafon for this difference feems to have been, that, in the firft cafe, though in years, he might have been hale and capable of taking his fhare in the duty of the convent; in the fecond, he was broken with age, and no longer equal to the functions of a canon.

Impreffed with this idea the bifhop very benevolently interceded in his favour, and laid his injunctions on the new-elected prior in the following manner.

Fol. 56. " In Dei nomine Amen. Nos *Willmus*, &c. confide-" rantes *Petrum Berne*," late prior " in adminiftratione fpiritualium " et temporalium prioratus laudabiliter vixiffe et rexiffe; ipfumque " fenio et corporis debilitate confractum; ne in opprobrium " religionis *mendicari cogatur*;—eidem annuam penfionem a Domino " *Johanne Sharp*, alias *Glaftonbury*, priore moderno," and his fuc-ceffors, and, from the Priory or church, to be payed every year during his life, " de voluntate et ex confenfu expreffis" of the faid *John Sharp*, " fub ea que fequitur forma verborum — affig-" namus :

1ft. That the faid prior and his fucceffors, for the time being, *honefte exhibebunt* of the fruits and profits of the priorfhip, " eidem " efculenta et poculenta," while he remained in the Priory " fub " confimili portione eorundem prout convenienter priori," for the time being, *miniftrari contigerit*; and in like manner *uni famulo*, whom he fhould chufe to wait on him, as to the *fervientibus* of the prior.

Item. " Invenient feu exhibebunt eidem unam honeftam came-" ram" in the Priory, " cum focalibus neceffariis feu opportunis " ad eundem.

Item.

Item. We will, ordain, &c. to the said *P. Berne* an annual pension of ten marks, from the revenue of the Priory, to be paid by the hands of the prior quarterly.

The bishop decrees farther, that *John Sharp*, and his successors, shall take an oath to observe this injunction, and that before their installation.

" Lecta et facta sunt hæc in quodam alto oratorio," belonging to the bishop at *Suthwaltham, May* 25, 1478, in the presence of *John Sharp*, who gave his assent, and then took the oath before witnesses, with the other oaths before the chancellor, who decreed he should be inducted and installed ; as was done that same day.

How *John Sharp*, alias *Glastonbury*, acquitted himself in his prior-ship, and in what manner he made a vacancy, whether by resig-nation, or death, or whether he was removed by the visitor, does not appear ; we only find that some time in the year 1484 there was no prior, and that the bishop nominated canon *Ashford* to fill the vacancy.

LETTER

LETTER XXIII.

THIS *Thomas Ashford* was most undoubtedly the last prior of *Selborne*; and therefore here will be the proper place to say something concerning a list of the priors, and to endeavour to improve that already given by others.

At the end of bishop *Tanner's Notitia Monastica*, the folio edition, among *Brown Willis's Principals* of *Religious Houses* occur the names of eleven of the priors of *Selborne*, with dates. But this list is imperfect, and particularly at the beginning; for though the Priory was founded in 1232, yet it commences with *Nich. de Cantia*, elected in 1262; so that for the first thirty years no prior is mentioned; yet there must have been one or more. We were in hopes that the register of *Peter de Rupibus* would have rectified this omission; but, when it was examined, no information of the sort was to be found. From the year 1410 the list is much corrected and improved; and the reader may depend on it's being thence forward very exact.

A LIST

A List *of the* Priors *of* Selborne Priory, *from* Brown Willis's Principals *of* Religious Houses, *with additions within* [   ] *by the author.*

[*John* - - - was prior, *fine dat.*] [y]

Nich. *de Cantia* el.   -   -   -   -   -   -   1262.

[*Peter* —— was prior in   -   -   -   1271.]

[*Richard* —— was prior in   -   -   1280.]

Will. Basing was prior in   -   -   -   1299.

Walter de Insula el. in   -   -   -   -   1324.

[Some difficulties, and a devolution; but the election confirmed by bishop *Stratford*.]

John de Wintōn   -   -   -   -   -   1339.

Thomas Weston   -   -   -   -   -   1377.

John Winchester, [*Wynchestre*]   -   -   1410.

[Elected by bishop *Beaufort* " per viam vel formam " simplicis compromissi.]

[*John Stype,* alias *Stepe,* in   -   -   -   -   1411.]

Peter Bene [alias *Berne* or *Bernes,* appointed keeper, and, by lapse to bishop *Wayneflete,* prior] in   -   -   1454.

[He resigns in 1468.]

John Morton, [Prior of *Reygate*] in   -   -   1468.

[The canons by compromise transfer the power of election to the bishop.]

---

[y] See, in Letter XI. of these Antiquities, the reason why prior *John* - - -, who had transactions with the *Knight's Templars,* is placed in the list before the year 1262.

Fff   *Will.*

Will. Winfor [*Wyndefor*, prior for a few days]   1471.
 [but removed on account of an irregular election.]

Thomas Farwill [*Fairwife*, vicar of *Somborne*]   1471.
 [by compromife again elected by the bifhop.]

[*Peter Berne*, re-elected by fcrutiny in   1472,]
 [refigns again in 1478.]

John Sharper [*Sharp*] alias *Glaftonbury*.   1478.
[Canon-reg. of *Bruton*, elected by the bifhop by compromife.]

[*Thomas Afhford*, canon of *Selborne*, laft prior elected by
 the bifhop of *Winchefter*, fome time in the year   -   -   1484,
 and depofed at the diffolution.]

LETTER

LETTER XXIV.

Bishop *Wayneflete*'s efforts to continue the Priory still proved un-successful; and the convent, without any canons, and for some time without a prior, was tending swiftly to it's dissolution.

When *Sharp*'s, alias *Glastonbury*'s, priorship ended does not appear. The bishop says that he had been obliged to remove some priors for male-administration: but it is not well explained how that could be the case with any, unless with *Sharp*; because all the others, chosen during his episcopate, died in their office, *viz.* *Morton* and *Fairwise*; *Berne* only excepted, who relinquished twice voluntarily, and was moreover approved of by *Wayneflete* as a person of integrity. But the way to shew what ineffectual pains the bishop took, and what difficulties he met with, will be to quote the words of the libel of his proctor *Radulphus Langley*, who appeared for the bishop in the process of the impropriation of the Priory of *Selborne*. The extract is taken from an attested copy.

" *Item*—that the said bishop—dicto prioratui et personis ejusdem
" pie compatiens, sollicitudines pastorales, labores, et diligentias
" gravissimas quam plurimas, tam per se quam per suos, pro re-
" formatione premissorum impendebat: et aliquando illius loci
" prioribus, propter malam et inutilem administrationem, et dis-
" pensationem bonorum predicti prioratus, suis demeritis exigen-

" tibus,

" tibus, amotis; alios priores in quorum circumſpectione et
" diligentia confidebat, prefecit: quos tamen male ſe habuiſſe ac
" inutiliter adminiſtrare, et adminiſtraſſe, uſque ad preſentia tem-
" pora poſt debitam inveſtigationem, &c. invenit." So that he
deſpaired with all his care—" ſtatum ejuſdem reparare vel reſtau-
" rare: et conſiderata temporis malicia, et preteritis timendo et
" conjecturando futura, de aliqua bona et ſancta religione ejuſdem
" ordinis, &c. juxta piam intentionem primevi fundatoris ibidem
" habend. deſperatur."

William Wainfleet, biſhop of *Wincheſter*, founded his college of
Saint *Mary Magdalene*, in the univerſity of *Oxford*, in or about the
year 1459; but the revenues proving inſufficient for ſo large and
noble an eſtabliſhment, the college ſupplicated the founder to
augment it's income by putting it in poſſeſſion of the eſtates belonging to the Priory of *Selborne*, now become a deſerted convent,
without canons or prior. The preſident and fellows ſtate the circumſtances of their numerous inſtitution and ſcanty proviſion, and
the ruinous and perverted condition of the Priory. The biſhop
appoints commiſſaries to inquire into the ſtate of the ſaid monaſtery; and, if found expedient, to confirm the appropriation of
it to the college, which ſoon after appoints attornies to take poſſeſſion, *September* 24, 1484. But the way to give the reader a
thorough inſight reſpecting this tranſaction, will be to tranſcribe
a farther proportion of the proceſs of the impropriation from the
beginning, which will lay open the manner of proceeding, and
ſhew the conſent of the parties.

IMPROPRIATIO

Impropriatio Selborne, 1485.

" Univerfis fan&te matris ecclefie filiis, &c. *Ricardus* Dei
" gratia *prior* ecclefie conventualis de *Novo Loco*, &c^z. ad uni-
" verfitatem veftre notitie deducimus, &c. quod coram nobis
" commiffario predi&to in ecclefia parochiali S^{ti}. *Georgii de Esfher*,
" di&t. *Winton.* dioc. 3°. die *Augufti*, A. D. 1485. Indi&tione tertia
" pontificat. *Innocentii* 8^{vi}. ann. 1mo. judicialiter comparuit venera-
" bilis vir *Jacobus Prefton*, S. T. P. infrafcriptus, et exhibuit
" literas commiffionis—quas quidem per magiftrum *Thomam*
" *Somercotes* notarium publicum, &c. legi fecimus, tenorem fe-
" quentem in fe continentes." The fame as N°. 103, but dated
—" In manerio noftro de *Esfher*, *Augufti*, 1mo. A. D. 1485, et
" noftre confec. anno 39." [N°. 103 is repeated in a book
containing the like procefs in the preceding year by the fame
commiffary, in the parifh church of St. *Andrew* the apoftle, at
Farnham, *Sept.* 6th, anno 1484.] " Poft quarum literarum le&tu-
" ram—di&tus magifter *Jacobus Prefton*, quafdam procuratorias
" literas mag. *Richardi Mayewe* prefidentis, ut afferuit, *collegii beate*
" *Marie Magdalene*, &c. figillo rotundo communi, &c. in cera
" rubea impreffo figillatas realiter exhibuit, &c. et pro eifdem
" dnis fuis, &c. fecit fe partem, ac nobis fupplicavit ut juxta

---

z *Ecclefia Conventualis* de *Novo Loco* was the monaftery afterwards called the *New
Minfter*, or *Abbey* of *Hyde*, in the city of *Winchefter.* Should any intelligent reader wonder to fee that the *prior* of *Hyde Abbey* was commiffary to the bifhop of *Wintōn*, and fhould
conclude that there was a miftake in titles, and that the *abbot* muft have been here
meant ; he will be pleafed to recolle&t that this perfon was the fecond in rank ; for,
" next under the abbot, in every abbey, was the prior." Pref. to Notit. Monaft.
p. xxix. Befides, abbots were great perfonages, and too high in ftation to fubmit to
any office under the bifhop.

" formam

" formam in eifdem traditam procedere dignaremur, &c." After thefe proclamations no contradictor or abjector appearing—" ad " inftantem petitionem ipfius mag. *Jac. Preston*, procuratoris, &c. " procedendum fore decrevimus vocatis jure vocandis; nec non " mag. *Tho. Somercotes*, &c. in actorum noftrorum fcribam nomi- " navimus. Confequenter et ibidem tunc comparuit magifter " *Michael Clyff*, &c. et exhibuit in ea parte procuratorium fuum," for the prior and convent of the cathedral of *Winton*, " et fecit " fe partem pro eifdem.—Deinde comperuit coram nobis, &c. " honeftus vir *Willmus Cowper*," proctor for the bifhop as patron of the Priory of *Selborne*, and exhibited his " procuratorium, &c." After thefe were read in the prefence of *Clyff* and *Cowper*, " Pref- " ton, viva voce," petitioned the commiffary to annex and ap- propriate the Priory of *Selborne* to the college—" propter quod " fructus, redditus, et proventus ejufdem coll. adeo tenues funt, " et exiles, quod ad fuftentationem ejus, &c. non fufficiunt."—The commiffary, " ad libellandum et articulandum in fcriptis"— ad- journed the court to the 5th of *August*, then to be held again in the parifh church of *Esfher*.

W. Cowper being then abfent, *Radulphus Langley* appeared for the bifhop, and was admitted his proctor. *Preston* produced his libel or article in fcriptis for the union, &c. " et admitti petiit " eundem cum effectu; cujus libelli tenor fequitur.—In Dei " nomine, Amen. Coram nobis venerabili in Chrifto patre " *Richardo*, priore, &c. de *Novo Loco*, &c. commiffario, &c." Part of the college of Magd. dicit. allegat, and in his " fcriptis " proponit, &c."

" *Imprimis*

" *Imprimis*—that said college consists of a president and eighty
" scholars, besides sixteen choristers, thirteen servientes inibi al-
" tissimo famulantibus, et in scientiis plerisque liberalibus, pre-
" sertim in sacra theologia studentibus, nedum ad ipsorum presi-
" dentis et scholarium pro presenti et imposterum, annuente deo,
" incorporandorum in eodem relevamen; verum etiam ad omnium
" et singulorum tam scholarium quam religiosorum cujuscunque
" ordinis undequaque illuc confluere pro salubri doctrina volen-
" tium utilitatem multiplicem ad incrementa virtutis fideique
" catholice stabilimentum. Ita videlicet quod omnes et singuli
" absque personarum seu nationum delectu illuc accedere volentes,
" lecturas publicas et doctrinas tam in grammatica in loco ad
" collegium contiguo, ac philosophiis morali et naturali, quam
" in sacra theologia in eodem collegio perpetuis temporibus con-
" tinuandas libere atque gratis audire valeant et possint, ad
" laudem gloriam et honorem Dei, &c. extitit fundatum et
" stabilitum."

For the first *item* in this process see the beginning of this letter.
Then follows *item* the second—" that the revenues of the college non
" sufficiunt his diebus." " *Item*—that the premisses are true, &c.
" et super eisdem laborarunt, et laborant publica vox et fama.
" Unde facta fide petit pars eorundem that the Priory be annexed
" to the college: ita quod dicto prioratu vacante liceat iis ex
" tunc to take possession, &c." This libel, with the express
consent of the other proctors, we, the commissary, admitted, and
appointed the sixth of *August* for proctor *Preston* to prove the
premisses.

Preston produced witnesses, *W. Gyfford*, S. T. P. *John Nele*, A. M.
John Chapman, chaplain, and *Robert Baron*, literatus, who were ad-
mitted and sworn, when the court was prorogued to the 6th of
August;

August; and the witneſſes, on the ſame 5th of *August*, were examined by the commiſſary, " in capella infra manerium de *Eſsher* ſituata, " ſecrete et ſingillatim." Then follow the " literæ procuratoriæ :" firſt that of the college, appointing *Preſton* and *Langport* their proctors, dated *August* 30th, 1484; then that of the prior and convent of the cathedral of *Wintōn*, appointing *David Huſband* and *Michael Cleve*, dated *September* 4th, 1484 : then that of the biſhop, appointing *W. Gyfford, Radulphus Langley*, and *Will. Cowper*, dated *September* 3d, 1484. Conſec. 38°.—" Quo die adveniente " in dicta eccleſia parochiali," appeared " coram nobis" *James Preſton* to prove the contents of his libel, and exhibited ſome letters teſtimonial with the ſeal of the biſhop, and theſe were ad- mitted; and conſequenter *Preſton* produced two witneſſes, viz. Dominum *Thomam Aſhforde nuper priorem dicti prioratus*, et *Willm. Rabbys* literatum, who were admitted and ſworn, and examined as the others, by the commiſſary; " tunc & ibidem aſſiſtente ſcriba ſecrete & ſingillatim ;" and their depoſitions were read and made public, as follows :

Mr. *W. Gyfford*, S. T. P. aged 57, of the ſtate of *Magd. Coll.* &c. &c. as before :

Mr. *John Nele*, aged 57, proves the articles alſo :

Robert Baron, aged 56 :

Johannes Chapman, aged 35, alſo affirmed all the five articles :

Dompnus Thomas Aſhforde, aged 72 years—" dicit 2^{dum}. 3um. 4um. " articulos in eodem libello contentos, concernentes ſtatum dicti " prioratus de *Selebourne*, fuiſſe et eſſe veros."

W. Rabbys, ætat 40 ann. agrees with *Gyfford*, &c.

Then follows the letter from the biſhop, " in ſubſidium pro- " bationis," abovementioned — " *Willmus*, &c. ſalutem, &c. " noverint univerſitas veſtra, quod licet nos prioratui de *Sele-* " *bourne*, &c. pie compacientes ſollicitudines paſtorales, labores,

" diligentias

" diligentias quam plurimas per nos & commiffarios noftros pro
" reformatione ftatus ejus impenderimus, jufticia id pofcente;
" nihilominus tamen," &c. as in the article—to " defperatur,"
dated " in manerio noftro de *Esfher*, *Aug.* 3d, 1485, & confec. 39."
Then, on the 6th of *Auguft*, *Prefton*, in the prefence of the other
proctors, required that they fhould be compelled to anfwer;
when they all allowed the articles " fuiffe & effe vera;" and the
commiffary, at the requeft of *Prefton*, concluded the bufinefs, and
appointed *Monday*, *Auguft* 8th, for giving his decree in the fame
church of *Esfher*; and it was that day read, and contains a recapi-
tulation, with the fentence of union, &c. witneffed and attefted.

As foon as the prefident and fellows of *Magdalen college* had obtained the decifion of the commiffary in their favour, they proceeded to fupplicate the *pope*, and to entreat his holinefs that he
would give his fanction to the fentence of union. Some difficulties were ftarted at *Rome*; but they were furmounted by the
college agent, as appears by his letters from that city. At length
pope Innocent VIII. by a *bull*[a] bearing date the 8th day of *June*, in the
year of our Lord 1486, and in the fecond year of his pontificate,
confirmed what had been done, and fuppreffed the convent.

Thus fell the confiderable and well-endowed Priory of *Selborne*
after it had fubfifted about two hundred and fifty-four years: about
feventy-four years after the fuppreffion of Priories alien by
Henry V. and about fifty years before the general diffolution of
monafteries by *Henry* VIII. The founder, it is probable, had

---

[a] There is nothing remarkable in this *bull* of *pope Innocent* except the ftatement of the
annual revenue of the Priory of *Selborne*, which is therein eftimated at 160 *flor. auri*;
whereas bifhop *Godwin* fets it at 337*l.* 15*s.* 6¼*d.* Now a *floren*, fo named, fays
Camden, becaufe made by *Florentines*, was a gold coin of king *Edward* III. in value 6*s.*
whereof 160 is not one feventh part of 337*l.* 15*s.* 6¼*d.*

G g g

fondly

fondly imagined that the facredness of the inftitution, and the pious motives on which it was eftablifhed, might have preferved it inviolate to the end of time—yet it fell,

    " To teach us that God attributes to *place*
    " No fanctity, if none be thither brought
    " By men, who there frequent, or therein dwell."

                                   Milton's Paradise Loft.

LETTER XXV.

WAINFLEET did not long enjoy the fatisfaction arifing from this new acquifition; but departed this life in a few months after he had effected the union of the Priory with his late founded college; and was fucceeded in the fee of *Winchefter* by *Peter Courtney*, fome time towards the end of the year 1486.

In the beginning of the following year the new bifhop releafed the prefident and fellows of *Magdalen College* from all actions refpecting the Priory of *Selborne*; and the prior and convent of Saint *Swithun*, as the chapter of *Winchefter* cathedral, confirmed the releafe [a].

N. 293. " Relaxatio *Petri* ēpi *Wintōn Ricardo Mayew*, Prefi-" denti omnium actionum occafione indempnitatis fibi debite pro " unione Prioratus de *Selborne* dicto collegio. *Jan.* 2. 1487. et " tranflat. anno 1°."

N. 374. " Relaxatio *prioris* et *conventus* S^{ti}. *Swithini Wintōn* " confirmans relaxationem *Petri* ep. *Winton*." 1487. *Jan.* 13.

    [a] The bifhops of *Winchefter* were patrons of the Priory.

                                     Affhforde,

Aſhforde, the depoſed prior, who had appeared as an evidence for the impropriation of the Priory at the age of ſeventy-two years, that he might not be deſtitute of a maintenance, was penſioned by the college to the day of his death ; and was living on till 1490, as appears by his acquittances.

REG. A. ff. 46.

" Omnibus Chriſti fidelibus ad quos preſens ſcriptum per-
" venerit, *Richardus Mayew*, preſidens, &c. et ſcolares, ſalutem in
" Domino."

" Noveritis nos prefatos preſidentem et ſcolares dediſſe, conceſ-
" ſiſſe, et hoc preſenti ſcripto confirmaſſe *Thome Aſhforde, capellano*,
" quendam annualem redditum ſex librarum treſdecim ſoli-
" dorum et quatuor denariorum bone et legalis monete
' Anglie—ad terminum vite prefati *Thome*"—to be paid from the poſſeſſions of the college in *Baſingſtoke.*—" In cujus rei teſtimo-
" nium ſigillum noſtrum commune preſentibus apponimus.
" Dat. Oxon. in coll. noſtro ſupra dicto primo die menſis *Junii*
" anno regis *Ricardi* tertii ſecundo." viz. 1484. The college, in their grant to *Aſhforde*, ſtyle him only *capellanus* ; but the annuitant very naturally, and with a becoming dignity, aſſerts his late title in his acquittances, and identifies himſelf by the addition of *nuper priorem*, or late prior.

As, according to the perſuaſion of the times, the depriving the founder and benefactors of the Priory of their maſſes and ſervices would have been deemed the moſt impious of frauds, biſhop *Wainfleet*, having by ſtatute ordained four *obits* for himſelf to be celebrated in the chapel of *Magdalen College*, enjoined in one of them a ſpecial collect for the anniverſary of *Peter de Rupibus*, with a particular prayer—" *Deus Indulgentiarum.*"

The

The college also sent *Nicholas Langrish*, who had been a chantry priest at *Selborne*, to celebrate mass for the souls of all that had been benefactors to the said Priory and college, and for all the faithful who had departed this life.

N. 356. *Thomas Knowles*, presidens, &c.—" damus et con-
" cedimus *Nicholao Langrish* quandum capellaniam, vel salarium,
" five alio quocunque nomine censeatur, in prioratu *quondam* de
" *Selborne* pro termino 40 annorum, si tam diu vixerit. Ubi
" dictus mag^r. *Nicholaus* celebrabit pro animabus omnium bene-
" factorum dicti prioratus et coll. nostri, et omnium fidelium
" defunctorum. Insuper nos, &c. concedimus eidem ibidem
" celebranti in sustentationem suam quandam annualem pensi-
" onem sive annuitatem octo librarum &c.—in dicta capella
" dicti prioratus—concedimus duas cameras contiguas ex parte
" boreali dicte capelle, cum *una coquina*, et cum uno stabulo
" conveniente pro tribus equis, cum pomerio eidem adjacente voc.
" le *Orcheyard*—Preterea 26s. 8d. per ann. ad inveniendum unum
" clericum ad serviendum sibi ad altare, et aliis negotiis necessariis
" ejus."—His wood to be granted him by the president on the progress.—He was not to absent himself beyond a certain time ; and was to superintend the coppices, wood, and hedges.——
" Dat. 5to. die *Julii*. an°. *Hen*. VIII^vi. 36°." [viz. 1546.]

Here we see the Priory in a new light, reduced as it were to the state of a *chantry*, without prior and without canons, and attended only by a priest, who was also a sort of bailiff or woodman, his assistant clerk, and his female cook. *Owen Oglethorpe*, president, and *Magd. Coll.* in the fourth year of *Edward* VI. viz. 1551, granted an annuity of ten pounds a year for life to *Nich. Langrish*, who, from the preamble, appears then to have been fellow of that society : but, being now superannuated for business, this pension is

granted

granted him for thirty years, if he fhould live fo long. It is faid of him—" cum jam fit provectioris etatis quam ut," &c.

Laurence Stubb, prefident of *Magd. Coll.* leafed out the Priory lands to *John Sharp*, hufbandman, for the term of twenty years, as early as the feventeenth year of *Henry* VIII.—viz. 1526 : and it appears that *Henry Newlyn* had been in poffeffion of a leafe before, probably towards the end of the reign of *Henry* VII. *Sharp's* rent was vi^{li}. per ann.—Regift. B. p. 43.

By an abftract from a leafe lying before me, it appears that *Sharp* found a houfe, two barns, a ftable, and a *duf*-houfe, [*dove*-houfe] built, and ftanding on the fouth fide of the old Priory, and late in the occupation of *Newlyn*. In this abftract alfo are to be feen the names of all the fields, many of which continue the fame to this day [b]. Of fome of them I fhall take notice, where any thing fingular occurs.

And here firft we meet with *Paradyfs* [Paradife] *mede*. Every convent had it's *Paradife* ; which probably was an enclofed orchard, pleafantly laid out, and planted with fruit-trees.—*Tyle-houfe* grove, fo diftinguifhed from having a *tiled houfe* near it [c].

[b] It may not be amifs to mention here that various names of *tithings, farms, fields, woods,* &c. which appear in the ancient deeds, and evidences of feveral centuries ftanding, are ftill preferved in common ufe with little or no variation :—as *Norton, Southington, Durton, Achangre, Blackmore, Bradfhot, Rood, Pleftor,* &c. &c. At the fame time it fhould be acknowledged that other places have entirely loft their original titles, as *le Buri* and *Trucftede* in this village ; and *la Liega,* or *la Lyge,* which was the name of the original fite of the Priory, &c.

[c] Men at firft heaped fods, or fern, or heath, on their roofs to keep off the inclemencies of weather ; and then by degrees laid ftraw or haum. The firft refinements on roofing were fhingles, which are very ancient. Tiles are a very late and imperfect covering, and were not much in ufe till the beginning of the fixteenth century. The firft tiled houfe at *Nottingham* was in 1503.

Butt wood

Butt-wood clofe; here the fervants of the Priory and the village-fwains exercifed themfelves with their long bows, and fhot at a mark againft a butt, or bank [d].—*Cundyth* [conduit] *wood:* the engroffer of the leafe not underftanding this name has made a ftrange barbarous word of it. *Conduit-wood* was and is a fteep, rough cow-pafture, lying above the Priory, at about a quarter of a mile to the fouth-weft. In the fide of this field there is a fpring of water that never fails; at the head of which a ciftern was built which communicated with leaden pipes that conveyed water to the monaftery. When this refervoir was firft conftructed does not appear, we only know that it underwent a repair in the epifcopate of bifhop *Wainfleet,* about the year 1462 [e]. Whether thefe pipes only conveyed the water to the Priory for common and culinary purpofes, or contributed to any matters of ornament and elegance, we fhall not pretend to fay; nor when artifts and mechanics firft underftood any thing of *hydraulics,* and that water confined in tubes would rife to it's original level. There is a perfon now living who had been employed formerly in digging for thefe pipes, and once difcovered feveral yards, which they fold for old lead.

There was alfo a plot of ground called *Tan-houfe garden:* and " *Tannaria fua,*" a *tan-yard of their own,* has been mentioned in Letter XVI. This circumftance I juft take notice of, as an inftance that monafteries had trades and occupations carried on within themfelves [f].

[d] There is alfo a *Butt-clofe* juft at the back of the village.

[e] N. 381. " Claufure terre abbatie ecclefie parochiali de *Seleburne.* ix *s.* iiii *d.* " Reparacionibus domorum predicti prioratus iiii *lib.* xi *s. Aque conduct.* ibidem. " xxiii *d.*"

[f] There is ftill a wood near the Priory called *Tanner's wood.*

Regiftr.

Regiftr. B. pag. 112. Here we find a leafe of the parfonage of *Selborne* to *Thomas Sylvefter* and *Miles Arnold*, hufbandmen—of the tythes of all manner of corne pertaining to the parfonage—with the offerings at the chapel of *Whaddon* belonging to the faid parfonage. Dat. *June* 1. 27th. *Hen.* 8th. [viz. 1536.]

As the chapel of *Whaddon* has never been mentioned till now, and as it is not noticed by bifhop *Tanner* in his *Notitia Monaftica*, fome more particular account of it will be proper in this place. *Whaddon* was a chapel of eafe to the mother church of *Selborne*, and was fituated in the tithing of *Oakhanger*, at about two miles diftance from the village. The farm and field whereon it ftood are ftill called *chapel farm* and *field*[g]: but there are no remains or traces of the building itfelf, the very foundations having been deftroyed before the memory of man. In a farm yard at *Oakhanger* we remember a large hollow ftone of a clofe fubftance, which had been ufed as a *hog-trough*, but was then broken. This ftone, tradition faid, had been the baptifmal font of *Whaddon chapel*. The chapel had been in a very ruinous ftate in old days; but was new-built at the inftance of bifhop *Wainfleet*, about the year 1463, during the *firft* priorfhip of *Berne*, in confequence of a fequeftration iffued forth by that vifitor againft the Priory on account of notorious and fhameful dilapidations[h].

[g] This is a manor-farm, at prefent the property of *Lord Stawell*; and belonged probably in ancient times to *Jo.* de *Venur*, or *Venuz*, one of the firft benefactors to the Priory.

[h] See Letter XIX. of thefe Antiquities.—" Summa total. folut. de novis edifica-" tionibus, et reparacionibus per idem tempus, ut patet per comput."

" Videlicet de nova edificat. *Capelle Marie* de *Wadden*. xiiii *lib.* v s. viii *d.*---Repara-" cionibus ecclefie Prioratus, cancellor. et capellar. ecclefiarum et capellarum de *Sel-*" *borne*, et *Eftworhlam*"---&c. &c.

The

The *Selborne* rivulet becomes of some breadth at *Oakhanger*, and, in very wet seasons, swells to a large flood. There is a bridge over the stream at this hamlet of considerable antiquity and peculiar shape, known by the name of *Tunbridge*: it consists of one single blunt *gothic* arch, so high and sharp as to render the passage not very convenient or safe. Here was also, we find, a bridge in very early times; for *Jacobus de Hochangre*, the first benefactor to the Priory of *Selborne*, held his estate at *Hochangre* by the service of providing the king one foot-soldier for forty days, and by building this bridge. " *Jacobus de Hochangre* tenet *Hochangre* in com. " *Southampton, per Serjantiam* [i], inveniendi unum valectum in exer- " citu Domini regis [scil. *Henrici* III[tii].] per 40 dies; et ad fa- " ciendum *pontem* de *Hochangre*: et valet per ann. C. s."

Blount's Ancient Tenures, p. 84.

A *dove-house* was a constant appendant to a manerial dwelling : of this convenience more will be said hereafter.

A *corn-mill* was also esteemed a necessary appendage of every manor ; and therefore was to be expected of course at the Priory of *Selborne*.

The prior had *secta molendini*, or *ad molendinum* [k]; a power of compelling his vassals to bring their corn to be ground at his mill, according to old custom. He had also, according to bishop *Tan-ner, secta molendini de Strete*: but the purport of *Strete*, we must confess, we do not understand. *Strete*, in old English, signifies a road or highway, as *Watling Strete*, &c. therefore the prior might

---

[i] *Sargentia*, a sort of tenure of doing something for the king.

[k] " Servitium, quo feudatorii grana sua ad Domini molendinum, ibi molenda per-" ferre, exconsuetudine, astringuntur."

have

have some mill on a high road. The Priory had only one mill originally at *Selborne*; but, by grants of lands, it came possessed of one at *Durton*, and one at *Oakhanger*, and probably some on it's other several manors[1]. The mill at the Priory was in use within the memory of man, and the ruins of the mill-house were standing within these thirty years : the pond and dam, and miller's dwelling, still remain. As the stream was apt to fail in very dry summers, the tenants found their situation very distressing, for want of water, and so were forced to abandon the spot. This inconvenience was probably never felt in old times, when the whole district was nothing but woodlands : and yet several centuries ago there seem to have been two or three mills between *Well-head* and the *Priory*. For the reason of this assertion, see Letter XXIX. to Mr. *Barrington*.

Occasional mention has been made of the many privileges and immunities enjoyed by the convent and it's priors; but a more particular state seems to be necessary. The author therefore thinks this the proper place, before he concludes these antiquities, to introduce all that has been collected by the judicious bishop *Tanner*, respecting the Priory and it's advantages, in his *Notitia Monastica*, a book now seldom seen, on account of the extravagance of it's price ; and being but in few hands cannot be easily consulted[m]. He also adds a few of it's many privileges from other authorities :—the account is as follows. *Tanner*, page 166.

---

[1] *Thomas Knowles*, president, &c. ann. *Hen.* 8vi. xxiii°. [viz. 1532.] demised to *J. Whitelie* their mills, &c. for twenty years. Rent. xxiii *s.* iiii *d.*---*Accepted Frewen*, president, &c. ann. *Caroli* xv. [viz. 1640.] demised to *Jo. Hook* and *Elizabeth*, his wife, the said mills. Rent as above.

[m] A few days after this was written a new edition of this valuable work was announced, in the month of *April* of the year 1787, as published by Mr. *Nasmith*.

Hhh SELEBURNE.

S E L E B U R N E.

A priory of *black canons*, founded by the often-mentioned *Peter de Rupibus*, bishop of *Winchester*, A. D. 1233, and dedicated to the blessed *Virgin Mary:* but was suppressed — and granted to *William Wainfleet*, bishop of *Winchester*, who made it part of the endowment of St. *Mary Magdalene College* in *Oxford*. The bishops of *Winchester* were patrons of it. [Pat. 17. *Edw*. II.] Vide in *Mon. Angl.* tom. II. p. 343. " Cartam fundationis ex ipso autogra-
" pho in archivis *Coll. Magd. Oxōn.* ubi etiam conservata sunt re-
" gistra, cartæ, rentalia et alia munimenta ad hunc prioratum
" spectantia.

" Extracta quædam e registro MSS. in bibl. Bodl. *Dodsworth,*
" vol. 89. f. 140.

" Cart. antiq. N. N. n. 33. P. P. n. 48. et 71. Q. Q. n. 40.
" plac. coram justit. itin. [Southampton] 20 *Hen*. rot. 25. De
" eccl. de *Basing*, & *Basingstoke*. Plac. de juratis apud *Winton.*
" 40 *Hen.* III. rot.—Profecta molendini de *Strete.* Cart. 54. *Hen.* III
" m. 3. [*De mercatu, & feria apud Seleborne*, a mistake.] Pat. 9.
" *Edw*. I. m.—Pat. 30. *Edw*. I. m.—Pat. 33. *Edw*. I. p. 1.
" m.—Pat. 35. *Edw*. I. m.—Pat. 1. *Edw*. II. p. 1. m. 9. Pat. 5.
" *Edw*. II. p. 1. m. 21. De terris in *Achanger*. Pat. 6. *Edw*. II.
" p. 1. m. 7. de eisdem. Brev. in Scacc. 6. *Edw*. II. Pasch.
" rot. 8. Pat. 17. *Edw*. II. p. 1. m.—Cart. 10. *Edw*. III. n. 24.
" Quod terræ suæ in *Seleburn, Achangre, Norton, Basings, Basing-*
" *stoke*, and *Nately*, sint de afforestatæ, and pro aliis libertatibus.
" Pat. 12. *Edw*. III. p. 3. m. 3.—Pat. 13. *Edw*. III. p. 1. m.—
" Cart. 18. *Edw*. III. n. 24."

" N. N. 33.

" N. N. 33. Rex conceſſit quod prior, et canonici de *Seleburn*
" habeant per terras ſuas de *Seleburne, Achangre, Norton, Brompden,*
" *Baſinges, Baſingſtoke,* & *Nately,* diverſas libertates.

" P. P. 48. Quod prior de *Seleburne,* habeat terras ſuas quietas
" de vaſto, et regardo."—*Extracts from Ayloffe's Calendars of
Ancient Charters.*

" Placita de juratis & aſſis coram *Salom* de *Roff,* & ſociis
" ſuis juſtic. itiner. apud *Wynton* in comitatu *Sutht.*—anno regni
" R. *Edvardi* filii reg. *Henr.* octavo.—Et *Por de Seleborn* ht
" in *Selebr.* ſure. thurſet. pillory, *emendaſſe panis,* & *ſuis."*
[cereviſiæ.]—*Chapter-houſe Weſtminſter.*

" Placita Foreſte apud *Wynton* in com. *Sutham*—Anno reg.
" *Edwardi* octavo coram Rog. de *Clifford.*—&c. Juſtic. ad eadem
" placita audienda et tminand. aſſigtis.

" Carta Pror de *Seleburn,* H. Dei gra. rex. angl. &c. Con-
" ceſſim. prior. ſce. *Marie* de *Seleburn.* et canonicis ibidem Deo
" ſervient. — — — q ipi et oes hoies ſui in pdcis terris ſuis et
" tenementis manentes ſint in ppetum quieti de ſectis Swanemotor.
" et omnium alior. placitor. for. et de *eſpeltamentis* canum. et de
" omnibus ſubmonitoibz. placitis querelis et exaccoibus et oc-
" coibz. ad for. et for. et viridar. et eor. miniſtros ptinentibz."—
Chapter-houſe, Weſtminſter.

" Plita Foreſtarum in com. *Sutht.* apud *Suthamton* — — anno
" regni regis *Edwardi tcii* poſt conqueſtum quarto coram *Johe
Mantvers.* &c. juſtic. itinand. &c.

De hiis qui clamant libtates infra Foreſtas in com. *Sutht.*

" Prior de *Selebourne* clamat eſſe quietus erga dnm regem de
" omnibus finibus et amerciamentis p tnſgr. et omnibus exaccoibz
" ad Dom. regem vel hered. ſuos ptinent. pret. plita corone reg.

" Item

" Item clamat q^d si aliquis hominum suorum de terris et
" ten. p. delicto suo vitam aut membrum debeat amittere vel
" fugiat, & judico stare noluerit vel aliud delictum fecit pro
" quo debeat catella sua amittere, ubicunq; justitia fieri debeat
" omnia catella illa sint ptci Prioris et successor. suor. Et liceat
" eidem priori et ballis suis ponere se in seisinam in hujusmodi
" catall. in casibus pdcis sine disturbacone ballivor. dni reg.
" quorumcunque.

" Item clam. quod licet aliqua libtatum p dnm regem con-
" cessar. pcessu temporis quocunq ; casu contingente usi non
" fuerint, nlominus postea eadm libtate uti possit. Et pdcus prior
" quesitus p justic. quo waranto clamat omn. terr. et ten. sua in
" *Seleburne, Norton, Basynges, Basyngestoke,* & *Nattele,* que prior
" domus pdte huit & tenuit X^{mo}. die *April* anno regni dni *Hen.*
" reg. pavi dni reg. nue XVIII. imppm esse quieta de vasto et
" regardo, et visu forestarior. et viridarior. regardator. et omnium
" ministrorum foreste." &c. &c.—*Chapter-house, Westminster.*

LETTER.

LETTER XXVI.

THOUGH the evidences and documents of the Priory and parish of *Selborne* are now at an end, yet, as the author has still several things to say respecting the present state of that convent and it's *Grange*, and other matters, he does not see how he can acquit himself of the subject without trespassing again on the patience of the reader by adding one supplementary letter.

No sooner did the Priory (perhaps much out of repair at the time) become an appendage to the college, but it must at once have tended to swift decay. *Magdalen College* wanted now only two chambers for the chantry priest and his assistant; and therefore had no occasion for the hall, dormitory, and other spacious apartments belonging to so large a foundation. The roofs neglected, would soon become the possession of daws and owls; and, being rotted and decayed by the weather, would fall in upon the floors; so that all parts must have hastened to speedy dilapidation and a scene of broken ruins. Three full centuries have now passed since the dissolution; a series of years that would craze the stoutest edifices. But, besides the slow hand of time, many circumstances have contributed to level this venerable structure with the ground; of which nothing now remains but one piece of a wall of about ten feet long, and as many feet high, which probably was part of an out-house. As early as the latter end of the reign of *Hen.* VII.

we

we find that a farm-houfe and two barns were built to the fouth of the Priory, and undoubtedly out of it's materials. Avarice again has much contributed to the overthrow of this ftately pile, as long as the tenants could make money of it's ftones or timbers. Wantonnefs, no doubt, has had a fhare in the demolition; for boys love to deftroy what men venerate and admire. A remarkable inftance of this propenfity the writer can give from his own knowledge. When a fchoolboy, more than fifty years ago, he was eye-witnefs, perhaps a party concerned, in the undermining a portion of that fine old ruin at the north end of *Bafingftoke* town, well known by the name of *Holy Ghoft Chapel*. Very providentially the vaft fragment, which thefe thoughtlefs little engineers endeavoured to fap, did not give way fo foon as might have been expected; but it fell the night following, and with fuch violence that it fhook the very ground, and, awakening the inhabit-ants of the neighbouring cottages, made them ftart up in their beds as if they had felt an earthquake. The motive for this dangerous attempt does not fo readily appear: perhaps the more danger the more honour, thought the boys; and the notion of doing fome mifchief gave a zeft to the enterprize. As *Dryden* fays upon an other occafion,

"It look'd fo like a fin it pleas'd the more."

Had the Priory been only levelled to the furface of the ground, the difcerning eye of an antiquary might have afcertained it's *ichnography*, and fome judicious hand might have developed it's dimenfions. But, befides other ravages, the very foundations have been torn up for the repair of the highways: fo that the fite of this convent is now become a rough, rugged pafture-field, full of

                                                            hillocks

hillocks and pits, choaked with nettles, and dwarf-elder, and trampled by the feet of the ox and the heifer.

As the tenant at the Priory was lately digging among the foundations, for materials to mend the highways, his labourers difcovered two large ftones, with which the farmer was fo pleafed that he ordered them to be taken out whole. One of thefe proved to be a large *Doric* capital, worked in good tafte; and the other a bafe of a pillar; both formed out of the foft freeftone of this diftrict. Thefe ornaments, from their dimenfions, feem to have belonged to maffive columns; and fhew that the church of this *convent* was a large and coftly edifice. They were found in the fpace which has always been fuppofed to have contained the fouth tranfept of the Priory church. Some fragments of large pilafters were alfo found at the fame time. The diameter of the capital was two feet three inches and an half; and of the column, where it had ftood on the bafe, eighteen inches and three quarters.

Two years ago fome labourers digging again among the ruins founded a fort of rude thick vafe or urn of foft ftone, containing about two gallons in measure, on the verge of the brook, in the very fpot which tradition has always pointed out as having been the fite of the convent kitchen. This clumfy utenfil [n], whether intended for holy water, or whatever purpofe, we were going to procure, but found that the labourers had juft broken it in pieces, and carried it out on the highways.

---

[n] A judicious antiquary, who faw this vafe, obferved, that it poffibly might have been a ftandard *meafure* between the monaftery and it's tenants. The priory we have mentioned claimed the affize of bread and beer in *Selborne* manor; and probably the adjuftment of dry meafures for grain, &c.

The

The Priory of *Selborne* had poffeffed in this village a *Grange*, an ufual appendage to manerial eftates, where the fruits of their lands were ftowed and laid up for ufe, at a time when men took the natural produce of their eftates in kind. The manfion of this fpot is ftill called the *Grange*, and is the manor-houfe of the convent poffef-fions in this place. The author has converfed with very ancient people who remembered the old original *Grange*; but it has long given place to a modern farm-houfe. *Magdalen College* holds a court-leet and court-baron° in the great wheat-barn of the faid *Grange*, annually, where the *Prefident* ufually fuperintends, attended by the *burfar* and *fteward* of the college. ᴾ

The following uncommon prefentment at the court is not un-worthy of notice. There is on the fouth fide of the king's field, (a large common-field fo called) a confiderable tumulus, or hillock, now covered with thorns and bufhes, and known by the name of *Kite's Hill*, which is prefented, year by year, in court as not ploughed. Why this injunction is ftill kept up refpecting this fpot, which is furrounded on all fides by arable land, may be a queftion not eafily folved, fince the ufage has long furvived the knowledge of the intention thereof. We can only fuppofe that as the prior, befides *thurfet* and *pillory*, had alfo *furcas*, a power of life and death, that he might have referved this little eminence as the place of execution for delinquents. And there is the more reafon to fuppofe fo, fince a fpot juft by is called *Gally* [Gallows] hill.

° The time when this court is held is the mid-week between *Eafter* and *Whitfuntide*.

ᴾ *Owen Oglethorpe*, prefident, &c. an. *Edw. Sexti*, primo [viz. 1547.] demifed to *Robert Arden Selborne Grange* for twenty years. Rent viii.- -*Index of Leafes*.

The

The lower part of the village next the *Grange*, in which is a pond and a ftream, is well known by the name of *Gracious-ftreet*, an appellation not at all underftood. There is a lake in *Surrey*, near *Chobham*, called alfo *Gracious-pond*: and another, if we miftake not, near *Hedleigh*, in the county of *Hants*. This ftrange denomination we do not at all comprehend, and conclude that it may be a corruption from fome *Saxon* word, itfelf perhaps forgotten.

It has been obferved already, that Bifhop *Tanner* was miftaken when he refers to an evidence of *Dodfworth* "*De mercatu et* FERIA *de Seleburne.*" *Selborne* never had a chartered fair; the prefent fair was fet up fince the year 1681, by a fet of jovial fellows, who had found in an old almanack that there had been a fair here in former days on the firft of *Auguft*; and were defirous to revive fo joyous a feftival. Againft this innovation the vicar fet his face, and perfifted in crying it down, as the probable occafion of much intemperance. However the fair prevailed; but was altered to the twenty-ninth of *May*, becaufe the former day often interfered with wheat-harveft. On that day it ftill continues to be held, and is become an ufeful mart for cows and calves. Moft of the lower houfe-keepers brew beer againft this holiday, which is dutied by the excifeman; and their becoming victuallers for the day without a licenfe is overlooked.

Monafteries enjoyed all forts of conveniencies within themfelves. Thus at the Priory, a low and moift fituation, there were ponds and ftews for their fifh: at the fame place alfo, and at the *Grange* [c]in *Culver-croft*, there were dove-houfes; and on the hill oppofite to the *Grange* the prior had a warren, as the names of *The Coney-crofts* and *Coney-croft Hanger* plainly teftify[d].

[c] *Culver*, as has been obferved before, is *Saxon* for a pigeon.
[d] A warren was an ufual appendage to a manor.

Nothing

Nothing has been said as yet respecting the tenure or holding of the *Selborne* estates. *Temple* and *Norton* are manor farms and freehold; as is the manor of *Chapel* near *Oakhanger*, and also the estate at *Oakhanger-house* and *Black-moor*. The *Priory* and *Grange* are leasehold under *Magdalen college*, for twenty-one years, renewable every seven: all the smaller estates in and round the village are copyhold of inheritance under the college, except the little remains of the *Gurdon-manor*, which had been of old leased out upon lives, but have been freed of late by their present lord, as fast as those lives have dropped.

Selborne seems to have derived much of it's prosperity from the near neighbourhood of the Priory. For monasteries were of considerable advantage to places where they had their sites and estates, by causing great resort, by procuring markets and fairs, by freeing them from the cruel oppression of forest-laws, and by letting their lands at easy rates. But, as soon as the convent was suppressed, the town which it had occasioned began to decline, and the market was less frequented; the rough and sequestered situation gave a check to resort, and the neglected roads rendered it less and less accessible.

That it had been a considerable place for size formerly appears from the largeness of the church, which much exceeds those of the neighbouring villages; by the ancient extent of the burying ground, which, from human bones occasionally dug up, is found to have been much encroached upon; by giving a name to the hundred; by the old foundations and ornamented stones, and tracery of windows that have been discovered on the north-east side of the village; and by the many vestiges of disused fish-ponds still to be seen around it. For ponds and stews were multiplied in the times of popery, that the affluent might enjoy some

variety

variety at their tables on faſt days; therefore the more they abounded the better probably was the condition of the inha- bitants.

More PARTICULARS *reſpecting the* OLD FAMILY TORTOISE, *omitted in the* Natural Hiſtory.

BECAUSE we call this creature an abject reptile, we are too apt to undervalue his abilities, and depreciate his powers of inſtinct. Yet he is, as Mr. *Pope* ſays of his lord,

— — — " Much too wiſe to walk into a well:"

and has ſo much diſcernment as not to fall down an haha; but to ſtop and withdraw from the brink with the readieſt precaution.

Though he loves warm weather he avoids the hot ſun; becauſe his thick ſhell, when once heated, would, as the poet ſays of ſolid armour—" ſcald with ſafety." He therefore ſpends the more ſultry hours under the umbrella of a large cabbage-leaf, or amidſt the waving foreſts of an aſparagus-bed.

But as he avoids heat in the ſummer, ſo, in the decline of the year, he improves the faint autumnal beams, by getting within the reflection of a fruit-wall: and, though he never has read that planes inclining to the horizon receive a greater ſhare of

warmth,

warmth *, he inclines his shell, by tilting it against the wall, to collect and admit every feeble ray.

Pitiable seems the condition of this poor embarrassed reptile: to be cased in a suit of ponderous armour, which he cannot lay aside; to be imprisoned, as it were, within his own shell, must preclude, we should suppose, all activity and disposition for enter-prize. Yet there is a season of the year (usually the beginning of June) when his exertions are remarkable. He then walks on tip-toe, and is stirring by five in the morning; and, traversing the gar-den, examines every wicket and interstice in the fences, through which he will escape if possible: and often has eluded the care of the gardener, and wandered to some distant field. The motives that impel him to undertake these rambles seem to be of the amo-rous kind: his fancy then becomes intent on sexual attachments, which transport him beyond his usual gravity, and induce him to forget for a time his ordinary solemn deportment.

* Several years ago a book was written entitled "Fruit-walls improved by inclining "them to the horizon:" in which the author has shewn, by calculation, that a much greater number of the rays of the sun will fall on such walls than on those which are per-pendicular.

ADDITIONS

ADDITIONS to LETTER XLI, page 236.

Of all the propensities of plants none seem more strange than their different periods of blossoming. Some produce their flowers in the winter, or very first dawnings of spring; many when the spring is established; some at midsummer, and some not till autumn. When we see the *helleborus fœtidus* and *helleborus niger* blowing at Christmas, the *helleborus hyemalis* in January, and the *helleborus viridis* as soon as ever it emerges out of the ground, we do not wonder, because they are kindred plants that we expect should keep pace the one with the other. But other congenerous vegetables differ so widely in their time of flowering, that we cannot but admire. I shall only instance at present in the *crocus sativus*, the vernal, and the autumnal crocus, which have such an affinity, that the best botanists only make them varieties of the same *genus*, of which there is only one *species*; not being able to discern any difference in the *corolla*, or in the internal structure. Yet the *vernal crocus* expands it's flowers by the beginning of March at farthest, and often in very rigorous weather; and cannot be retarded but by some violence offered:—while the *autumnal* (the *Saffron*) defies the influence of the spring and summer, and will not blow till most plants begin to fade and run to seed. This circumstance is one of the wonders of the creation, little noticed, because a common occurrence; yet ought not to be overlooked on account

of

of it's being familiar, since it would be as difficult to be explained
as the most stupendous phænomenon in nature.

> Say, what impels, amidst surrounding snow
> Congeal'd, the *crocus*' flamy bud to glow?
> Say, what retards, amidst the summer's blaze,
> Th' *autumnal bulb*, till pale, declining days?
> The GOD of SEASONS; whose pervading power
> Controls the sun, or sheds the fleecy shower:
> He bids each flower his quick'ning word obey;
> Or to each lingering bloom enjoins delay.

APPENDIX.

APPENDIX,

NUMBER I.

II.

Carta Petri et conventus ecclesie Winton. pro fundatione prioratus de Seleburne, &c. dat. 1233.

Omnibus Christi fidelibus ad quos presens scriptum pervenerit. P. divina miseracione *Wintōn* ecclesie minister humilis salutem in Domino : Ex officio pastorali tenemur viros religiosos, quī pauperes spiritu esse pro Christo neglectis lucris temporalibus elegerunt ; spirituali affectu diligere, fovere pariter et creare, eorumq; quieti sollicite providere ; ut tanto uberiores fructus de continua in lege Dei meditatione percipiant, quanto a conturbationibus malignorum amplius fuerint ex patroni provisione et ecclesiastica defensione securi. Hinc est quod universitati vestre notificamus, nos divine caritatis instinctu, de assensu conventus ecclesie nostre *Wintōn*, fundasse domum religiosam, ordinis magni patris *Augustini*, in honore Dei et gloriose semper virginis ejusdem Dei genetricis

Marie,

Marie, apud *Seleburne*; ibidemque canonicos regulares inftituiffe:
ad quorum fuftentationem et hofpitum et pauperum fufceptionem,
dedimus, conceffimus, et prefenti carta noftra confirmavimus
eifdem canonicis, totam terram quam habuimus de dono *Jacobi*
de *Acangre* : et totam terram, curfum aque, bofcum et pratum que
habuimus de dono *Jacobi* de *Nortone*; et totam terram bofcum et
redditum que habuimus de dono domini *Henrici* regis *Anglie*; cum
omnibus predictarum poffeffionum pertinentiis. Dedimus etiam
et conceffimus in proprios ufus eifdem canonicis ecclefiam
predicte ville de *Seleburne*, et ecclefias de *Bafing*, et de *Bafinge-
ftok*, cum omnibus earundem ecclefiarum capellis, libertatibus, et aliis pertinenciis; falva honefta et fufficienti fuf-
tentatione vicariorum in predictis ecclefiis miniftrantium;
quorum prefentatio ad priorem predicte domus religiofe de *Seleburne* et canonicos ejufdem loci in perpetuum pertinebit. Preterea
poffeffiones et redditus, ecclefias five decimas, quas in epifcopatu
noftro adempti funt, vel in pofterum, Deo dante, juftis modis
poterunt adipifci, fub noftra et *Winton* ecclefie protectione fuf-
cepimus, et epifcopalis auctoritate officii confirmavimus; eadem
auctoritate firmiter inhibentes, ne quis locum, in quo divino funt
officio mancipati, feu alias eorum poffeffiones, invadere vi vel
fraude vel ingenio malo occupare audeat, vel etiam retinere, aut
fratres converfos, fervientes, vel homines eorum aliqua violentia
perturbare, five fugientes ad eos caufa falutis fue confervande a
feptis domus fue violenter prefumat extraere. Precipimus autem
ut in eadem domo religiofa de *Seleburne* ordo canonicus, et regularis converfatio, fecundum regulam magni patris *Auguftini*,
quam primi inhabitatores profeffi funt, in perpetuum obfervetur;
et ipfa domus religiofa a cujuflibet alterius domus religiofe
fubjectione libera permaneat, et in omnibus abfoluta; falva in

omnibus

omnibus epifcopali auctoritate, et *Winton* ecclefie dignitate. Quod
ut in nofterum ratum permaneat et inconcuffum, prefenti fcripto
et figilli noftri patrocinis duximus confirmandum. His teftibus
domino *Waltero* abbate de *Hyda*. Domino *Walters* Priore de
fancto *Swithuno*, domino Stephano priore de Motesfonte, magiftro
Alano de *Stoke*; magiftro Willo de fancte *Marie* ecclefia, tunc
officiali noftro; *Luca* archidiacon' de furr'. magiftro *Humfrido* de
Millers, *Henrico* & *Hugone* capellanis, *Roberto* de *Clinchamp*, et *Petro*
Roffinol clericis, et multis aliis. Datum apud *Wlnes* [a] per manum
P. de cancellis. In die fanctorum martirum *Fabiani* et *Sebaftiani*.
Anno Domi millefimo ducentefimo tricefimo tercio.

Seal, two faints and a bifhop praying :
Legend: SVI M. SITE. BONI. PETR' PAVL' E PATRONI.

---

[a] Probably *Wolvefey-houfe* near *Winchefter*.

NUMBER II.

III.

(Ni 108.)

Carta petens licentiam eligendi prelatum a Domino Episcopo Wintoniensi.

Defuncto prelato forma petendi licentiam eligendi.

DOMINO et patri in Christo reverendo domino & P. Dei gratia Wintoniensi episcopo, devoti sui filii supprior monasterii de S. Wintoniensis diocefeos salutem cum subjectione humili, reverentiam, et honorem. Monasterio nostro de S. in quo sub protectione vestra vivimus, sub habitu regulari, Prioris solacio destituto per mortem bone memorie, &c. quondam Prioris nostri, qui tali hora in aurora diem clausit extremum, vestre paternitati reverende et dominationi precipue istum nostrum et nostri monasterii casum flebilem cum merore nunciamus; ad vestre paternitatis refugium fratres nostres A. et C. canonicos destinantes, rogando et petendo devote quatenus nobis dignemini licenciam tribuere, ut monasterio predicto, Prioris regimine destituto, providere possimus, invocata Spiritus sancti gratia, per electionem canonicam de Priore. Actum in monasterio predicto 5 kalend. &c. anno Domini, &c. Valeat reverenda paternitas vestra semper in Domino.

Forma

Forma licencie conceſſe.

P. Dei gratia Wintonienſis epiſcopus dileċtis in Chriſto filiis
ſuppriori et conventui talis loci ſalutem, gratiam, et benediċtionem.
Viduitatem monaſterii veſtri vacantis per mortem quondam R.
Prioris veſtri, cujus anime propicietur altiſſimus, paterno compacientes affeċtu, petitam a nobis eligendi licenciam vobis concedimus, ut patronus. Datum apud, &c. 3 kalend. Jul. anno
conſecrationis noſtre tertio.

Forma decreti poſt eleċtionem conficiendi.

In nomine Domini noſtri Jheſu Chriſti, Amen.   Monaſterio
beate Marie talis loci Winton. dioc. ſolacio deſtituto per mortem
R. quondam Prioris ipſius; ac corpore ejus, prout moris eſt, ec-
cleſiaſtice ſepulture commendato; petita cum devocione licentia
per fratres K. etP . canonicos a ven: in Chriſto patre et domino
domino P. Dei gratia Wintonienſi epiſcopo ejuſdem monaſterii
patrono, eligendi priorem, et optenta; die dato, a toto capitulo
ad eligendum vocati fuere evocandi, qui debuerunt, voluerunt,
et potuerunt comode eleċtioni prioris in monaſterio prediċto intereſſe: omnes canonici in capitulo ejuſdem eccleſie convenerunt
tali die, anno Dom. &c. ad traċtandum de eleċtione ſui prioris
facienda; qui, invocata Spiritus Sanċti gratia, ad procedendum
per formam ſcrutinii concencientes.

(N. 108.)

Modus procedendi ad electionem per formam scrutinii.

OMNIBUS in capitulo congregatis qui debent volunt et poſſunt comode intereſſe electioni eligendi ſunt tres de capitulo ᵃ *non noſtro obediencias ores* ᵇ, qui erunt ſcrutatores, et ſedebunt in angulo capituli; et primo requirent vota ſua propria, videlicet, duo re- quirent *tertium* et duo *alterum,* &c. dicendo ſic, " Frater P. in- " quem concentis ad eligendum in prelatum noſtrum ?" quibus examinatis, et dictis eorum per vicem ex ipſis in ſcriptura redactis, vocabunt ad ſe omnes fratres ſingillatim, primo ſuppriorem, &c. Et unus de tribus examinatoribus ſcribet dictum cujuſlibet. Celebrato ſcrutinio, *publicare db coram omnibus. Facta ptmodū concenſum* collectione apparebit in quem pars major capituli et ſanior concentit; quo viſo, major pars dicet minori, " Cum " major pars et ſanior capituli noſtri concenciat in fratrem R. " ipſe eſt eligendus, unde, ſi placet, ipſum communiter eliga- " mus;" ſi vero omnes acquieverint, tunc ille qui majorem vocem habet in capitulo ſurgens dicet, " Ego frater R. pro toto " capitulo eligo fratrem R. nobis in paſtorem;" et omnes dicent; " Placet nobis." Et incipient, " TE DEUM LAUDAMUS." Si vero in unum concordare nequiverint, tunc hiis, qui majorem vocem habet inter illos qui majorem et ſaniorem partem capituli conſtituerint, dicet, " Ego pro me et illis qui mecum concenciunt " in fratrem R. eligo ipſum in," &c. Et illi dicent, " Placet " nobis," &c.

ᵃ Fratres canonicos. See *Forma decreti, &c.*
ᵇ Obedientiores ſc. more regular. In virtute obedientiæ occurs in *Not. Viſit.*

Forma

Forma riĉte prefentandi eleĉtum.

Reverendo in Chrifto patri et domino domino P. Dei gratia Winton. epifcopo devoti sui filii frater R. Supprior conventualis beate Marie de tali loco, et ejufdem loci Conventus, cum fubjeĉtione humili, omnem obedienciam, reverenciam, et honorem. Cum conventualis ecclefia beate Marie talis loci, in qua fub proteĉtione veftra vivimus fub habitu regulari, per mortem felicis recordationis R. quondam prioris noftri deftituta ecclefia priore, qui 6to kalend. Jul. in aurora anno Dom. &c. diem claufit extremum; de corpore ejus, prout moris eft, ecclefiaftice tradito fepulture; petita a vobis, tanquam a Domino, et vero ejufdem ecclefie patrono et paftore, licencia eligendi priorem et optenta; convenientibus omnibus canonicis prediĉte ecclefie in capitulo noftro, qui voluerunt debuerunt et potuerunt comode electioni noftre intereffe, tali die anno Dom. fupradiĉto, invocata Spiritus Sanĉti gratia, fratrem R. de C. ejufdem ecclefie canonicum unanimi affenfu et voluntate in priorem noftrum, ex puris votis fingulorum, unanimiter eligimus. Quem reverende paternitati veftre et dominacioni precipue Priorem vero patrono noftro et paftore confirmandum, fi placet, tenore prefentium prefentamus; *dignitatem* veftram humiliter et devote rogantes, quatenus, diĉte electioni felicem prebere volentes affenfum, eidem R electo noftro *nunc* confirmabitis, et quod veftrum eft paftorali *folicitudine* impendere dignemini. In cujus rei teftimonium prefentes litteras figillo capituli noftri fignatas paternitati veftre *tranfmittimus.* Valeat reverenda paternitas veftra femper in Domino. Datum tali loco die et anno fupradiĉtis. Omnes et finguli, per fratres.

A. B.

A. B. et C. ejuſdem ecclefie canonicos de voluntate tocius conventus ad inquirenda vota fingulorum conſtitutos, fecreto et fin-
gillatim requifiti; tandem publicato fcrutinio et faĉta votorum
coleĉtione inventum eſt, majorem et feniorem partem tocius
capituli diĉte ecclefie in fratrem S. de B. diĉte ecclefie canonicum
unanimiter et concorditer concenciſſe; vel fic, quando inventum
omnes canonicos diĉte ecclefie preter duos in fratrem, A. D.
quibus ſtatim majori parti eligendum adquiefcenter: frater k.
fupprior ecclefie memorate, juxta poteſtatem fibi a toto conventu
traditam, vice confociorum fuorum et fua ac tocius conventus,
diĉtum fratrem S. de B. in priorem ejuſdem ecclefie elegit, fub
hac forma; " Ego frater fupprior conventualis ecclefie beate
" Marie talis loci, poteſtate et auĉtoritate mihi a toto conventu
" diĉte ecclefie tradita et commiſſa, quando, puplicato fcrutinio
" et omnibus circa hoc rite peraĉtis, inveni majorem et partem
" feniorem tocius capituli noſtri in fratrem S. de B. virum
" providum unanimiter concenfiſſe, ipfum nobis et ecclefie noſtre,
" vice tocius conventus, in priorem eligendum; et eidem elec-
" tioni fubfcribo; cui electioni omnes canonici noſtri concence-
" runt, et fubfcripferunt."—" Ego frater de C. prefenti elec-
" tioni concencio, et fubfcribo." Et fic de fingulis electoribus;
in *cujus rei* teſtimonium figillum capituli noſtri apponi fecimus ad
prefentes.

WILLMUS

N U M B E R III.

Vifitatio Notabilis de Seleburne.

1387.

Willmus permiffione divina Winton Epifcopus dilectis filiis Priori et Conventui Prioratus de Selborne Ordinis S^{ti}. Auguftini, noftræ diocefeos, Salutem, gratiam, et ben. Sufcepti regiminis cura paftoralis officii nos inducit invigilare folicite noftrorum remediis fubjectorum, et eorum obviare periculis ac fcandala removere; ut fic de vinea domini per cultoris providi farculum vicia extirpentur inferantur virtutes, exceffus debite corrigantur, et fubditorum mores in nimium prolapforum per appoficionem moderaminis congrui reformentur: Hanc nempe folicitudinem noftris humeris incumbentem affidua meditacione penfantes, ne fanguis vefter de manibus noftris requiratur, ad vos et veftrum Prioratum fupradictum, prout noftro incumbebat officio paftorali, nuper ex caufa defcendimus vifitandi; et dum inter vos noftre vifitacionis officium iteratis vicibus actualiter exercuimus, nonnulla reperimus que non folum obviant regularibus inftitutis, verum eciam que religioni veftre non congruunt, nec conveniunt honef-

tati ;;

tati; ad que per noſtrum antidotum debite reformanda opem et operam prout expedit et oportet apponimus, quas credimus efficaces, infra ſcripta ſiquidem precepta noſtra pariter et decreta, ſanctorum patrum conſtitucionibus editis et debite promulgatis canoniciſque ac regularibus inſtitutis fulcita, vobis noſtri ſigilli roborata munimine tranſmittimus, inter vos futuris temporibus efficaciter obſervanda, quatinus ad Dei laudem, divini cultus ac veſtræ religionis augmentum, ipſis mediantibus, per viam ſalutis feliciter incedatis; mores et actus veſtri abſtrabantur a noxiis, et ad ſalutaria dirigantur.

No. I. In primis ut Domino Deo noſtro, a quo cuncta bona procedunt, et omnis religio immaculata ſumpſit exordium, in Prioratu veſtro predicto ſerviatur laudabiliter in divinis; Vobis, in virtute ſancte obediencie ac ſub majoris excommunicationis ſententie pena, firmiter injungendo mandamus, quatinus hore canonice, tam de nocte quam de die, in choro a conventu cantentur; miſſe quoque de beata Maria et de die, necnon miſſe alie conſuete horis et devocione debitis et cum moderatis pauſacionibus celebrentur: nec liceat alicui de conventu ab horis et miſſis hujuſmodi ſe abſentare, aut, poſtquam incepte fuerint, ante complecionem earum ab ipſis recedere quoviſmodo; niſi ex cauſa neceſſaria vel legitima per priorem vel ſuppriorem aut alium preſidentem loci, ut convenit, approbanda; in quo caſu ipſorum omnium conſciencias apud altiſſimum arctius oneramus; contrarium vero facientes in proximo tunc capitulo celebrando abſq accepcione qualibet perſonarum regularem ſubeant diſciplinam; acrius inſuper puniendi ſi contumacia vel pertinacia delinquencium hoc expoſcat; ſi quis vero poſt trinam correpcionem debite ſe non correxerit in premiſſis, pro ſingulis vicibus quibus contrarium fecerit ipſum ſingulis ſextis feriis in pane et aqua dumtaxat precipimus jejunare.

No. II. Item

No. II. Item quia in visitacione nostra predicta comperimus evidenter quod silencium, quasi in exilio positum, ad quod juxta regulam Sti. Augustini efficaciter estis astricti, locis et temporibus debitis inter vos minime observatur contra observancias regulares; Vobis omnibus et singulis firmiter injungendo mandamus, quatinus silencium, prout vos decet, regula supradicta, de cetero locis et temporibus *hujusmodi* observetis; a vanis et frivolis colloquiis, sicut decet, vos penitus abstinendo: illos vero, qui silencium *hujusmodi* in locis predictis non observaverint, animadversione condigna precipimus castigari; et, si quis tercio super hoc legitime convictus fuerit, preter regularem disciplinam, die, quo debite silencium non tenuerit, pane et servicia dumtaxat et legumine sit contentus.

No. III. Item quia nonnulli concanonici et confratres prioratus vestri predicti validi atq; sani et in sacerdocio constituti celebracionem missarum absq; causa legitima indebite ac minus voluntarie multociens, ut dicitur, negligunt et omittunt; fundatorum aliorumq; benefactorum suorum animas, pro quibus sacrificia offerre tenentur, suffragiis nequiter defraudando; Vobis, ut supra, firmiter injungendo mandamus, quatinus vos omnes et singuli Prioratus predicti concanonici et confratres in sacerdocio constituti frequenter confiteamini confessoribus per Priorem deputandis; quos quidem confessores discretos et idoneos, prout numerus personarum dicti conventus exigit, per vos dominum Priorem predictum precipimus deputari; missasque, impedimento cessante legitimo, tam pro vivis quam pro defunctis, pro quibus orare tenemini, de cetero, quanto frequencius poteritis, celebretis devocius, sicut decet; impedimentum vero predictum cum contigerit Priori vel Suppriori Prioratus predicti per illud pacientes infra triduum declarari volumus et exponi, ac per eorum alterum

prout

prout juftum fuerit approbari, vel eciam reprobari; 'in quo cafu ipforum omnium tam exponencium quam approbancium apud altiffimum confciencias diftrictius oneramus; contrarium vero facientes, primo fuper hoc convicti, proxima quarta feria fequenti in pane, fervifia, et legumine; fecundo vero convicti feria quarta et fexta fequentibus modo confimili; tercio vero convicti dictis feriis extunc fequentibus in pane et aqua jejunent, quoufque judicio prioris fe correxerint in premiffis; ftatuentes preterea quod Prior et Supprior Prioratus predicti contra *hujufmodi* delinquentes femel fingulis menfibus diligenter inquirant, et quos culpabiles invenerint in premiffis modo predicto ftudeant caftigare.

No. IV. Item quia tranfitus communis fecularium perfonarum utriufque fexus per clauftrum Prioratus veftri in congruis temporibus minime exercetur, et potiffime horis illis quibus fratres de conventu in contemplacione fancta ftudiis quoque ac lectionibus variis inibi occupantur; unde diffoluciones plurime provenerunt, et poterunt in futuro verifimiliter provenire, ac ipforum fratrem quieti et religionis honeftati plurimum derogatur: Vobis ut fupra arcius injungendo mandamus, quatinus, cum fecundum regulam fancti Auguftini converçacio veftra debeat effe a fecularibus *hujufmodi* feparata, ad animarum ac' eciam rerum pericula, que poffent et folent ex concurfu *hujufmodi* provenire, caucius evitanda; tranfitum communem predictum per prefatum clauftrum de cetero fieri nullatenus permittatis, per quem veftra devocio et religionis honeftas vulneram vel eciam impediri valeant quovifmodo, fub pena excommunicacionis majoris quam in contravenientes intendimus canonice fulminare: illum vero, ad quem oftiorum clauftri cuftodia pertinet, fi propter illius negligenciam five culpam tranfitus *hujufmodi* fuftineatur indebite, ut

prefertur;

prefertur; pro fingulis vicibus, quibus hoc factum fuerit, fingulis quartis feriis in pane, fervifia, et legumine dumtaxat jejunet; et, fi nec fic fe correxerit debite in hac parte, ab officio deponatur, ac alius, magis providus, loco fuo celeriter fubrogetur.

No. V. Item quia oftia ecclefie atq; clauftri prioratus veftri predicti non fervantur nec ferantur temporibus debitis, nec modo debito, ut deceret; fed cuftodia eorundem agitur et omittitur multociens necgligenter; adeo quod *fufpecte perfone* et *alie inhonefte* per ecclefiam et clauftrum *hujufmodi* incedunt frequenter in tenebris atq; umbris, temporibus eciam fufpectis et illicitis, indecenter; unde dampna et fcandala varia pluries provenerunt, et im pofterum verifimiliter poterunt provenire; Vobis, ut fupra, mandamus, firmiter injungentes, quatinus dicta oftia de cetero claudi faciatis, et claufa per miniftros idoneos cuftodiri temporibus debitis, prout decet; vobis inhibentes expreffe, ne oftia ecclefie veftre predicte, (illa videlicet que inter navem ipfius ecclefie et chorum ejufdem exiftunt) nec oftia clauftri que ducunt ad extra, et per que introitus fecularium in ipfum clauftrum patere poterit, de mane, antequam prima incipiatur in choro; aut commeftionis tempore; nec eciam de fero, poftquam conventus collationem inceperit; nifi in caufa utili vel neceffaria per priorem vel fuppriorem, ut convenit, approbanda, aperiantur de cetero quovis modo: ad que fideliter exequenda facriftam, qui pro tempore fuerit, ad cujus officium premiffa pertinent fub pena amocionis ab officio fuo arcius oneramus, acrius per nos puniendum prout nobis videbitur expedire.

No. VI. Item quia nonnulli concanonici et confratres prioratus veftri minus fapiunt in lectura, non intelligentes quid legant, fed literas quafi prorfus ignorantes, dum pfallunt vel legunt, accentum brevem pro longo ponunt pluries, et e contra; et per invia gradi-

entes

entes sanum scripturarum intellectum adulterantur multociens, et
pervertunt; sitque, ut dum scripturas sacras non sapiant, ad per-
petrandum illicita proniores reddantur: Vobis Domino Priori in
virtute obedientie, firmiter injungendo mandamus, quatinus,
cum legere et non intelligere sit necgligere, noviciis et aliis minus
sufficienter literatis idoneus de cetero deputetur magister, qui ipsos
in cantu et aliis primittivis scienciis instruat diligenter juxta regularia instituta; quatinus, in eisdem perfectius eruditi, cecitatis
squamis et ignorancie nebulis depositis, que legant intelligant et
agnoscant, et contemplandum clarius misteria Scripturarum effici-
antur, ut convenit, promciores.

No. VII. Item quia constituciones sive decretales Romanorum
Pontificum vestrum ordinem concernentes, (ille videlicet de quibus
in constitucionibus recolende memorie Domini Ottoboni, quondam
sedis Apostolice in Anglia legate, sit mensio specialis) inter vos nullatenus recitantur, prout per constituciones ejusdem legati recitari
mandantur; unde, dum decretales ipsas et contenta in eis penitus
ignoratis, committitis multociens que prohibentur expressius per
easdem in vestrarum periculum animarum: Vobis firmiter injungendo mandamus, quatinus, ne ignoranciam aliquam pretendere
poteritis in hac parte, decretales predictas, prout in prefatis domini constitucionibus Ottoboni plenius recitantur, in quodam qua-
terno seu volumine absque more dispendio faciatis conscribi; ipsas bis singulis annis in vestro capitulo, juxta formam constitutio-
num dictarum, recitari clarius facientes, ad informacionem rudium
et perfectionem eciam provectorum; adjicientes preterea, ut magis-
tri noviciorum presencium et eciam futurorum ipsos in regula
S^{ti}. Augustini diligentur instruant et informant, ipsam regulam eis
vulgariter exponendo; quodque iidem novicii per frequentem reci-
tacionem ejusdem illam sciant quasi cordetenus, sicut in dictis

constitucionibus

conftitucionibus plenius continetur, per quam incedere poterint via recta et errorum tenebras caucius evitare : fuper execucione vero premifforum debite facienda dominum priorem prioratus vef-tri predicti arcius oneramus quatinus ea que premifimus in hoc cafu fub pena fufpenfionis ab ipfius officio per menfem diligencius exequatur.

No. VIII. Item quia canonici et confratres prioratus veftri predicti, ipforum propriam voluntatem pocius quam utilitatem communem fectantes, non veftes neceffarias, cum opus fuerit,. fed certam et limitatam ac determinatam quantitatem peccunie, velut annuum redditum, pro veftibus *hujufmodi* percipiunt an-nuatim, contra regulam Sti. Auguftini ac domini Ottoboni et aliorum fanctorum patrum canonica inftituta; fitque, ut, dum effrenis illa religioforum cupiditas, aliena fpecie colorata, vetita concupifcat, fancta religio, folutis conftantie frenis, in luxum labentem ad latitudinis tramites que ducunt ad mortem, mifera-biliter nofcitur declinare : cui quidem morbo peftifero, ne pu-trefcat et vermes generet corruptivos, mederi cicius cupientes nichil novi ftatuendo fed fanctorum patrum veftigiis inherendo, volumus ac eciam ordinamus, quod canonicis et confratribus me-moratis prefentibus et futuris de bonis et facultatibus communibus prioratus veftri predicti veftris ufibus deputatis veftes et calci-amenta, cum indiguerint, neceffaria, juxta facultates predictas, et nullo modo peccuniam, pro eifdem, per eos qui fuper hiis miniftrandi gerent officium de cetero miniftrentur; veftes vero inveteratas et ineptas *hujufmodi* canonicorum camerario communi tradi volumus pauperibus erogandas juxta regulam Sti. Auguftini, et alias canonicas fanctiones contrarium vero facientes, fi camera-rius fuerit, penam fufpenfionis ab officio ipfum incurrere volumus ipfo facto ; fi vero alius canonicus de conventu exiftat, preter

alias

alias pmas regulares tam peccunia quam eciam indumentis novis careat illo anno.

No. IX. Item quia nonnulli canonici et confratres Prioratus vesti predicti opportunitate captata, extra septa Prioratus absque societate honesta, evagandi causa, nulla super hoc optenta licencia, se transferunt pluries indecenter; alii preterea provectiores certis officiis deputati ad maneria et loca alia officiis *hujusmodi* assignata equitant, quando placet, ibidem manentes pro eorum libito volantatis, nullo canonico ipsis in socium assignato, contra ordinis decenciam et religionis eciam honestatem, constitucionesque Sanctorum Patrum editas in hac parte: Cum igitur religiosos extra eorum Prioratum sic vagari aut in eorum maneriis vel ecclesiis eis appropriatis soli manere expresse prohibeant canonica instituta; nos, premissa fieri de cetero prohibentes. Vobis firmiter injungendo mandamus, quatinus, cum aliquis Prioratus vestri canonicus vel confrater super vel pro negociis propriis vel eciam communibus exire contigerit, prius ad hoc a Priore vel Suppriore, si presentes in Prioratu fuerint, alioquin, ipsis absentibus, ab ipso qui protunc conventui preesse contigerit, licenciam habeat specialem; cui assignari volumus unum canonicum in socium, ne suspicio sinistra vel scandalum oriatur; qui, associata eisdem juxta qualitatem negocii cometiva honesta, in eundo et eciam redeundo gravitate servata modestius semper incedant, et expletis negociis ad Prioratum cicius revertantur, que regularibus conveniunt institutis devocius impleturi: contrarium vero facientes, absque remissione seu accepcione qualibet personarum, regularem subeant disciplinam; super quo presidencium conventus consciencias arcius oneramus, ipsosque nichilominus pro singulis vicibus, quibus excesserint in premissis, singulis sextis feriis in pane et aqua

jejunent;

jejunent; et fi officiarius fuerit, ipfo facto, fi aliquod canonicum non obfiftat, ab ipfius officio fit fufpenfus.

No. X. Item quia comperimus evidenter, quod nonnulli canonici domus veftre, fecundum carnem pocius quam fecundum fpiritum diffolute viventes, nulla caufa racionabili fubfiftente, nudi jacent in lectis abfque femoralibus et camifiis contra eorum obfervancias regulares; Vobis igitur fermiter injungendo mandamus, quatinus vos omnes et finguli canonici Sti. Auguftini regulam et in ea parte ordinis veftri canonica inftituta de cetero efficaciter obfervetis: contrarium vero facientes fingulis quartis feriis in pane, fervifia, et legumine tantummodo fint contenti; fi quis vero poft trinam correptionem reus inventus fuerit in hac parte pro fingulis vicibus fingulis extunc feriis fextis in pane et aqua hunc precipimus jejunare; Priorem vero ac Suppriorem domus predicte fub pena fufpenfionis ab officiis eorundem arcius onerantes, quatinus fuper premiffis fepius et diligenter inquirant, et quos culpabiles invenerint eos penis predictis percellere non poftponant.

No. XI. Item quia nonnullos canonicos et confratres Prioratus veftri predicti publicos reperimus venatores ac venacionibus *hujufmodi* fpreto jugo regularis obfervancie, publice intendentes, ac canes tenentes venaticos, contra regularia inftituta; unde diffolutiones quamplures, animarum pericula corporumque, ac rerum difpendia multociens oriuntur; nos volentes hoc frequens vicium a Prioratu predicto radicitus extirpare; Vobis omnibus et fingulis tenore prefencium inhibemus, vobis *nichilominus* firmiter injungentes, ne quifquam canonicorum Prioratus veftri predicti publicis venacionibus vel clamofis ex propofito intendere de cetero, vel eciam intereffe; canefve venaticos per fe vel alios tenere prefumat, publice vel occulte, infra Prioratum vel extra, contra

formam

formam capituli, " Ne in agro dominico," et alias canonicas fanctiones; per hoc autem Prioratus veftri predicti nec juri
vel confuetudini, quod vel quam habere dinofcitur, in ea parte
non intendimus in aliquo derogare: contrarium vero facientes
preter difciplinas et penas alias canonicas pro fingulis vicibus fingulis quartis et fextis feriis in pane et fervifia jejunando precipimus
caftigari.

No. XII. Item quia canonici Prioratus veftri predicti quibus officia forinfeca et intrinfeca committuntur, fingunt fe, cum poffent et
deberent in choro divinis officiis intereffe, in officiis *hujufmodi* fibi
commiffis multociens occupari, que poffent ante vel poft horas
hujufmodi commode fieri, et eciam exerceri; propter quod cultus
divinus minuitur, et alii clauftrales nimium onerantur; Vobis
in virtute fancte obedientie et fub pena excommunicacionis majoris firmiter injungendo mandamus, quatinus officiarii quicunque
ecclefie veftre predicte in choro ejufdem divinis officiis a modo
perfonaliter interfint, nifi ex caufa legitima officiorum fuorum et
per prefidentem conventus, qui pro tempore fuerit, approbanda,
eos contigerit abfentare; in quo cafu de et fuper abfencia fua
legalitateque caufarum pretenfarum in hac parte ipforum prefi-
dencium et officiariorum confciencias apud altiffimum diftrictius
oneramus.

No. XIII. Item, quia juxta fapientis doctrinam ubi majus
iminet periculum, ibi caucius eft agendum, volumus et eciam
ordinamus, quod duo canonici difcreti et idonei de conventu
Prioratus veftri predicti per ipfum conventum vel majorem partem
ejufdem annis fingulis de cetero eligantur, qui bis in anno ad
maneria, tam Priori quam eciam pro reftentacione conventus
hujufmodi ceteris que officiariis affignata, perfonaliter fe transferant

et

et accedant, ſtatum maneriorum ipſorum tam in edificiis quam eciam in ſtauro vivo vel mortuo plenarie ſupervifuri; quique ſuper hiis que invenerunt in eiſdem conventui ſupradicto relacionem fidelem in ſcriptis, ut convenit, facere teneantur; ut, ſi mors alicujus officiarii vel caſus ilius fortuitus evenerit, de ſtatu officii *hujuſmodi* cujuſcumque conventum non lateat memoratum; premiſſa vero vobis precipimus efficaciter obſervanda ſub pena noſtro arbitrio limitanda, vobis, ſi in hiis necgligentes fueritis vel remiſſi, acrius infligenda.

No. XIV. Item quia ſolitus et antiquus numerus canonicorum in Prioratu veſtis predicto, quod dolenter referimus, adeo jam decrevit, ac eciam minuitur in preſenti, quod ubi xiiii. canonici vel circiter in habitu et obſervanciis regularibus in dicto Prioratu ſolebant Altiſſimo devocius famulari, (quibus de bonis et poſſeſſionibus ipſius Prioratus veſtri communibus que poſſidetis in victu et veſtitu juxta decenciam ordinis regularis honorifice ac debite fuerat miniſtratum) modo vero undecim canonici dumtaxat exiſtunt et ſerviunt in eodem; quo fit, ut dum regis regum cultum attenuet cohabitancium paucitas, contra multiformis nequitie hoſtem minuatur exercitus bellatorum : Cum igitur juxta prefati domini Ottoboni conſtitutiones aliorumque ſanctorum patrum canonica inſtituta, canonicorum antiquus numerus ſit ſervandus, ac juxta ſapientis doctrinam " In multitudine " populi ſit dignitas regis, et in paucitate plebis ignominia " principis accendatur;" Vobis in virtute ſancte obedientie ac ſub pena majoris excomm. firmiter injungendo mandamus, quatinus, cum omni diligentia et celeritate debitis, de viris idoneis religioni diſpoſitis, et honeſtis vobis abſque more diſpendio providere curetis; ipſos in ordinem veſtrum regularem in ſupple-

cionem

cionem majoris numeri requifiti, feu faltem illius numeri canonicorum ad quorum fuftentacionem congruam, aliis oneribus vobis
incumbentibus debite fupportatis, veftre jam habite fuppetunt
facultates; fuper quibus veftram et cujuflibet veftrum confcienciam arcius oneramus; celerius admittentes, ad augmentum cultus
divini et perfectionem majorem ordinis regularis, pro fundatoribus
et benefactoribus veftris devocius, ut convenit, interceffuros.

No. XV. Item quia comperimus evidenter quod vos, domine
Prior, cui ex debito veftri officii hoc incumbit, de proprietariis
canonicis Prioratus veftri predicti, juxta conftitutiones domini
legati editas in hac parte, inquificionem debitam hactenus non
feciftis, minifterium vobis creditum in ea parte necgligentius omit-
tendo; quo fit, ut ille peftifer hoftis antiquus paftoris confiderans
continuatam defideam oves miferas et errantes, ipfius hoftis ne-
quiffimi fraude deceptas in fitim avaricie prolabentes laqueo pro-
prietatis feduxit, contra fanctorum patrum canonica inftituta, in
fuarum grave periculum animarum; Vos igitur requirimus et
monemus, vobifque in virtute obediencie firmiter injungendo
mandamus, quatinus dicti legati conftitutiones, ut convenit, imi-
tantes fuper proprietariis *hujufmodi* faltim bis in anno inquificionem
faciatis de cetero diligentem; ipfos, fi quos inveneritis, animad-
verfione condigna juxta regularia inftituta canonice punientes; fi
vero id adimplere necglexentis, adminiftracione veftra, ipfo facto
noveritis vos privatum, donec premiffa fueritis diligenter executi,
prout in conftit. homini ottoboni legati predicti plenius continetur.

No. XVI. Item, cum fecundum conftit. dicti legati et aliorum
fanctorum patrum canonica inftituta, abbates et priores, proprios
abbates non habentes, nec non officiarii quicunque teneantur bis
faltim fingulis annis prefente toto conventu vel aliquibus ex fe-
nioribus ad hoc a capitulo deputatis de ftatu Prioratus et de ad-

miniftracione

miniſtracione ſua plenariam reddere rationem, quod tum in Prioratu veſtro predicto invenimus hactenus non ſervatum, unde plura ſecuntur incommoda, et veſtre utilitati communi plurimum derogatur; Vobis in virtute obediencie firmiter injungendo mandamus, quatinus prefati domini legati, domini videlicet Ottoboni, necnon bone memorie domini Stephani quondam Archiepiſcopi Cant. conſtit. editas in hac parte, faciatis inter vos de cetero firmiter obſervari, ſub pena ſuſpenſionis officiariorum ipſorum ab eorum *hujuſmodi* officiis, dictique Prioris ab adminiſtracione ſua, quam, ſi premiſſa necglexerint obſervare, ipſo facto, donec id perfecerint, ſe noverint incurriſſe, prout in dictis conſtit. dicti Ottoboni plenius continetur.

No. XVII. Item quia in Prioratu veſtro predicto et eccleſia ejuſdem ac in nonnullis domibus, edificiis, muris et clauſuris eccleſie veſtre prelibate, necnon maneriorum ipſius Prioratus certis diverſis officiis deputatorum, quas et quæ preceſſorum et predeceſſorum veſtrorum induſtria ſumptuoſe conſtruxerat, quamplures enormes et notabiles ſunt defectus, reparatione neceſſaria indigentes; unde ſtatum ipſius Prioratus ac maneriorum predictorum deformitas occupat, et multa incommoda inſecuntur; Vobis igitur in virtute obedientie firmiter injungendo mandamus, quatinus defectus *hujuſmodi*, pro veſtra utilitate communi abſque dilationis incommodo, quamcicius poteritis, juxta vires reparari debite faciatis; alioquin Priorem ceteroſque officiarios quoſcumque, qui in premiſſis necgligentes fuerint vel remiſſi, niſi infra ſex menſes poſt notificacionem preſencium ſibi factam ad debitam reparationem defectuum *hujuſmodi* ſe preparaverint, cum effectu, ipſo facto ab officiis ſuis *hujuſmodi* ſint ſuſpenſi.

No. XVIII. Item, quia per vendiciones et conceſſiones liberacinoum et *corrodiorum* hactenus per vos factas, reperimus dictum Pri-

M m m 2   oratum

oratum multipliciter fore gravatum, adeo quod ea, que ad divini cultus augmentum, suftentacionem pauperum, et infirmorum, pia devocio fidelium erogavit, mercenariorum ceca cupiditas jam abforbet; fitque, ut dum bona ejufdem Prioratus in alios ufus quam debitos, ne dixerimus in prophanos, nepharie convertantur, altiffimo famulancium in eadem numerus minuitur, pauperes et infirmi fuis porcionibus, ac ipfa ecclefia divinis obfequiis inequiter defraudantur, contra intencionem piiffimam fundatorum, in veftrarum periculum animarum; Indempnitati igitur ipfius ecclefie veftre in hac parte debite providere, dictum quoque tam frequens incommodum ab eadem radicitus extirpare volentes, bone memorie domini Ottoboni legati predicti aliorumque fanctorum patrum veftigiis inherentes; Vobis tenore prefencium diftrictius inhibemus, eciam fub pena excomm. majoris, ne *corrodia*, liberaciones, aut penfiones perfonis aliquibus imperpetuum vel ad tempus vendatis de cetero, vel aliqualiter concedatis, abfque noftro confenfu et licencia fpeciali; prefertim cum vendiciones *hujufmodi*, que fpecies alienacionis exiftunt, Prioratus veftri predicti detrimentum procurent et enormem eciam generat lefionem; fi quis vero contra hanc noftram inhibicionem aliquid attemptare prefumpferit, nifi id quod fic prefumpferit revocaverit, ab officio fit fufpenfus prout in conftit. domini Ottoboni clarius continetur.

No. XIX. Item quia quedam certe perpetue cantarie pro fundatoribus et aliis benefactoribus veftris tam in genere quam in fpecie antiquitus conftitute per diverfos prefbyteros in Prioratu veftro predicto debite celebrande, pro quibus plura donaria recipiftis a multis retro actis temporibus, ac eciam de prefenti, ut afferitur, funt fubftracte, contra piam intencionem ac ordinacionem eciam fundatorum, in veftrarum grave periculum animarum; Vobis igitur, in virtute fancte obedientie ac fub majoris excom.

fentencie

fentencie pena, firmiter injungendo mandamus, quatinus cantarias predictas juxta formam inftitucionum et ordinacionum earum faciatis de cetero debite celebrari, ac eifdem congrue deferviri, fi redditus et proventus ad *hujufmodi* cantarias antiquitus affignati ad hoc fufficiant hiis diebus, alioquin prout redditus et proventus earum, aliis omnibus eifdem incumbentibus debite fupportatis, fufficiunt de prefenti, dolo et fraude ceffantibus quibufcunque; fuper quo veftram confcienciam arcius oneramus, a modo deferviri debite faciatis.

No. XX. Item vobis et omnibus et fingulis in virtute fancte obediencie ac fub majoris excom. fentencie pena firmiter injungendo mandamus, quatinus elemofinas in Prioratu veftro predicto antiquitus fieri confuetas, et eas ad quas tenemini ex ordinacione antiqua pro animabus fundatorum et aliorum benefactorum veftrorum juxta facultates veftras fuper quibus veftras confciencias arctius oneramus, prout divinam effugere volueritis ulcionem diftribui de cetero faciatis; precipientes preterea quod fragmenta feu reliquiæ tam de aula Prioris quam eciam de refectorio proveniencià, abfque diminucione qualibet, per elimofinarium vel ipfius locum tenentem integre colligantur, pauperibus fideliter eroganda; alioquin, fi elimofinarius *hujufmodi* remiffus vel negligens fuerit in premiffis, penam fufpenfionis ab officio fe noverit incurfurum.

No. XXI. Item quia debilibus et infirmis humanitatis preberi fubfidium jubet caritas, et pietas intelpellat; Vobis domino Priori ceteris obedienciariis Prioratus veftri predicti, quorum intereft in hac parte in virtute fancte obediencie firmiter injungendo mandamus, quatinus confratribus veftris debilibus et infirmis, ipforum infirmitate durante, in efculentes et poculentes eorum infirmitatibus congruentibus, necnon in medicinis et aliis juxta infirmitatis *hujufmodi* qualitatem et Prioratus facultates, de bonis veftris com-

munibus

munibus et ficut antiquitus fieri confueverat de cetero faciatis debite procurari, fub pena fufpenfionis ab officiis veftris fi circa
premiffa necgligentes fueritis vel remiffi, ipfo facto, quoufq; id quod
necgligenter omiffum fuerit perfeceritis, incurrenda; prout in
conftit. domini Ottoboni plenius continetur; ftatuentes preterea
quod camere infirmaria veftra, cum opus fuerit, infirmis canonicis
fint communes, ne, quod abfit, aliquis fibi retineat in eifdem vel
vendicet proprietatem, contra fancti Auguftini regulam et conftit.
fanctorum patrum editas in hac parte.

No. XXII. Item cum necgligencia five remiffio in perfonis
precidencium fit plurimum deteftanda, facilitas quoq; venie
incentivum prebeat delinquendi; Vobis domino Priori, Suppriori,
aliifq; conventus predicti prefidentibus quibufcumq; prefentibus
et futuris, in virtute fancte obediencie firmiter injungendo mandamus, Quatinus, cum correctiones in perfonis ipfius conventus
imineant faciende, ipfas, prout ad vos pertinet, abfq; acceptione
qualibet perfonarum juxta quantitatem delictorum et perfonarum
qualitatem veftrafq; obfervancias regulares cum maturitate debita,
et difcretione previa, facere ftudeatis; alioquin vos fuppriorem
ceterofq; prefidentes predictos, fi necgligentes vel remiffi aut culpabiles fueritis in premiffis, canonica noftra monicione premiffa
penam fufpencionis ab officiis veftris extunc incurrere volumus
ipfo facto, donec *hujufmodi* negligenciam, remiffionem, culpam,
vel defidiam a vobis excufferitis in hac parte; pena prefacto domino Priori in hoc cafu, ut convenit, infligenda nobis fpecialiter
refervata.

No. XXIII. Item cum confuetudines laudabiles Prioratus cujuf-
cumq; ordinacionefque ac ftatuta que ufus longevi temporis appro-
bavit merito fint fervandæ; Vobis domino Priori ac fingulis officiariis Prioratus veftri predicti prefentibus et futuris in virtute fancte

obediencie,

obediencie, et sub penis infra scriptis, firmiter injungendo mandamus; Quatinus pitancias et alias distribuciones quascunque, in quibuscunque rebus consistant et quocunque nomine censeantur, in obitibus, anniversarius festivitatibus, aut aliis diebus, conventui, aut ab uno officio alii officio ex ordinacione antiqua debitas et consuetas, in canonicum aliquod non obsistat a modo faciatis persolvi, sub pena porcionis duple, cujus partem unam conventui predicto, alteram vero partem certis piis usibus nostro arbitrio limitandis debite persolvendam specialiter reservamus.

No. XXIV. Item cum vendiciones boscorum, firme maneriorum vel eciam ecclesiarum, aut alia domus vestre ardua negocia imineant facienda, illa, sine tractatu ac deliberacione provida cum conventu predicto ac eorum consensu expresso vel majoris et sanioris partis ejusdem, de eetero fieri prohibemus; aliter autem *hujusmodi* negocia ardua facta nullius existunt firmitatis; et nichilominus Priorem aliosque officiarios quoscumq; qui contra presentem prohibitionem nostram quicquam attemptaverint in premissis, penam suspensionis ab officiis eorundem ipso facto se noverint incursuros, cum ex *hujusmodi* factis privatis ecclesiis dispendia multociens provenerunt; illa quoque que omnes tangunt ab omnibus merito debeant approbari.

No. XXV. Item volumus ac eciam ordinamus, quod sigillum vestrum commune sub quinque clavibus ad minus de cetero custodiatur; quarum unam penes Priorem, secundam penes suppriorem, terciam penes precentorem, et reliquas duas claves penes confratres alios per conventum ad hoc nominandos decrevimus remanere, per ipsos fideliter custodiendas; inhibentes preterea sub pena excom. majoris ne quicquam cum dicto sigillo communi a modo sigelletur, nisi litera *hujusmodi* sigellanda primitus legatur, inspiciatur, et eciam intelligatur a majore et saniore parte tocius

conventus,

conventus, et ad ipfam figillandam communis vefter prebeatur confenfus, cum ex facto *hujufmodi* plura poffunt difpendia verifimiliter provenire; ad hac vobis omnibus et fingulis tenore prefencium inhibemus, ne *compatres alicujus pueri* de cetero fieri prefumatis, noftra fuper hoc licencia non obtenta, cum ex *hujufmodi* cognacionibus religiofis domibus difpendia fepius invenire nofcuntur; contrarium vero facientes, preter difciplinas alias regulares, fingulis fextis feriis per menfem proxime tunc fequentem in pane et aqua jejunando precipimus caftigari.

No. XXVI. Item quia nonnulli canonici domus veftre predicte, freno abjecto obfervancie regularis, *caligis* de *Burneto* et fotular*ium* bafp. in *ocrearum* loco ad modum fotularium uti publice non verentur, contra confuetudinem antiquam laudabilem ordinis fupradicti, in perniciofum exemplum et fcandalum plurimorum; nos igitur honeftatem dicti ordinis obfervare volentes, Vobis domino Priori in virtute fancte obediencie firmiter injugendo mandamus, Quatinus quofcumq; veftros canonicos et confratres ad utendun de cetero ocreis feu botis fecundum antiquas veftri ordinis obfervancias regulares per quafcumq; cenfuras ecclefiafticas, et, fi opus fuerit, par incarceracionis penam canonice compellatis, fub pena fufpenfionis ab officio veftro predicto.

No. XXVII. Item quia tres vel due partes conventus domus veftre non comedunt cotidie in refectorio, prout conftitutiones fanctorum patrum fanxerunt provide in hac parte; Vobis dicti Prioratus conventui firmiter injungendo mandamus, Quatinus tres vel faltem due partes veftrum cotidie in refectorio hora prandii de cetero comedant et remaneant debite, ficut decet; vobis arcius injungentes, quod nullus veftrum in manfiunculis aut locis aliis privatis eciam cum hofpitibus fuis regularibus vel fecularibus vel confratibus fuis comedat; hoftilaria cum hofpitibus, refectorio in in communi mifericordia, caufa recreacionis, et aula Prioris dumtaxat exceptis; hanc tamen Prior apponat providenciam diligentem, ut, fine perfonarum accepcione, nunc hos nunc illos ad refectionem convocet, quos magis noverit indigere; fuper execucione vero debita premifforum Priorem ac alios conventui prefidentes fub pena fufpenfionis ab eorum officiis arctius oneramus.

No. XXVIII. Item, cum fecundum fanctorum patrum conftituciones, juniores canonici a fuis prelatis vivendi normam habeant affumere, ac iidem prelati fuper fua converfacione teftium copiam debeant obtinere; Vobis domino Priori in virtute obedientie diftricte precipiendo mandamus, Quatinus capellanum veftrum canonicum fingulis de cetero mutetis annis, juxta conftitutiones fanctorum patrum editas in hac parte; ut fic, qui vobifcum fuerint in officio predicto, per doctrine laudabilis exercicium plus valeant in religione proficere, ac eos innocencie teftes, fi vobis, quod abfit, crimen aliquod feu fcandalum per aliquorum invidiam imponatur, prompte poteritis invocare.

No. XXIX. Item, cum communis exquifitus ornatus prefertim in religiofis perfonis a jure fit penitus interdictus; Vobis tenore prefencium inhibemus, ne quivis veftram de cetero in fuis veftibus *furruris* preciofis aut *manicis* nodulatis *zonisve* fericis auri vel argenti ornatum habentibus utatur de cetero quovis modo, cum abufus *hujufmodi* ad pompam et oftentacionem ac fcandalum ordinis manifefte tendere dinofcatur.

No. XXX. Item, quia fingula officia funt fingulis committenda perfonis; Vobis in virtute obediencie et fub excom. fententie pena firmiter injungendo mandamus, ut officia fingula veftri Prioratus, que per canonicos officiarios gubernari folebant, per officiarios *hujufmodi*, per vos communiter vel divifim juxta Prio-

ratus

ratus predicti morem folitum eligendos, quibus ipfa officia, ut olim, committi volumus exercenda, fingulariter de cetero gubernentur.

XXXI. Item, cum plus timeri foleat id quod fpecialiter injungitur quam quod generaliter imperatur; Vobis omnibus et fingulis inhibemus, ne aliquis veftrum, ad curam animarum non admiffus, clericis aut laicis facramentum unctionis extreme vel euchauriftia miniftrare, matrimonia ve folempnizare, non habita fuper hiis parochialis prefbyteri licencia, quomodolibet prefumatis, fub pena excom. majoris fententie in hac parte a canone fulminate.

XXXII. Item quia comperimus in noftris vifitacionibus fupradictis vafa et pallas altaris, necnon et veftimenta facra ecclefie veftre, atque corporalia, tam immunda relinqui, quod interdum aliquibus funt horrori; ut igitur honor debitus divinis impendatur; Vobis firmiter injungendo mandamus, Quatinus vafa, corporalia, pallas, et veftimenta predicta, ac cetera ecclefie ornamenta munda nitida et honefta decetero conferventur hoc quoq; infuper injungentes, ut in ecclefiâ veftra celebrantibus *vinum bonum*, purum, et incorruptum ad facramentum altaris conficiendum per eum qui fuper hoc gerit officium, et non *corruptum*, et *acetofum*, prout fieri confuverit, impofterum miniftretur; nimis enim videtur abfurdum in facris fordes necgligere, que dedecerunt in prophanis.

XXXIII. Item licet fanctorum reliquias, vafa, aut veftimenta facra feu libros ecclefie in *vadem dari*, aut pignori obligari canonica prohibeant inftituta, a vobis tamen in dictis vifitacionibus comperimus contrarium effe factum; Vobis igitur domino Priori, tenore prefencium, firmiter injungendo mandamus, quatinus ab *hujufmodi* impignoracionibus extra cafus a jure permiffos vos decetero penetus abftinentes, *hujufmodi* pignori obligata curetis recolligere, et ea

ecclefie

ecclefie veftre reftituere, abfq; more difpendio, ficut decet; ftatuentes preterea ut omnes carte ac munimenta quecumq; ftatum bona
et poffeffiones domus veftre qualitercumq; contingentes, fub tribus
feruris et clavibus remaneant, futuris temporibus fideliter confervande.

XXXIV. Item cum religiofi de bono in melius continue debeant proficifci, ac ex facre fcripture lectione et infpectione qualiter
id faciant plenius inftrui valeant; Vobis firmiter injungendo mandamus, ut, completis hiis, que ad veftri ordinis et regularis difcipline obfervanciam pertinent atq; fpectant, in clauftro fedentes
fcripture facre lectioni fancteq; contemplacioni devocius infiftatis,
ficq fecundum regule veftre exigenciam taliter codices infpiciendos
requiratis, ut in eis quid fugiendum quid fubfequendum ac cujuf-
modi premium inde confequendum fuerit agnofcere valeatis.

XXXV. Item vobis Domino Priori injungimus, quod cum pa-
rentes vel confanguinei alicujus confratris veftri ad eum acceffe-
rint, caufa vifitandi eundem, liberaliter fecundum ftatum fui
exigentiam per vos vel illum qui fuper hoc miniftrandi gerit officium infra Prioratum honefte et debite procurentur; fed videant
fratres ne nimis fint in talibus Prioratui onerofi.

XXXVI. Item quia parum eft jura condere nifi executioni debita demandentur, ea quoque folent labili memorie eo tenacius
commendari quo veraciter audientium auribus fuerint fepius incul-
cata; et, ne veftrum quis piam ignorantiam pretendere valeat
premifforum; Vobis firmiter injungendo mandamus, quatinus
has noftras injunctiones et decreta pariter fupradicta in aliquo volumine competenti abfque more difpendio confcribi plenius faciatis, eaque omnia et fingula bis annis fingulis de cetero coram toto
conventu plenius recitari; vos nichilominus omnes et fingulos
monemus primo fecundo et tercio peremptorie, vobis infuper in

virtute

virtute obediencie arctius injungentes, quatinus ipfas injunctiones noftras et decreta predicta omnia et fingula prout ad vos et veftrum quemlibet pertinent et fingulariter vos concernunt, teneatis de cetero ac eciam obfervetis, fub penis et cenfuris ecclefiafticis fupradictis, et aliis penis canonicis in contravenientes quofcumque, prout contumacia delinquencium exegerit, per nos impofterum canonice infligendis. Poteftatem autem premiffa corrigendi, mutandi in toto vel in parte, interpretandi, declarandi et eifdem addendi, et eciam detrahendi, ac penas adjiciendi, fufpendendi, necnon fuper compertis aliis in vifitatione noftra predicta procedendi, criminaque et defectus ac exceffus in ipfa comperta et delata corrigendi, ac canonice puniendi, et fuper ipfis novas injunctiones infuper faciendas, ficut et prout opus fuerit et nobis videbitur expedire, nobis eciam fpecialiter refervamus. In quorum omnium teftimonium figillum noftrum fecimus hiis apponi. Dat. apud Wynton vicefimo feptimo die menfis Septembris anno Domini millefimo ccc° octogefimo feptimo et noftre confecrationis anno, vicefimo.

(L. S.)

NUMBER

NUMBER IV.

(No. 50.)

INDENTURA PRIORIS *de* SELBORNE *quorundam tradit. Petro Barnes sacristæ ibidem ann. Hen.* 6. - - - *una cum confiss. ejusdem Petri script.*

Hec indentura facta die lune proxime post ffestum natalium Dni anno *regis* Henrici sexti post conquestum anglie v. - - - - - inter ffratrem Johannem Stepe priorem ecclesie beate Marie de Selborne & Petrum Bernes sacrist. ibidem videlicet quod predictus prior deliveravit prefato Petro omnia subscripta In primis XXII amit XXXI aubes vid. v. sine parura pro quadragesima XXII manicul. Item XXII stole Item VIII casule vid. III albe pro quadragesima Item XI dalmatic. vid. I debit. Item XVI cape vid. IIII veteres Item unam amittam I albam cum paruris unum manipulum I stolam I casulam et duas dalmaticas de dono Johannis Combe capellani de Cicestria pro diebus principalibus Item I amittam I aubam cum paruris I manupulum I stolam I casulam de dono ffratris Thome Halybone canonicis Item I amittam I aubam cum paruris I manupulum I stolam I casulam pertinentem ad altare sancte Catherine virginis pro priore Item I amittam II aubas cum paruris II manipul II stolas et II casulas pertinentes ad altare sancti Petri de dono patris Ricardi holte Item de dono
ejusdem

ejufdem ıı tuella vid. ı cum fruictello et ı canvas pro eodem altare Item ı tuellum pendentem ad terram pro quadragefima Item vı tuell cum ffruictibus xv tuell fine ffruictell Item ıııı tuell pro lavatore Item v corporas Item ıı ffruictell pro fummo altare fine tuellis Item ıı coopertor pro le cefte Item ıı pallias de ferico debili Item ı velum pro quadragefima Item ı tapetum viridi coloris pro fummo altare ıı ridell cum ıııı ridellis parvis pertinent. ad dict. altare Item vıı offretor vid. v debit. Item ıııı vexilla Item ıııı pelves ııı queffones vid. ı de ferico Item ıı fuper altaria Item quinq; calices vid. ıııı de auro Item ıı cruettes de argento de dono dni Johannis Combe capellani de Ciceftre Item vııı cruettes de peuter Item ı coupam argent. et deaur. Item ıı *ofculator argent.* Item ı ofculatorium cum offe *digiti auricular Sti. Johannis Baptifte* Item ı crux argent. et deaur. non radicat. Item turribulum argent et deaur. Item ı anulum cum faphiro Item ı aliud anulum ı politum aureum Item ı anulum argent. et deauratum *Sti. Edmundi.* Item ı concha cum pereo infixo Item ı ciftam argent. et deaur. Item ı imaginem beate Marie argent. et deaurat. Item ı parvam crucem *cum v reliquiis* Item ı *junctorium Sti. Ricardi* Item ı tecam pro reliquiis imponend Item ı *calefactor Sti. Ricardi* Item ıııı candelebra vid. ıı de ftagno et ıı de ferro Item ı *pecten Sti. Ricardi* Item ıı viell de criftall In parte fract Item ı pelvim de coper ad lavator Item ıı *ofculat. de coper* Item ı parvum terribulum de latyn Item ı vas de coper pro frank et fence confecrand Item ı pixidem de juery pro corpore Chrifti Item ıı vafa de plumbo pro oleo confervando Item ı patellam eneam ferro ligat. Item tripodem ferr. Item ı coftrell contum ıı lagen et ı potrell. Item ıı babyngyres Item ıı botelles de corio vid. ı de quarte et ı de pynte Item ııı anul. arg. et ı pixidem Ste Marie de Waddon Item ( )

Inftrumenta

Inftrumenta pro Sandyng Item i ledbnyff Item i fhafshobe Item
i fecurim Item ii fcabell. de ferro pro cancell Item i plane Item
i ciftam fine cerura Item xiiii fonas Item xix taperes ponder
xiii ℔ et dimid. Item ii torches ponder xx ℔ Item xii ℔
cere et dimid. Item de candelis de cera ponder vi ℔ Item i ℔ de
frank et fence Item i lagenam olei Item ix pondera de plumbo
(Vide de ftauro in tergo) et in tergo fcribuntur
hæc, " ii vacce i fus iiii hoggett et iiii porcell"

NUMBER V.

(N. 381.)

A PAPER *conteyning the value of the* MANORS *and* LANDS *pertayning to the* PRIORIE *of* SELBORNE. iv. Edw. 3. *With a note of charges yſſuing out of it.*

SELEBORNE PRIORATUS.

SUMMA totalis valoris maneriorum terrarum tenementorum et premiſſorum ejuſdem Prioratus in ffeſto Sᵗⁱ. Michaelis Archang. anno ſecundo Regis Edvardi 4ᵗⁱ. ut patet Rotul. de valoribus liberat.

ˣˣ
IIII VI li. (i. e. LXXXVI li.) X s. VI d.

Inde in redditibus reſolutis domino pape domino Archiepiſcopo et in diverſis ffeodis certis perſonis conceſſis ac aliis annualibus repriſis in eiſdem Rotul. de valoribus annotatis per annum XIIII li. XIX s. V d.

Et remanet de claro valore LXXI li. X s. VIII d.

Videlicet Aſſignantur pro
{
Quatuor canonicis et quatuor ffamulis deo et eccleſie ibìd. ſervientibus pro eorum vadiis veſtur. et diet. ut patet per bill inde faꝗ. per annum XXX li.

Diverſis creditoribus pro eorum debitis perſolvendis ut patet per parcell inde faꝗ. XV li. XV s. IIII d.

Reparacionibus Eccleſiarum domorum murorum et clauſurarum ejuſdem Prioratus per annum XV li. XV s. IIII d.

Annua pencione Domini Prioris ei aſſignata per annum quouſque remanet X li.
}

SELEBORNE

SELBORNE PRIORATUS.

Modo sequitur de Reformatione premiſſorum.

Redditus omn. ffirmis et Pencionibus.

Summa total. valorum ibid. miſis et deſperatis inde deductis prout patet per declaracionem Dni Petri Prioris de Seleborne ad man. Dni noſtri Wynton apud Palacium ſuum de Wolſley preſentat. per ipſum ultimo die ffebr. Ann. Domini MCCCCLXII. et penes ipſum remanet

LXXI li. x s. VIII d. unde per ipſum Dnum noſtrum Wynton aſſignantur in fforma ſequente videlicet.

Aſſignantur ut ſupra

Pro quatuor canonicis et quatuor ffamulis deo et ecclesie ibid. ſervientibus pro eorum Diet. vadiis et veſtur. ut patet per bill inde ſact.

xxx li.

Pro annua pencione Prioris quouſque *remanet.*

x li.

Pro diverſis creditoribus pro eorum debitis perſolvendis ut patet per bill inde ſact.

xv li. xv s. IIII d. per II annos ad xxxI li. x s. VIII d. ultra LV li. XIIII d. de vendit. ſtauri.

Pro diverſis reparacionibus ecclesiarum domorum murorum et clauſurarum ut patet per bill.

xv li. xv s. IIII d. per II annos ad xxxI li. x s. VIII d. Summa total. valoris pro debitis et reparacionibus aſſignat. cum LV li. XIIII d. de vendit. Stauri ut ſupra CXVIII li. II s. VI d.

Debita

Debita que debentur ibid per diverſos tenentes et ffirmarios ad feſtum
Sᵗⁱ. Michaelis anno tertio Regis Edvardi 4ᵗⁱ. videlicet.

Abbas de Derford de ffeod ffirme ſua ad ɪx li. vɪ s. ⎱
vɪɪɪ d. per annum a retro   ⎰ xx li. vɪɪ s. xɪ d..

Thomas Perkyns armig. ffirmarius Rectorie de ⎱
Eſtworlam pro uno anno finiente ad ffeſtum Sᵗⁱ ⎬ ʟx s.
Mich. anno ɪɪ. Regis Edvardi 4ᵗⁱ.   ⎰

Johannes Shalmere ball. de Selborne debet   ʟxxv s.
Ricardus Cawry debet de eodem anno   vɪ s.

                          Summa   xxvɪɪ li. vɪɪɪ s. xɪ d.

Thomas Perkyns armig. debet de ffirme ſua pre- ⎱
dicta ad feſtum Sᵗⁱ. Mich. ann. vɪɪ et ultra feod. ⎬ vɪɪ li. vɪ s vɪɪɪ d.
ſuum ad xx s. per annum   ⎰

Thomas lusſher debet pro ffirme ſua ad xʟ s. per ⎱
annum cum feod. ſuis ad xx s. per annum   ⎰ c. s.

Hugo Pakenham debet de reddit. ſuo ad xx s. ⎱
per ann.   ⎰ c. s.

Abbas de Derford debet de ffeod ffirme ſua ultra ⎱
xx li. vɪɪ s. xɪ d. ut ſupra pro annis ɪɪɪ. ɪɪɪɪ. et ⎬ xxvɪɪɪ li.
Regis Edvardi   ⎰

Walterus Berlond ffirmarius de *Shene* debet   ɪx li. v s. ɪɪd.
Henr. Shafter ffirmarius *ffeod* de Baſynſtoke   xɪɪ li. ɪɪɪɪ d.
Henr. lode nuper ffirmarius manerii de *Chede* debet xx li.

        Total ʟxxxxɪv li. xɪɪ d.

                         Summa   ʟxvɪ li. xɪɪ s. vɪ d

I N D E X.

INDEX.

O o o 2

INDEX.

INDEX.

INDEX.

INDEX.

I N D E X.

INDEX.

INDEX.

Sedge-

INDEX.

* For the amazing ravages committed on turnips, wheat, clover, field cabbage-seeds, &c. by *slugs*, and a rational and easy method of destroying them, see a sensible letter by Mr. *Henry Vagg,* of *Chilcompton,* in the county of *Somerset,* lately made public at the request of the gentlemen of that neighbourhood.

Sudington,

I N D E X.

INDEX.

W

Yard,

INDEX.

Y

L I S T of the P L A T E S

to the

H I S T O R Y of S E L B O R N E,

from the

DRAWINGS of S. H. GRIMM.

ERRATA in the HISTORY of SELBORNE.

Page.	Line.	
11	13	for *scences* read *scenes*.
31	15	for *teems* read *teams*.
91	7	dele comma, and for *or* read *of*.
115	21	for *plantive* read *plaintive*.
163	22	for *dilation* read *dilatation*.
197	13	for *effusus* read *conglomeratus*.
228	7	for *nurtered* read *nurtured*.
234	14	for *setter-worth* read *setter-wort*.
235	8	for *blea-berries* read *bil-berries*.
237	2	after *incessus* dele colon, and insert it after *est*.
237	13	over *vera* dele ʌ
240	7	for *winded* read *winged*.
246	21	for *Wordleham* read *Ward le ham*, & passim.
258	2	for *fresh-mowed* read *fresh-moved*.
280	23	for *peak-nosed* read *peaked-nosed*.
309	8	for *Woolmer* read *Wolmer*, & passim.
352	12	for *esteemed* read *deemed*.
423	18	for *founded* read *found*.

Printed in the United States
By Bookmasters